Legionellosis

Diagnosis and Control in the Genomic Era

Editor:

Prof. Jacob Moran-Gilad, MD MPH FESCMID
Dept. of Health Systems Management
School of Public Health
Faculty of Health Sciences
Ben Gurion University of the Negev
Beer Sheva, Israel

Assistant Editor:

Rachel E. Gibbs, BA
Medical School for International Health
Ben Gurion University of the Negev
Beer Sheva, Israel

Caister Academic Press **www.caister.com**

Caister Academic Press, UK
www.caister.com

ISBN: 978-1-913652-53-1 (paperback)
ISBN: 978-1-913652-54-8 (ebook)

DOI: https://doi.org/10.21775/9781913652531

Contents

Editor Biographies

Prof. Jacob Moran-Gilad is a physician board-certified in clinical microbiology and in public health and the Principal Investigator of the Microbiology, Advanced Genomics and Infection Control Applications Laboratory (MAGICAL group) at the Ben Gurion University of the Negev. Prof. Moran-Gilad is particularly interested in harnessing diagnostic technologies to inform public health policy and has published extensively on the topic. He is the chair of the National Advisory Committee for Microbiology and Member of the Epidemic Management Team in Israel, the past chairperson of the ESCMID Study Group for Genomic and Molecular Diagnostics (ESGMD) and current Programme Director for the European Congress on Clinical Microbiology and Infectious Diseases (ECCMID), the world's premiere conference in the field of infection. Over the last years Prof. Moran-Gilad made several notable contributions to the field of Legionella, as a board member of the European Study Group for Legionella Infections (ESGLI), organiser of the ESGLI conference, chair of the national advisory committee for legionellosis and author of scientific papers focusing on outbreak investigation, genomic typing tools and molecular diagnostics for legionellosis.

Rachel E. Gibbs is a medical student at the Medical School for International Health at Ben Gurion University of the Negev and a research associate at the Microbiology, Advanced Genomics and Infection Control Applications Laboratory (MAGICAL). Rachel has an interest in infectious diseases and particularly relating to outbreak investigations and public health.

Current books of interest

Chapter 1

Introduction to Legionellosis Diagnosis and Control in the Genomic Era

Rachel E. Gibbs and Jacob Moran-Gilad*

MAGICAL group, Department of Health Systems Management, School of Public Health, Ben Gurion University of the Negev, Israel

*giladko@post.bgu.ac.il

DOI: https://doi.org/10.21775/9781913652531.01

Abstract

Legionella bacteria are a leading cause of infectious disease and mortality worldwide. Moreover, from the perspectives of bacteriology, public health, epidemiology and clinical medicine, *Legionella* is an exceptional and unique micro-organism. Historically, the study of *Legionella* is inherently multidisciplinary as leading up to human infection, the bacteria interacts with freshwater and other various ecosystems, protozoa, complex molecular secretion systems, man-made water systems and finally, human colonization and antibiotic treatment. These multifaceted interactions often render the studies of *Legionella* as an applicable model for the prevention, assessment and control of other infectious diseases causing outbreaks. Since the discovery of the organism in 1977, the techniques which specialists use for research on *Legionella* have rapidly transformed. With chapters written by a diverse array of specialists, this book exemplifies the dynamic nature of *Legionella* while new methods such as genomics revolutionize this domain.

Summary of chapters

Legionella was only discovered after its role in causing human disease was identified, a phenomenon which was enhanced by the installation of infrastructural technology involving the aerosolization of fresh water. Nevertheless, the study of *Legionella* in its original fresh-water ecology remains relevant to disease control initiatives today. In the following chapter, Rafael A. Garduño delves deeply into the freshwater ecology of *L. pneumophila* with emphasis on interactions with freshwater amoebae, the biotic and abiotic components of freshwater, pleomorphic variations of *L. pneumophila's* developmental forms, and a brief discussion of *L. longbeachae* ecology in soil environments. This chapter presents *Legionella* as an adaptive and intracellular parasite with specific requirements and complex mechanisms employed to survive and replicate within live, freshwater amoebae.

After the ecological components of *L. pneumophila* as an intracellular parasite are established, Elisabeth Kay et al. describe in Chapter 3 the molecular biological features which it uses to effectively invade, survive within and control host cells from protozoa to human alveolar macrophages. Of these mechanisms, the most studied and essential are the intricate secretion systems (T1SS – T4SS) and particularly the highly conserved Dot/Icm T4BSS. In addition to secretion systems, this chapter explains the importance of the secreted proteins and genetic regulation which manipulate and respond to the environment, effect virulence and contribute to the extraordinary infective versatility of *Legionella*. These complex molecular invasive properties are important for the control and prevention of *Legionella* outbreaks in humans and suggest excellent models for the studies of intracellular pathogens.

Chapter 4 examines the developing understanding of *L. pneumophila* reference strains in a historical context. After a general introduction, Kozak-Muiznieks et al. describes the clinical picture and epidemiology of

Legionnaire's Disease (LD) and Pontiac Fever (PF). They outline modes of transmission and as well as outcomes and trends of diagnostic studies. Then the chapter provides a detailed description of how each reference strain was discovered. This includes the epidemiology of their respective outbreaks, and how the investigation of each strain contributed to the overall understanding of *Legionella* and legionellosis. The story behind each reference strain demonstrates the importance and usefulness of genomics in outbreak investigations and beyond. Not only has sequencing analysis offered a plethora of game-changing information, but genomics also helps point to remaining gaps in understanding to inform further investigation.

Subsequently, Chapters 5 and 6 go into detail about the clinical features of legionellosis from symptoms to diagnostics and treatment. In Chapter 5 Giancarlo et al., take on the clinical picture of both LD and PF, including how the nonspecific clinical picture can lead to difficulties in prompt diagnosis and treatment, as well as tools and protocols developed to improve care and determine the severity of the illness. The antibiotic treatment regimens are also present with explanations of the research showing the superiority of fluoroquinolones and macrolides. The chapter also brings forward the unique presentation and treatment requirements of immunocompromised patients. Overall, prompt diagnosis and treatment with correct antibiotic coverage is very important for improving clinical outcomes. Due to the ambiguity of legionellosis' clinical symptoms, Chapter 6 outlines the importance of laboratory diagnostics. From culture, urine antigen, antibody detection, PCR and direct immunofluorescence, this section explains the development and shortcomings of each diagnostic method and how to use them appropriately.

Most of our chapters emphasize on *L. pneumophila* and especially *L. pneumophila* serogroup 1, as it is the culprit organism in the majority of legionellosis cases in humans. However, in Chapter 7 Diane S.J. Lindsay describes the clinical and epidemiological features of non-*L. pneumophila* species of the *Legionella* genus. While not found in abundance, the non-*L. pneumophila* species have unique environmental characteristics, and some have been implicated in cases of both LD and PF. This chapter describes each species' historical significance as well as important flaws in the current diagnostic strategies which has led to these species being widely undetected.

Chapter 8 discusses the essential process of prevention, risk assessment and risk management of legionellosis outbreaks. The prevention of legionellosis is indeed a complicated topic necessitating more support and attention from scientific efforts and public health initiatives. Surman-Lee et al. describes the governing bodies responsible for such regulation along with the shortcomings of these processes. The specific components of the risk assessment and management process require in-depth knowledge of the chain of causation of an LD outbreak. This includes an understanding of the pathogen and methods of transmission as well as the physical and chemical components of the industrial habitats of the buildings and systems which it can proliferate. This chapter outlines the various control strategies as well as various lesser known areas of risk requiring more research.

After the details of risk assessment and prevention are laid out, de Jong et al. describe in Chapter 9 the European approach to the systematic response of reported cases and outbreaks. As legionellosis is an international phenomenon, and cases of travel-associated LD (TALD) are on the rise, the European Legionnaires' Disease Surveillance Network (ELDSNet) as a standardized and multinational effort of surveillance is an important established model. This chapter outlines the evolution of the organizations in charge of surveillance in the EU along with their strategies, shortcomings, partnerships, definitions and patterns from the valuable data compiled by the system over time.

Chapter 10 marks the return to the theme of *Legionella* diagnosis in the genomic era. Here, Fry et al. lays out the various genotyping methods used for identification of *L. pneumophila*. The chapter explains the background and overall importance of genotyping while emphasizing *L. pneumophila* due to its widespread nature with many travel-associated cases and the homogeneity of the *L. pneumophila* serogroup 1. From the first use of plasmid analysis in the 1980's to the more current and widely accepted sequence-based typing (SBT) scheme, Fry et al. points out the historical context, benefits and disadvantages of each genotyping method. Leading up to the final chapter, chapter 10 sets up whole genome sequencing (WGS) as the answer to the imperfections of the SBT scheme.

Lastly, in the concluding chapter, Wüthrich et al. explain the typing of *Legionella* isolates at the frontier of phylogenetic analysis, next generation sequencing and genomics. The chapter outlines the explicit need for typing strategies with higher resolution, among them being the prominent level of genetic conservation of *L. pneumophila*, and especially *L. pneumophila* ST1, as a result of its intracellular lifecycle. The chapter reviews the methods of *Legionella* species determination along with serotyping and classical molecular typing. Afterwards, the use of genome-based typing in the context of other pathogens sets the stage for application of *Legionella* typing. They explain the basic workflow of WGS before providing a detailed review of the important methods; cgMLST, wgMLST and SNP-based analysis. Examples of various applications of WGS to the study of *Legionella* are offered along with emphasizing that while NGS and WGS contribute the most comprehensive genotypic resolution on isolates to date, epidemiological data remains essential in conjunction. Historically, *Legionella* has presented researchers with unique scenarios due to its fascinating pathogenic patterns. Today in the era of genomics with the boundless applications of WGS, it appears that our understanding of *Legionella* has only just scratched the surface.

Chapter 2

Freshwater Ecology of *Legionella pneumophila*

Rafael A. Garduño*

Department of Microbiology and Immunology, Dalhousie University, Halifax, Nova Scotia, Canada

*rafael.Garduño@dal.ca

DOI: https://doi.org/10.21775/9781913652531.02

Abstract

Legionella pneumophila is a Gram-negative freshwater bacterium that emerged in the mid-1970s as an opportunistic human pathogen and the causal agent of Legionnaires' disease. Legionnaires' disease is an atypical pneumonia which is always acquired from the environment. *L. pneumophila* is principally not transmitted from person-to-person which indicates that *L. pneumophila* is not adapted to the human host. In addition, *L. pneumophila* is an intracellular pathogen of amoebae, and infection of humans is merely accidental. This chapter discusses *L. pneumophila* as a highly adapted intracellular parasite whose existence depends on its ability to replicate intracellularly in free-living freshwater amoebae. The complex ecology of *L. pneumophila* is explored through its interactions with both the abiotic and the biotic components of the freshwater environment. Since *L. pneumophila* is a pleomorphic organism with many developmental, morphological and(or) physiological forms, a comparative analysis of the unique ecologies of relevant forms will be presented. Control measures used to mitigate the transmission of Legionnaires' disease and the risk management of water systems will be discussed in further chapters, though the information provided in this chapter contributes to the improvement of control initiatives.

Introduction

Legionella pneumophila, the main focus in this chapter, is the micro-organism which the bulk of scientific literature revolves around. Additionally, *Legionella* spp can live in different environments (e.g. freshwater, moist soil). This chapter will focus on the freshwater environment with discussion of *L. pneumophila*'s interaction with the abiotic freshwater environment and with other freshwater microorganisms (Figure 1). Lastly, *Legionella longbeacheae* which thrives in the soil environment will be mentioned.

Freshwater environments are quite diverse in physicochemical characteristics and nutrient levels. In oligotrophic freshwater bodies, organic substrates and inorganic ions are diluted and typically present in low concentrations. While eutrophic freshwater ecosystems are characterized by high influxes of organic carbon and nitrogen (e.g. wastewaters) and inorganic compounds (e.g. agricultural runoff). Laboratory experimentation has shown that free planktonic cells of *L. pneumophila* remain viable for long periods of time, but do not multiply in either oligotrophic tap water or deionized water devoid of nutrients (Skaliy and McEachern, 1979; Schofield, 1985; Lee and West, 1991; Al-Bana et al., 2014 and references within). *L. pneumophila* is adapted to survive in oligotrophic freshwater environments, but it does not depend on this environment directly for growth-supporting nutrients. However, it is clear that *L. pneumophila* requires high concentrations of select nutrients due to its fastidious dependence on iron and amino acids and efficient mechanism for obtaining nutrients from the intracellular environment of eukaryotes. It is unknown how these specialized nutrient requirements are

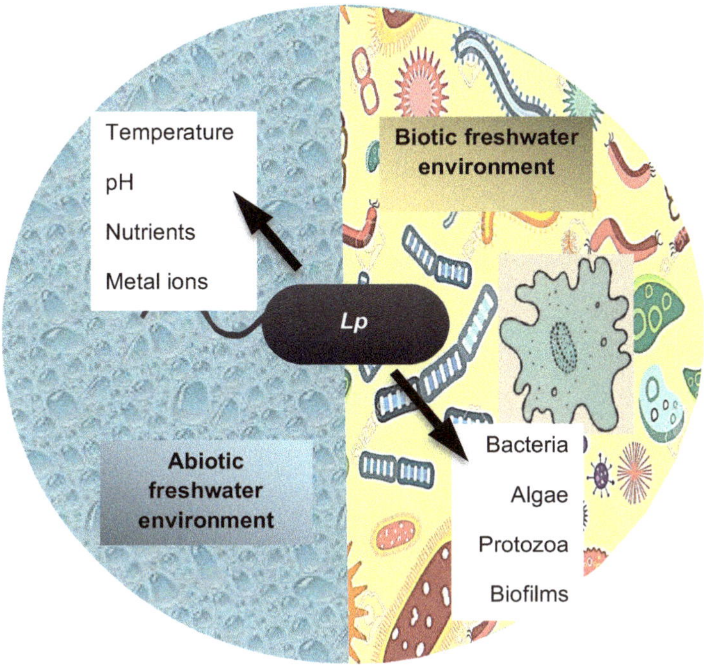

Figure 1. Schematic representation of the overall microbial ecology of *L. pneumophila* in the freshwater environment (background oval). The left half of the oval (blue background) is the abiotic component of freshwater. *L. pneumophila* interacts with the abiotic components (left arrow) which include physicochemical characteristics of water itself plus the chemicals dissolved in it. The right half of the oval (microbial background) is the biotic component of freshwater. *L. pneumophila* interacts with microorganisms (right arrow) primarily bacteria, algae and protozoa either as free planktonic organisms suspended in water or within biofilms. This chapter is structured according to this scheme.

congruent with the fact that *L. pneumophila* is ubiquitous in freshwater environments and thrives in man-made water systems. The vigorous growth of *L. pneumophila* in freshwater ecological niches where select nutrients are plentiful answer this question.

Predatory freshwater protozoa are the main consumers of bacterial biomass (Pernthaler, 2005). Consequently, this strong selective pressure lead freshwater bacteria to develope anti-predation strategies sometimes at the expense of optimizing growth (Greub and Raoult, 2004; Jousset, 2012). When a freshwater bacterium must choose between nutrient acquisition in an oligotrophic freshwater environment and surviving predation, and it will likely choose the latter (Baumgartner et al., 2016). This could potentially cause the bacteria to enter the evolutionary pathway towards pathogenesis (Harb et al., 2000; Greub and Raoult, 2004; Matz and Kjelleberg, 2005). *L. pneumophila* has evolved as an intracellular parasite of freshwater amoeba (Garduño 2008). Therefore, it is not surprising that *L. pneumophila* survives intravacuolar digestion inside protozoa. *L. pneumophila* then grows in a specialized intracellular membrane-bound compartment, the *Legionella*-containing vacuole (LCV), and derives nutrients from the eukaryotic cytosol. The relationship between *L. pneumophila* and protozoa is essential in the ecology of *L. pneumophila*, and this chapter will explore related key issues and recent developments

Legionella spp.

The genus *Legionella* includes more than 60 species of Gram-negative rod-shaped bacteria that primarily thrive in freshwater environments. Some species are less understood, especially those that are are non-culturable such as *Legionella drancourtii* (La Scola et al., 2004) or obligate symbionts of ill-known amoeba such as the X-bacteria of *Amoeba proteus* (Jeon, 2004; Newsome et al., 2001; Jacquier et al., 2013). Among the legionellae, *L. pneumophila* is the most studied, simply because it is the species that causes the most humans disease. The introduction of technology involving the aerosolization of freshwater increased the infective potential of *L. pneumophila*. When contaminated water is aerosolized, *L. pneumophila* can enter the lungs of susceptible individuals and infect alveolar macrophages and lung epithelial cells resulting in disease. The enhanced interaction between *L. pneumophila* and humans is therefore a recent phenomenon.

As an intracellular parasite of amoeba, *L. pneumophila* has not evolved to lose its ability to grow extracellularly and still can be isolated from its natural environment on agar plates. Although this ability could allow *L. pneumophila* to occupy a quite unique trophic position in the freshwater environment, it cannot be assumed that *L. pneumophila* replicates extracellularly in the freshwater environment by *in vitro* results (Spriggs, 1987). The evolutionary preservation of *L. pneumophila*'s ability to grow extracellularly could result from selective pressure or the remnant of a disappearing trait. Since *L. pneumophila*'s interactions with amoebae are facultative and pathogenic, it is plausible that *L. pneumophila* occupies an intermediate position between a free-living freshwater bacterium and a non-culturable intracellular endosymbiont in the evolutionary spectrum of *Legionella* as endobionts of amoeba (Garduño, 2008). It is also likely that *L. pneumophila*'s ability to grow extracellularly is not an essential survival strategy but rather an accessory trait that could be lost as the species behaves increasingly as an obligate pathogen of amoebae.

L. pneumophila is a highly pleomorphic bacterium with many different morphological forms linked through a developmental network regulated by a complex signaling and transcriptional regulatory system coordinating *L. pneumophila* differentiation (Garduño et al., 2008; Robertson et al., 2014). Why *L. pneumophila* produces so many different morphological forms is not clear. It could be that each form is adapted to optimally survive, colonize and thrive in particular ecological niches which again places *L. pneumophila* in a unique position to exploit different habitats (Spriggs, 1987). This implies that the various *L. pneumophila* forms interact differently with both the abiotic environment and other freshwater organisms.

Interaction of *L. pneumophila* with the abiotic freshwater environment

From the early years of *L. pneumophila* research, *Legionella* was initially studied in natural environments as opposed to within man-made water distribution systems (Fliermans, 1996). Given the mode of transmission of legionellosis and *L. pneumophila*, it is understandable that scientific efforts rapidly shifted to man-made systems (Fields, 1997). However, it is important that the understanding of *L. pneumophila* as a freshwater bacterium must begin in nature, and studies of *L. pneumophila* in natural water bodies should not be neglected.

It is widely accepted that *Legionella* is ubiquitous to the freshwater environment (Fliermans and Tyndall, 1992; Fields, 1997). *Legionella* has been found in almost every eutrophic or oligotrophic natural freshwater body where it has been searched for. This includes lakes (Fliermans et al., 1979 and 1981a), thermal and volcanic ponds (Fliermans, 1996 and references within), rivers (Fliermans et al., 1981a; Parthuisot et al., 2010; Li et al., 2015b), hot springs (Shen et al., 2015), groundwater (Riffard et al., 2001; Brooks, 2004; Costa et al., 2005) and surface river waters (Hsu et al. 2015).

The number of *Legionella* species and their relative abundance change depending upon the freshwater body. In most oligotrophic bodies the levels of *Legionella* are quite low and require the concentration of large volumes of water obtained by filtration, centrifugation or immunomagnetic separation for detection. In other cases, there may be non-culturable *Legionella* species present, and detection requires culture-independent and molecular methods. Although initially developed for clinical samples, amoebae enrichment is successfull for *Legionella* detection in environmental samples (Rowbotham, 1983; Berk et al., 2006; Hsu et al. 2015). Due to their widespread presence in freshwater environments, it is accepted that all man-made water systems are colonized by one or more

Legionella species. Furthermore, *Legionella* could be found in any water system fed either by urban potable water systems or stored natural freshwater. To provide examples of this, *Legionella* spp., including *L. pneumophila*, have been found in treated urban wastewaters (Brissaud et al., 2008), dental unit waterlines (Barbeau et al., 1998), storage water tanks in RVs (Litwin et al., 2013), reservoirs of collected rain water (Simmons et al., 2008) and even rain water puddles (Sakamoto et al., 2009; van Heijnsbergen et al., 2014).

Since planktonic *L. pneumophila* does not typically derive nutrients directly from freshwater, it is surmisable that free planktonic *L. pneumophila* cells cope with starvation with physiological strategies to survive until an ecological niche favourable for replication is found. This notion is supported by the finding that when suspended in oligotrophic freshwater, *L. pneumophila* shuts down the processes of genome and cell replication as well as protein synthesis, transcription and translation of mRNAs through the combined actions of the alarmone ppGpp and the global stress response factor RpoS, alternative sigma factor σS (Trigui et al., 2015). This process may represent a differentiation mechanism underlying the developmental network within which *L. pneumophila* produces quiescent cellular forms as an overall survival strategy against starvation (Garduño, 2008). Quiescent *L. pneumophila* forms include stationary phase forms (SPFs) and mature infectious forms (MIFs) (Garduño et al., 2008; and Robertson et al., 2014) as well as viable but non-culturable cells (VBNCCs) (Steinert et al. 1997, Ohno et al., 2003; Al-Bana et al., 2014; Li et al., 2014). These forms have the potential to differentiate back into metabolically active replicative forms when nutrients are available, or a new host is found.

When SPFs are suspended in water, before losing culturability they undergo morphological and physiological changes becoming like MIFs (Garduño et al., 2008). It is possible that the morphological changes of

MIF-like forms reflect the transcriptional changes recently reported by Li et al. (2015a). These include the active induction of genes involved in resistance to antibiotics and oxidative stress. As well as *bdhA*, a gene encoding an enzyme involved in the metabolism of poly-β-hydoxybutyrate (PHB), which is a reserve polymer grossly accumulated in SPFs and MIFs. Only in tap water, not deionized water, can *L. pneumophila* cells in transition become VBNCCs consume the PHB accumulated in their originating SPFs and MIFs (Al-Bana et al. 2014). This suggests that use of PHB is both a strategy to extend cell viability and depends on an uncharacterized component of tap water that is absent in deionized water (Al-Bana et al. 2014). Since the addition of salts to buffered deionized water restores culturability and viability to the levels observed in tap water (see subsection "Metal ions"), metal cations could be linked to the usage of PHB (Al-Bana et al. 2014).

LasM is a putative *L. pneumophila* membrane protein regulated by RpoS which is highly expressed in water (Li and Faucher, 2016). Deletion of LasM-encoding gene, *lpg1659*, results in early loss of culturability of *L. pneumophila* in water, but the Δ*lpg1659* mutant recovered wild-type levels of culturability in the presence of excess trace metals (Li and Faucher, 2016). Collectively, this data suggests that LasM's role in maintaining *L. pneumophila* culturability in water is related to the uptake of metal cations. Therefore, it seems reasonable to hypothesize that (i) metal cations are the factor needed by *L. pneumophila* to utilize PHB, (ii) that LasM is directly or indirectly involved in the uptake of metal cations, and (iii) PHB utilization could be LasM-dependent, pointing at a plausible mechanism by which LasM influences the survival of *L. pneumophila* in water. However, these hypotheses require further experimentation.

MIFs and MIF-VBNCCs are the most abundant planktonic forms of *L. pneumophila* in freshwater environments, but much remains unkown (Garduño, 2008; Robertson et al., 2014). Despite the technical challenges, it is important that studies of *L. pneumophila* in natural freshwater bodies are not neglected and should be conducted under standaridized conditions. The use of a defined freshwater medium, like Fraquil, is a step in the right direction to harmonize studies done with *L. pneumophila* in freshwater (Mendis et al., 2015).

Physicochemical factors

If considering *L. pneumophila* as a bacterial parasite adapted to grow intracellularly within amoeba or other freshwater protozoan hosts rather than a planktonic bacterium adapted to grow as a free-living organism in the freshwater environment, the effect of physicochemical factors might be more direct on *L. pneumophila*'s hosts than on *L. pneumophila* itself.

Temperature

Temperature among the most crucial factors in determining bacterial density in aquatic environments (White et al., 1991; Fliermans, 1996). *L. pneumophila* grows best at warm temperatures between 30 and 40 °C (Yee and Wadowsky 1982). *In vitro*, *L. pneumophila*'s optimal temperature for growth on Buffered Charcoal Yeast Extract (BCYE) agar plates is 36 ± 1 °C (International Organization for Standardization 1998, 2017). However, during *L. pneumophila* isolation, exposure to 50 °C for 30 minutes enhances recovery suggesting that *L. pneumophila* tolerates elevated temperatures (Groothuis and Veenendaal 1983). In fact, it was found that hot water tanks are an excellent reservoir for *L. pneumophila* in man-made water systems where *L. pneumophila* thrives in sediments (Fisher-Hoch and Smith 1982, Wadowsky et al. 1982, Joly et al. 1985). Further confirming *L. pneumophila's* heat tolerance is a study in which a total of 46 strains representing 26 non-*pneumophila* species and seven *L. pneumophila* serogroups were grown on BCYE plates incubated at high temperatures up to 42 °C (Veenendaal et al. 2017). Although most *Legionella* spp. were inhibited at 42 °C, *L. pneumophila* and *L. jordanis* were not affected by the high temperature.

Legionella spp. are not present in waters with temperatures above 63 °C (Fliermans, 1996). When the electron transport activity of *L. pneumophila* underwent assessment with metabolic dye *in vitro*, *L. pneumophila* remained metabolically active over temperatures between 25 and 60 °C (Fliermans et al., 1981b). A study of the growth and respiration of two clinical and two environmental strains through the production of CO_2 indicated that *L. pneumophila* remained metabolically active up to a temperature of 51.6 °C although at a reduced rate (Kusnetsov et al. 1996). However, cell multiplication stopped at temperatures between 48.4 and 50 °C implying that *L. pneumophila* exists in a VBNC state at temperatures ≥50 °C.

Regarding lower temperatures, there is controversy as to whether *L. pneumophila* can replicate at 25 °C or lower. *L. pneumophila* strains have grown in the laboratory at temperatures as low as 12 °C, but it is not clear whether low temperatures in nature are conducive (Söderberg et al., 2004). *Legionella* spp. have been isolated from waters with temperatures as low as 6 °C (Fliermans, 1996). However, this does not imply that *L. pneumophila* can grow at such temperatures. It is possible that in nature where *L. pneumophila* growth depends on the presence of amoebal hosts, that low temperatures limit growth of both amoebal hosts and *L. pneumophila*. This also applies to elevated temperatures where the lack of thermotolerant amoebal hosts could limit the *L. pneumophila* growth (van der Kooij et al., 2016).

The temperature growth range of *L. pneumophila* is likely strain-dependent (van der Kooij et al. 2016). A study on the growth *in vitro* of dominant *L. pneumophila* MLVA-8 genotypes showed that 20 environmental strains isolated from a water distribution system in Israel could grow at temperatures of 25-30 °C. On the other hand, 12 *L. pneumophila* strains isolated from clinical samples showed maximum growth rate and cell densities at temperatures of 37 °C or higher. Although, the 20 environmental strains peaked at temperatures below 37 °C. For strains belonging to a common genotype, Gt4, the clinical Gt4 strains were able to grow significantly better at 37 °C (Sharaby et al. 2017). When naturally occurring *Legionella* was followed from two surface water reservoirs to a potable water distribution system through a processing plant, it was observed that *Legionella* in the source water was more diverse and cold adapted whereas *Legionella* in the potable water system was thermotolerant and less diverse (Lesnik et al., 2016). The study also suggests that differences between cold reservoir and warm distribution systems were due to a selection of different strains or an adaptation of the same strains through the process.

Tolerances to high or low temperatures in *L. pneumophila* have distinct molecular basis. Thermotolerance depends on heat-stress proteins including chaperonins and other chaperone proteins like Hsp70 and Hsp90 (Lema et al. 1988, Hoffman et al. 1989, Fernandez et al. 1996). Growth at low temperatures seems to rely on a different set of proteins including LspDE, LspF and PilD, which are components of the type II secretion system of *L. pneumophila* (Söderberg et al. 2004) as well as PpiB, a secreted peptidyl-prolyl *cis-trans* isomerase different from Mip (Söderberg and Cianciotto 2008). Growth at higher temperatures is associated with filamentation presumbaly due to increased *L. pneumophila* HtpB chaperonin levels in response to elevated temperatures (Konishi et al., 2006; Piao et al., 2006). Over-expression of recombinant HtpB induces filamentation in *L. pneumophila* and other Gram-negative bacteria in a temperature-independent manner (Garduño et al., 2011). In turn, filamentation at higher growth temperatures has been proposed to aid in biofilm formation (Piao et al., 2006).

pH

In the isolation of *L. pneumophila* from water samples an acid treatment of pH 2.2 for five minutes enhances recovery suggesting that *L. pneumophila* survives exposures to solutions of low pH (Bopp et al., 1981; Kusnetsov et al., 1994). This contrasts the finding that a Philadelphia-1 strain of *L. pneumophila* cannot survive exposure to a pH between 2 and 3 in tap water under laboratory conditions (Katz and Hammel, 1987). This also contradicts the findings that when both SPFs grown *in vitro* and MIFs grown intracellularly in HeLa cells do not survive exposures to pH of 2 (Garduño et al., 2002b). These differences could be attributed to the presence of protozoan hosts in crude water samples which protect *L. pneumophila* from low pH.

Alkaline enviornments are not well tolerated by *L. pneumophila*, since treatment of SPFs at pH 11 rendered them unable to grow on agar plates (Garduño et al., 2002b). Survival in tap water was also improved at pH 4-7 rather than at pH 8 (Katz and Hammel 1987). Naturally occurring *L. pneumophila* in water samples are able to grow in batches of filter sterilized tap water in the presence of an associated microbiota at pH's between 5.5 and 9.5 and optimally around a pH of 6 (Wadowsky et al., 1985). However, the growth of the naturally occurring *L. pneumophila* was dependent on the associated uncharacterized microbiota potentially with amoeba. The pH range reported could be reflecting the pH requirement of the associated microbiota more than that of *L. pneumophila* (Wadowsky et al., 1985).

In vitro, *L. pneumophila*'s growth in nutrient-rich or chemically defined media shows a narrow optimal pH range of 6.3-6.9 (Feeley et al. 1979, Pine et al. 1979, Warren and Miller 1979, Ristroph et al. 1980 and 1981). This suggests that although survival is possible at acidic or alkaline pH's, *L. pneumophila* alone has not adapted to grow in a broad pH range. The pH of culture media raises as *L. pneumophila* grows, and the growth of *L.*

pneumophila is favoured in BCYE plates initially adjusted to pH 7.3. Thus, the combination of BCYE at pH 7.3 and incubation at 40 °C was optimal for the isolation of *L. pneumophila* from water samples also containing other *Legionella* spp. (Veenendaal et al. 2017).

Nutrients

As previously stated, *L. pneumophila* is adapted to survive, not grow, in oligotrophic freshwater environments. For growth, *L. pneumophila* requires high concentrations of select nutrients obtained from other freshwater microorganisms or directly from surrounding abiotic milieu. The nutrients critical for the growth of *L. pneumophila* are amino acids, iron, other metal cations of which sodium seems to play a unique role. Carbohydrates are not used by *L. pneumophila in vitro*, but recent studies have shown that they might play a role in the intracellular growth of *L. pneumophila* in protozoan hosts. Finally, the notion that metabolism changes as *L. pneumophila* differentiates between its developmental forms (Manske and Hilbi, 2014; Eisenreich and Heuner, 2016) is in agreement with previous observations suggesting that these forms have unique physiologies linked to virulence (Garduño et al., 2002b; Robertson et al., 2014).

Amino acids

The dependence of *L. pneumophila*'s growth on amino acids was established in studies aimed at formulating chemically defined media to grow *L. pneumophila* extracellularly *in vitro* (Pine et al., 1979; Warren and Miller, 1979; Ristroph et al., 1981). With the advent of whole genome sequencing and the definition of the *L. pneumophila* core genome, it was predicted that *L. pneumophila* cannot synthesize the amino acids Arg, Cys, Ile, Leu, Met, Thr, and Val (Price et al., 2014b). In addition to these seven essential amino acids, *L. pneumophila* seems to also be an auxotroph for the amino acid, Ser (George et al., 1980; Tesh and Miller, 1981). This suggests that Ser could be a major carbon, nitrogen and(or) energy source for *L. pneumophila* (George et al. 1980). In addition, glutamic acid is a non-essential amino acid that stimulates *L. pneumophila* growth (Tesh and Miller, 1981).

Isotopologue profiling is the study of metabolic pathways using metabolites and precursors labeled with ^{13}C, followed by tracking the relative abundances of ^{13}C-labeled compounds or ^{13}C isotopologues with quantitative NMR spectroscopy and(or) mass spectrometry. This approach is a powerful approach to elucidate the utilization of nutrients under defined conditions (Zamboni et al., 2009; Heuner and Eisenreich, 2013). By supplying [U-$^{13}C_3$] serine to *L. pneumophila* strain Paris grown in the defined medium of Ristroph et al. (1981), it was shown that Ser indeed is a major carbon and energy source for *L. pneumophila*. Isotopologue profiling supplied evidence of a major carbon flux from Ser to pyruvate which was metabolized to acetyl CoA and incorporated into the tricarboxylic acid cycle (Eylert et al. 2010).

When purified amino acids from the labeled cells were analyzed there was no detection of ^{13}C in Arg, Ile, Leu, Met and Val which thus confirms *L. pneumophila*'s auxotrophy for these amino acids (Eylert et al. 2010). However, isotopologue profiling did not highlight auxotrophies for Cys or Thr and indicated that *L. pneumophila* Paris could have auxotrophies for Pro, Phe and Tyr which were not previously shown to be essential. The case of Cys is uncertain, as in some studies Cys appears as a dispensable amino acid (Warren and Miller 1979, Tesh and Miller 1981). Though another study demonstrated that *L. pneumophila* Philadelphia-1, Knoxville-1, Paris and Lens are Cys auxotrophs. The Cys added to BCYE is used as a carbon source, which could explain why it is essential despite *L. pneumophila* not being a Cys auxotroph (Ewann and Hoffman 2006). In summary, although *L. pneumophila* has auxotrophies for several amino acids in a strain-dependent fashion, the confirmed core *L. pneumophila* auxotrophies are Arg, Ile, Leu, Met and Val. Although *L. pneumophila* does synthesize Ser *de novo*, it is a major carbon and energy source for *L. pneumophila* essential for growth intra- and extracellularly.

L. pneumophila must get amino acids from its environment, and essential protein building blocks are not readily available in freshwater as soluble nutrients. However, they could be from proteins released by nearby microorganisms through either natural secretion processes or lysis. Therefore, *L. pneumophila* produces extracellular proteases whose presumed function is to digest available extracellular proteins to generate free amino acids (Price et al. 2014b). Intracellularly, *L. pneumophila* obtains amino acids through a multistage mechanism resulting in a direct flux of amino acids into the LCV and their incorporation into *L. pneumophila* replicating cells (Price et al. 2014a; Schunder et al. 2014). The uptake of amino acids from freshwater may be mediated by several membrane transporters belonging to 12 different classes of ABC transporters and amino acid permeases as well as the 16 major facilitator protein genes of the *L. pneumophila* genome (Cazalet et al., 2004; reviewed by Hoffman, 2008). In addition, uptake from the intracellular compartments of *L. pneumophila* host cells requires transporters hypothesized to be on the membrane of the LCV (e.g. SLC1A5, identified by Wieland et al., 2005) (Price et al. 2014b).

Finally, amino acids may impact the ecology of *L. pneumophila* relates through regulation of *L. pneumophila* differentiation. As proposed by Molofsky and Swanson (2004), the regulation of *L. pneumophila* development is linked to concentrations of amino acids in the environment. Briefly, as levels of amino acids decrease the enzyme RelA is activated and ppGpp is synthesized. As the concentration of ppGpp increases, several regulators that mediate the differentiation of *L. pneumophila* into the transmissive forms (RpoS, LetA/S and CsrA) become activated. The RelA/ppGpp-independent loop of gene regulation linked to intracellular growth which is triggered by low levels of Arg sensed by the arginine repressor ArgR which in turn is regulated by RpoS (Hovel-Miner et al., 2009). It has been proposed that the ArgR repressor does more than simply participate in Arg metabolism. The ArgR repressor takes part in the regulation of many factors in stationary phase forms, such as 60 upregulated genes and 58 downregulated genes which are required for intracellular replication (Hovel-Miner et al., 2010; Fonseca and Swanson, 2014). The differentiation of transmissive forms of *L. pneumophila* to replicative forms is also regulated by amino acids. In this case, differentiation is triggered by high amino acid concentrations and mutants unable to sense the presence of amino acids or take up amino acids from the surrounding environment cannot initiate replication (Sauer et al., 2005; Hovel-Miner et al., 2010; Fonseca and Swanson 2014).

Metal ions

As many freshwater bacteria, *L. pneumophila* has specific requirements for metal cations. Compared to *Pseudomonas aeruginosa*, *L. pneumophila* requires more and higher concentrations of the metal cations common to both species (Reeves et al., 1981). Investigators developed a metal cation supplement designed to promote optimal growth in the DM tested containing calcium, cobalt, copper, iron, magnesium, manganese, molybdate, nickel, vanadate and zinc. The metals that had a major effect on growth were iron, zinc, magnesium and calcium (Reeves et al., 1981). In the defined medium of Warren and Miller (1979), several strains of *L. pneumophila* representing 6 serogroups required potassium phosphates and magnesium sulfate to achieve optimal growth (Tesh and Miller 1982).

Metal ions also serve non-nutritional purposes in *L. pneumophila*'s physiology. Calcium and magnesium mediate adherence of *L. pneumophila* cells to surfaces which can have implications in the formation of biofilms (Koubar et al., 2013). The *L. pneumophila* strain JR-32 in tap water and buffered deionized water, which both have a pH of 6.6, loses culturability and shows poor survival in deionized water. In contrast, survival and preservation of culturability was significantly better in tap water where *L. pneumophila* could use PHB inclusions (Al Bana et al. 2014). The addition of calcium nitrate, iron sulfate or sodium fluoride to buffered deionized water at pH 6.6 restored the culturability and survival levels of *L. pneumophila* to levels comparable to those of tap water (Al Bana et al. 2014). Since *L. pneumophila* does not grow in these experimental conditions, it appears that metal cations play a non-nutritional physiological role preserving the viability of *L. pneumophila*. The accompanying anions, such as fluoride or nitrate, may also promote survival of *L. pneumophila* in freshwater.

Iron is essential for the growth of *L. pneumophila*, as core element of *L. pneumophila* enzymes and iron supplementation is mandatory in the formulation of media intended for the isolation of *L. pneumophila* (Mengaud and Horwitz, 1993; Sadowsky et al., 1994). It is likely that in man-made water systems, *L. pneumophila* has access to large amounts of soluble iron and other metals derived from pipes and hot water tanks particularly in corroded systems (States et al. 1985). The mechanisms by which *L. pneumophila* acquires iron from the environment have been thoroughly studied (Cianciotto 2007, 2015). *L. pneumophila* can use ferrous iron (Fe^{2+}) by a direct uptake mediated by FeoB, an integral inner membrane protein, which actively transports Fe^{2+} into the cytoplasm. Access of Fe^{2+} to the periplasm must move via passive transport across the outer membrane through porins or outer membrane channels. As an obligate aerobic bacterium, *L. pneumophila* also acquires ferric iron (Fe^{3+}) via siderophores which are small organic molecules with high affinity for free Fe^{3+}. Siderophores are synthesized and released by bacteria and later recovered through the binding of iron-loaded siderophores to surface receptors. Most *L. pneumophila* strains synthesize legiobactin, a polycarboxylate-type siderophore, structurally identical to rhizoferrin which is a siderophore produced by many fungal species and bacteria (Burnside et al., 2015). LbtA, an enzyme with homology to siderophore synthetases of other bacteria, mediates the synthesis of legiobactin. LbtA likely assembles legiobactin by linking two citric acid moieties via a putrescine bridge (Burnside et al., 2015). The secretion of legiobactin to the extracellular space is mediated by LbtB, and the internalization of iron-loaded legiobactin is mediated by the outer membrane receptor LbtU and the inner membrane transporter LbtC (Cianciotto, 2015).

Further developments in the role of iron in *L. pneumophila* physiology are related to intracellular uptake. After the gene *lpg2959* was found to be important for intracellular growth, functional characterization of the Lpg2959 gene product showed that it is a second receptor of legiobactin, different from LbtU, and was renamed as LbtP (O'Connor et al., 2016). LbtP is important for intracellular growth in macrophages while LbtU is dispensable. LbtP mutants are also more sensitive to iron deprivation in culture media than LbtU mutants which indicates that these receptors have different roles in Fe^{3+} acquisition. Interestingly, neither LbtU nor LbtP mutants show growth defects in the amoebal host, *Acanthamoeba castellanii*, suggesting that intracellular iron acquisition mechanisms used by *L. pneumophila* in the environment differ from those used in mammalian hosts. In addition, iron limitation is linked to a pre-programmed exit strategy from mammalian cells which implies that *L. pneumophila* uses iron levels as an environmental cue to abandon the host cell after nutrients have been exhausted (O'Connor et al., 2016).

A transcriptomics study with *L. pneumophila* strain Paris found a gene encoding a protein with a conserved domain found in membrane transporters which was shown to be important for intracellular growth of *L. pneumophila* in macrophages and amoeba through iron transport (Portier et al., 2015). The gene was named *iroT* for iron transporter, but it had been previously named *mavN* in the Philadelphia-1 strain Lp02 after being identified as a factor for intracellular replication secreted by the Dot/Icm virulence-related type IV secretion system (T4SS) (Isaac et al., 2015). It was concluded that IroT/MavN is indeed a transmembrane transport protein translocated in strain Lp02 by the Dot/Icm T4SS, and it inserts in the LCV membrane to facilitate passage of Fe^{2+} into the lumen (Isaac et al., 2015).

Only Fe^{2+} seems to play an important yet unknown role in biofilm formation (Portier et al., 2016). Attached *L. pneumophila* cells within a monospecies biofilm highly express genes involved in iron acquisition (Hindré et al., 2008). Planktonic *L. pneumophila* cells undergo physiological changes when attaching to a biofilm resulting in the modulation of iron metabolism, induction of iron acquisition mechanisms and a stabilization of *L. pneumophila* in a complex biofilm when Fe^{2+} is chelated (Hindré et al., 2008; Portier et al., 2016).

Sodium tolerance in *L. pneumophila* is related to low virulence, but this mechanism has not been elucidated. Multiple passages of *L. pneumophila* in supplemented Mueller-Hinton agar selects for non-virulent mutants, and NaCl is the implicated selective medium component (Catrenich and Johnson, 1989). Genetic characterization of Na^{+}-tolerant non-virulent mutants aided in the discovery of the Dot/Icm type IV secretion system (T4SS), as the mutants were defective in T4S function (Marra et al., 1992; Vogel et al., 1996 and 1998). Sodium inhibits the growth of virulent *L. pneumophila* only when *L. pneumophila* has differentiated from the actively replicating exponential-phase form (EPF) to the non-replicative SPF (Byrne and Swanson, 1998). The highly infectious MIFs in which T4S is upregulated are also inhibited by sodium (Garduño et al., 2002b). It is plausible that either sodium is an important regulator of Dot/Icm function, the Dot/Icm T4SS is hyper-permeable to sodium or that sodium impacts the structural integrity of the T4SS and the *L. pneumophila* cell envelope. Although, the mechanism by which sodium inhibits growth of differentiated *L. pneumophila* transmissive forms or attenuates virulence has not yet been resolved. Mutations in a variety of *L. pneumophila* genes, different from the *dot* and *icm* genes, result in sodium tolerance with *nudA* encoding a Nudix hydrolase, *lqsR* encoding a response regulator and *clpP* encoding a protease (Edelstein et al., 2005; Tiaden et al., 2007; Li et al., 2010). However, deletion of *potD* encoding a polyamine binding protein results in sodium hypersensitivity (Nasrallah et al., 2014). The functionalities of NudA, LqsR, ClpP and PotD are so diverse that finding a common link to sodium tolerance or sensitivity is difficult. *L. pneumophila* must use sodium in a nutritional or metabolic capacity because, it enhances survival in water when present in low concentrations (Heller et al., 1997).

The inhibitory growth effect of sodium is temperature-dependent, as *L. pneumophila* seems to be more tolerant to sodium at low temperatures. While *L. pneumophila* loses culturability in saline solutions, the apparent "die-off" at temperatures ≥30 °C were not evident in sterile sea water (Heller et al., 1997). Therefore, it was concluded that *L. pneumophila* can survive in marine environments. This is in agreement with a case where *Legionella* spp. were found in sewage-contaminated, euthrophic sea water in Puerto Rico and a study in which *L. pneumophila*-infected amoeba species were isolated from estuarine sediments in both Massachusetts and the Great Salt Lake, Utah USA (Ortiz-Roque and Hazen, 1987; Gast et al., 2011).

Lastly, low sodium serum levels are seen in patients infected with *L. pneumophila* due to increased levels of vasopressin precursors correlating with disease severity and overrulling the osmoregulatory effect of anti-diuretic hormone (Schuetz et al., 2013). Perhaps, this an *L. pneumophila*-induced response in infected humans attempting to establish a favourable environment with low sodium concentration. Many questions about the role of sodium in the biology and ecology of *L. pneumophila* have yet to be answered.

Carbohydrates and other organic substrates

In early studies with DM, several compounds enhanced *L. pneumophila* growth without being essential. These compounds included rhamnose, choline, vitamins, pyruvate, glutathione and α-ketoglutaric acid (Pine et al. 1979, Ristroph et al. 1981). Similarly, metabolic precursors of amino acids such as N-acetylglutamate, a precursor of Arg, were used by *L. pneumophila* to synthesize essential amino acids (Tesh and Miller, 1983). It is plausible that *L. pneumophila* utilizes these organic substrates when available in the freshwater environment. Ketoacids like pyruvate and α-ketoglutaric acid protect *L. pneumophila* from oxygen radicals showing that some organics that favouring *L. pneumophila* growth mat not be used as nutrients (Pine et al., 1986). This may also be the case for pyruvate and glutamate which act as resuscitation enhancers of *L. pneumophila* VBNCCs in BCYE plates (Ducret et al., 2014).

Polyamines are multifunctional compounds, and it is not known how exactly they contribute to the growth of *L. pneumophila*. However, *L. pneumophila* lacks most of the canonical biosynthetic enzymes required to synthesize common biogenic polyamines such as spermidine and putrescine. Putrescine seems to be needed for the synthesis of legiobactin (Burnside et al., 2015). The addition of exogenous polyamines stimulates the intracellular growth of *L. pneumophila* in mammalian cells as well as the extracellular growth of *L. pneumophila* in the DM of Pine et al. (1979) without choline (Nasrallah et al., 2011). Therefore, *L. pneumophila*'s genome encodes a putative polyamine transporter system, PotABCD, to obtain polyamines from the environment. Similar to ArgR, which does more than regulate arginine metabolism, the spermidine-binding protein PotD participates in functions beyond polyamine transport (Nasrallah et al., 2014).

Although it is known that sugars, starch and other carbohydrates are not used for *L. pneumophila* growth *in vitro*, evidence has proved that carbohydrates can be metabolized by *L. pneumophila* during the late stages of intracellular growth. The first evidence that *L. pneumophila* scavenges carbohydrates from natural amoebal hosts came from a transcriptomics study of *L. pneumophila* strains Paris, Lens and Philadelphia-1 in *A. castellanii* (Brüggemann et al., 2006). The genome of *L. pneumophila* has genes involved in carbohydrate metabolism. This was the showed that genes of the Entner-Doudoroff pathway are expressed by *L. pneumophila* when growing inside amoeba. Eylert et al. (2010) demonstrated experimental evidence for the use of glucose by *L. pneumophila* with an isotopologue profiling study. As mentioned above, PHB is accumulated by SPFs and MIFs and is then used to maintain culturability or differentiate into VBNCCs when forms face starvation. PHB is also synthesized from acetyl-CoA derived from serine suggesting that use of carbohydrates may occur late in the intracellular growth cycle after amino acids are exhausted and *L. pneumophila* has begun to differentiate into MIFs. Intracellular glycogen might be a major source of glucose for *L. pneumophila*, and the genome of *L. pneumophila* encodes a variety of secreted amylases (Herrmann et al., 2011; Robertson et al., 2014).

Finally, fatty acids are actively metabolized by *L. pneumophila* and seem to play a role similar to amino acids in the regulation of *L. pneumophila* differentiation (Fonseca and Swanson, 2014). *L. pneumophila* synthesizes a variety of enzymes involved in the metabolism of fatty acids despite that fatty acids are not essential for *L. pneumophila* growth. One explanation is that these enzymes act as virulence factors and help establish *L. pneumophila* in the intracellular environment (Flores-Diaz et al., 2016).

Interaction of *L. pneumophila* with freshwater organisms

Protozoa

Amoebae

L. pneumophila is an intracellular parasite of freshwater amoeba and the literature describing the interactions of amoeba with *L. pneumophila* is abundant (Rowbotham, 1980; Harb et al., 2000; Borella et al., 2005; Molmeret et al., 2005; Albert-Weissenberger et al., 2007; Lau and Ashbolt, 2009; Taylor et al., 2009; Garduño, 2008; Richards et al., 2013; Escoll et al. 2013).

Interactions between *L. pneumophila* and amoebae in the natural environment seem to be sporadic, whereas those occurring in man-made environments such as cooling towers are frequent (Berk et al, 2006). Amoebae isolated from cooling towers are 16 times more likely to be infected with organisms, in general, than amoebae isolated from natural freshwaters. In addition, *L. pneumophila* shows low densities in natural waters, and Legionnaires' disease cases are seldom contracted from natural water reservoirs. Thus, aside from being a portal for exposing humans to *L. pneumophila*, cooling towers and similar man-made systems are "hot-spots" for *L. pneumophila*-amoebae interactions and may contribute to the evolution of *L. pneumophila*, other *Legionella* species and newly emerging pathogens (Berk et al., 2006). *L. pneumophila* strain Lens is equally capable of infecting 12 strains of amoeba. Four strains from each of the following genera: *Acanthamoeba, Naegleria* and *Vermamoeba* were isolated from cooling circuits of various man-made water systems (Dupuy et al. 2016). *Legionella* spp. including *L. pneumophila* have also been found in cultures of marine amoebae of the genera *Vannella* and *Platyamoeba* suggesting that *L. pneumophila* is capable of infecting a wide variety of amoebal hosts (Gast et al., 2011). In addition, once sheltered in the intracellular compartment, *L. pneumophila* can enter ecological niches beyond freshwater.

Not all *L. pneumophila* strains are equally capable of infecting freshwater amoebae which typically colonize man-made systems, namely lab strains of *Acanthamoeba, Naegleria* and *Vermamoeba/Hartmanella vermiformis* (Buse and Ashbolt, 2011). The ability of some *L. pneumophila* strains to infect various amoebal species is temperature-dependent (Buse and

Ashbolt, 2011). In addition, there seems to be particular species and strains of amoebae, such as *Naegleria lovaniensis*, *Willaertia magna*, *Acanthamoeba astronyxis* and close relatives of *Hartmanella* that are resistant to infection by *L. pneumophila* (Declerck et al., 2005; Dey et al. 2008; Amaro et al., 2015). Two mechanisms of amoebal resistance to *L. pneumophila* infection are avoidance in which *L. pneumophila* is not taken up and the second is expulsion which features pelleting of live undigested *L. pneumophila* (Amaro et al., 2015). In the avoidance mechanism, the selective uptake of certain bacterial species has been observed in feeding experiments with protozoa. For instance, *A. castellanii* and the ciliate *Tetrahymena tropicalis* preferentially ingested *L. pneumophila* over other bacteria (Declerck et al., 2005; Berk et al.,2008). Under certain conditions, common freshwater bacteria do not compete with *L. pneumophila,* and may even share food vacuoles with *L. pneumophila* or influence the intracellular replication process of *L. pneumophila* in a species-dependent manner (Declerck et al., 2005). However, the underlying mechanism by which amoebae selectively take up *L. pneumophila* or take up other bacteria at the expense of ignoring *L. pneumophila* is unknown. There are, however, specific receptors on different amoebal species that mediate the capture and trigger internalization of *L. pneumophila* (Venkataraman et al., 1997; Abu Kwaik et al. 1998b; Declerck et al., 2007a). These receptors interact with specific ligands on the surface of *L. pneumophila* cells such as RtxA or the 16-kDa protein promoting uptake in amoebae (Cirillo et al., 2002; Steudel et al., 2002). The mechanism of pelleting of live *L. pneumophila* by amoebae also remains unexplained except for the notion that amoebal strains capable of packaging *L. pneumophila* may have faster turnover rates of their food vacuoles particularly before encystation (Berk et al., 1998; Schuster, 1979).

A third group of protozoa including the amoeba, *Solumitrus palustris* which is a heterolobosean amoeba related to *Naegleria,* thrive in the presence of *L. pneumophila,* because they can efficiently ingest and digest *L.*

pneumophila (Amano et al., 2015). Aside from *S. palustris*, other non-identified heteroloboseans and several amoeboflagellates of the genera *Paracercomonas* and *Cercomonas* thrive when consuming *L. pneumophila* as food (Amaro et al., 2015). These protists that consume *L. pneumophila* could mediate the evolution of *L. pneumophila*'s pathogenic traits resulting in the high redundancy of virulence effectors in *L. pneumophila*. While some species of amoeba are permissive and support the vigorous intracellular replication of *L. pneumophila*, other species or strains are resistant to infection and can even use *L. pneumophila* as food. Collectively, in any freshwater ecosystem, a complex mosaic of *Legionella* strains differentially interact with a variety of amoebae resulting in varying outcomes depending on environmental variables such as temperature, pH, nutrient concentrations and the presence of other aquatic bacteria and non-amoebic protozoa.

Lastly, as *L. pneumophila* interact with amoebae, single-membraned vesicles holding live *L. pneumophila* are produced which are distinct from multi-lamellar structured pellets (Bouyer et al., 2007). In early references to multi-membranous pellets they were named "vesicles" (Berk et al., 1998; Berk et al., 2008; Denoncourt et al., 2014). Instead, the single-membraned vesicles seem to be morphologically equivalent to abundant free-floating vesicles produced by *L. pneumophila*-infected HeLa cells which contain live *L. pneumophila* MIFs surrounded by a continuous single-membrane (Bouyer et al., 2007; Garduño et al., 1998). These vesicles may be equivalent to those hand-drawn by Rowbotham (1983) as a possible outcome of *Acanthamoebae polyphaga* infection with *L. pneumophila* strain Leeds-1A SAP or an *L. pneumophila* serogroup six strain. Vesicles produced by HeLa cells may infact be non-ruptured LCV's that emerge from lysed host cells (Garduño et al., 1998). Vesicles produced by amoeba might contain *L. pneumophila* cells that did not fully activate a normal amoebal exit mechanism perhaps as a consequence of forced rapid turnover of intracellular vacuoles by a starving or prematurely dead host (Gao and Abu Kwaik, 2000). However, since the *L. pneumophila* cells contained in amoeba vesicles are protected from environmental stress and retain culturability better than free planktonic cells, it seems reasonable that these vesicles represent an alternate *L. pneumophila*-induced exit strategy (Bouyer et al., 2007). The formation of vesicles is an outcome of *L. pneumophila*-amoeba interactions that may be yet another planktonic form in which *L. pneumophila* can exist in freshwater.

Ciliates

Regarding non-amoebic protists, *L. pneumophila* grows in ciliates of the genus *Tetrahymena* in a temperature-dependent manner restricted to >30 °C (Berk et al., 2008 and references within). Instead of acting as hosts supporting the intracellular growth of *L. pneumophila*, ciliates seem to provide a sheltered yet transient intracellular environment relying on *L. pneumophila*'s ability to survive digestion inside food vacuoles. Ciliates also mediate the packaging of planktonic *L. pneumophila* forms into spherical pellets. The packaging of bacterial pathogens by protozoa enhances the environmental fitness of the packaged pathogens. This seems to occur more often than previously thought implying that bacterial survival in food vacuoles is a common adaptation towards pathogenesis (Denoncourt et al., 2014). In the presence of abundant bacteria, ciliates enter a feeding frenzy producing excessive amounts of pellets to store food for future use (Hojo et al., 2012). Since whole pellets are too large to be ingested by ciliates, the consumption of stockpiled pellets involves the gradual degradation of pellet structure likely aided by secreted protozoal enzymes. Therefore, the material wrapping the packaged bacteria would be partly of protozoal origin and when needed, it can be degraded and reutilized (Denoncourt et al., 2014).

When *Tetrahymena* ciliates ingest *L. pneumophila* SPFs, these SPFs differentiate into a form morphologically indistinguishable from MIFs produced in amoebae or in HeLa cells. This differentiation is rapid, and the newly formed MIFs are expelled by the ciliates packaged in pellets (Faulkner et al., 2008). Whether these MIFs and those produced in amoebae have a similar physiology remains to be determined. However, at least the MIFs produced in ciliates are more infectious and resistant to antibiotics than SPFs as are the MIFs produced in amoeba (Koubar et al., 2011; Garduño, 2008). However, *Tetrahymena* pellets MIFs produced in amoebae as well as *L. pneumophila* VBNCCs produced from SPFs or from amoebae-grown MIFs (McNealy et al., 2002; Al-Bana et al., 2014). Thus, what emerges from the interaction of *L. pneumophila* with ciliates is predominantly packaged MIFs or VBNCCs. Lastly, ciliates of the genera *Oxytricha*, *Stylonychia* and *Ciliophrya* all ingest *L. pneumophila,* although intracellular growth or pelleting has not been observed (Rasch et al., 2016). It is important that ciliates other than *Tetrahymena* can provide an intracellular environment for *L. pneumophila*.

Flagellates

It previously was unclear if flagellates would play a significant role in the ecology of *L. pneumophila* (Amaro et al., 2015). Now it is known that these protists can consume *L. pneumophila* as food. However, there is not indication that amoeboflagellates support the intracellular replication of *L. pneumophila*. Therefore, it should be assumed that the protozoa supporting the intracellular replication of *L. pneumophila* in water environments are primarily amoebae with some ciliates.

Freshwater bacteria and algae

It is uncertain whether *L. pneumophila* multiplies extracellularly in natural freshwater environments (Taylor et al., 2009). Undoubtedly, *L. pneumophila* can grow extracellularly in nutrient-rich, solid or liquid media. However, whether single ecological niches of freshwater environment could meet these requirements is unknown.

L. pneumophila can grow extracellularly in association with green unicellular algae or cyanobacteria and may even physically attach to these organisms (Hume and Hann, 1984a; Tison et al., 1980; Pope et al., 1982; Bohach and Snyder, 1983b; Hume and Hann, 1984b). An insoluble slime produced by the cyanobacterium *Fisherella* was purified and fractionated, and two of the fractions stimulated O_2 consumption by a clinical *L. pneumophila* serogroup 1 isolate as measured in a respirometer (Bohach and Snyder, 1983a). Furthermore, *L. pneumophila* assimilates radiolabelled compounds present in exudates produced by cyanobacterium *Fisherella* grown in the presence of $^{14}CO_2$ (Tison, 1987). *L. pneumophila* respired $^{14}CO_2$ which constitutes up to 40% of the total CO_2 production suggesting that *L. pneumophila* used unidentified algal compounds as a source of energy. However, these organic exudates alone were not sufficient to support the growth of *L. pneumophila* as the experiments were done in media supplemented with yeast extract (Tison, 1987). Cyanobacterial exudates and products, however, also protect *L. pneumophila* against desiccation in a nutrient-independent fashion (Berendt, 1981).

Other nutritional associations between *L. pneumophila* and microorganisms are nutritional symbioses and necrotrophic growth. Prokaryotes such as *Flavobacterium* and pseudomonads from freshwater environments, man-made systems and hot water tanks can stimulate *L. pneumophila* growth. A bacterial community could stimulate growth of naturally occurring *L. pneumophila* in sterile tap water (Wadowsky and Yee, 1985). In particular, using BCYE plates without cysteine, *L. pneumophila* colonies can grow as satellites around colonies of freshwater bacteria (Wadowsky and Yee, 1983; Stout et al. 1985). With the same approach, inhabitants of the human respiratory tract such as *Haemophilus influenza* and *Neisseria meningitidis* can also be nutritional symbionts for *L. pneumophila* (Stout et al. 1986). This proves that live bacteria growing in proximity to *L. pneumophila* can supply essential nutrients including amino acid cysteine. These reports, however, cannot determine if nutrients are secreted by live cells or released by lysis of dead cells.

The necrotrophic ability of *L. pneumophila* have been described (Temmerman et al., 2006). Heat killed organisms including bacteria, amoebae and yeast were added to a *L. pneumophila* serogroup 1 strain ATCC 33152 in sterile tap water, and *L. pneumophila* growth was monitored by culture and by qPCR of the *mip* gene. *L. pneumophila* replicated to various extents depending on the killed organism used to stimulate growth. However, the Gram-positive bacteria, *Bacillus* and *Lactobacillus*, were poor stimulators. A heat killed complex biofilm originally obtained from coupons installed in a cooling tower water circuit stimulated necrotrophic growth of *L. pneumophila* showing that *L. pneumophila* is capable of deriving essential nutrients from a variety of dead cells (Temmerman et al. 2006). Necrotrophic growth was less vigorous than intracellular growth in *A. castellanii* in general. While LCVs were being established, necrotrophic growth resulted in *L. pneumophila* numbers larger than those in *A. castellanii* and with a ratio of 100 dead stimulating bacteria for every *L. pneumophila* cell (Temmerman et al., 2006). Although the major contributor to *L. pneumophila*'s numbers in the experimental system used was growth in amoebae, necrotrophic growth was not negligible. Similar experiments in freshwater environments are necessary with distinguished *L. pneumophila* inputs resulting from intracellular and extracellular growth. Two questions remain regarding whether growth requirements of *L. pneumophila* can be met in a single freshwater ecological niche and whether *L. pneumophila* replicates extracellularly in freshwater to meaningful levels have yet to be answered.

Various microorganisms, particularly Gram-positive cocci, inhibit the growth of *L. pneumophila* (Flesher et al., 1980). The growth inhibitory factor of a *Streptococcus* strain was concentrated from a culture filtrate, and it was found to be smaller than 1-kDa and resistant to heat and proteolysis (Flesher et al., 1980). Although the factor was not identified, it was likely a bacteriocin or a bacteriocin-like peptide, as these compounds have been found to have anti-*L. pneumophila* activity (Verdon et al., 2008; Marchand et al., 2011 and 2015). Inhibition of *L. pneumophila* growth by bacteria isolated from man-made water systems has also been observed (Toze et al., 1990). Two bacterial strains that inhibited growth of several *Legionella* species on BCYE plates including *L. pneumophila* also stimulated growth of the same legionellae on plates lacking cysteine (Toze et al. 1990). This suggests that some bacteria stimulating *L. pneumophila* growth could be inhibitory under certain conditions (Wadowsky and Yee, 1983,1985; Stout et al., 1985, 1986). The genus, *Aeromonas,* are common inhabitants of freshwater systems, and many *Aeromonas* spp. strains inhibit *L. pneumophila* growth. Though, all *Aeromonas* strains do not uniformly affect different *Legionella* spp. (Toze et al., 1994). This shows that the interactions leading to growth inhibition are highly heterogeneous. It is not uncommon to find sporadic bacterial or fungal contaminants on BCYE plates, and they can produce a zone of inhibition within which colonies of *L. pneumophila* cannot grow. Therefore, it seems reasonable that in natural environments there must be freshwater microorganisms capable of inhibiting the extracellular growth of *L. pneumophila.*

Biofilms

Biofilms are complex and adherent structures usually colonized with multiple species built entirely by microorganisms on solid-water interfaces. When biofilms are formed at water-air interfaces they have a different structural organization and are referred to as floating biofilms. In the laboratory, mono-species biofilms can be developed and studied, but these have limitations when applying findings to natural environments. The interactions of *L. pneumophila* with biofilms or floating biofilms have been thouroughly reviewed (e.g. Lau and Ashbolt, 2009; Declerck, 2010; Abdel-Nour et al., 2013). Despite that *L. pneumophila* replication depends on the presence of amoeba, planktonic *L. pneumophila* can colonize pre-existing biofilms and survive within the complex microbial community. Biofilms, in turn, provide a protective haven from which *L. pneumophila* evades environmental stresses, in particular, those in man-made water systems such as disinfection. This section will focus on the extracellular growth of *L. pneumophila* in freshwater environments

L. pneumophila can form mono-species biofilms under two models. In the first model, the biofilm forms as *L. pneumophila* grows extracellularly in a nutrient-rich medium (Piao et al., 2005; Mampel et al., 2006). In the second model, a large inoculum of *in vitro* grown cells forms the biofilm in filtered tap water by becoming sessile and without necessarily replicating (Andreozzi et al., 2014). Regarding the second model, *L. pneumophila* does not always form monospecies biofilms in tap water, particularly in flow reactors (Fields and Lucas, 2006). Collectively, this means that *L. pneumophila* has the capabilities to (i) attach to surfaces, (ii) change physiology and gene expression patterns to adjust from a planktonic survival lifestyle to a sessile lifestyle and (iii) build mature structures through quorum sensing in a temperature-dependent manner (Hindré et al., 2008).

Planktonic *L. pneumophila* can actively multiply in the nutrient-rich broth while sessile *L. pneumophila* does not replicate despite the ample available nutrients (Mampel et al., 2006). The physiological change from a planktonic to a sessile lifestyle is linked to differentiation towards quiescence. However, sessile *L. pneumophila* cells might replicate within a monospecies biofilm if fed by a medium that poorly supports the growth of planktonic *L. pneumophila* (Pécastaings et al., 2010). It is possible that the physiological state of sessile *L. pneumophila* cells in monospecies biofilms fed by a nutrient-rich medium is an intermediate between replicating planktonic cells and stationary phase forms (SPFs) as determined by gene expression patterns (Hindré et al., 2008; Andreozzi et al., 2014). In this intermediate state, resistance to oxidative stress and iron acquisition seem to be functionally important (Hindré et al., 2008). Upon introduction of amoebae into a static, non-growing biofilm of sessile cells, *L. pneumophila* can replicate and the biofilm grows despite amoebae not being part of the biofilm (Mampel et al., 2006). This suggests that amoebae-grown *L. pneumophila* cells, MIFs, contribute significantly to biofilm build-up by adhering to them. It is not clear, however, whether MIFs underwent some form of differentiation upon attachment to the biofilm. It has been suggested that upon attachment to an existing biofilm, MIFs produce exopolysaccharides and a chemotactic factor (Bigot et al., 2013). Finally, differentiation of *L. pneumophila* into multinucleated filaments in a monospecies biofilm system and the formation of mat biofilms was obvious at temperatures higher than 37 °C (Piao et al., 2006). The formation of multinucleated filaments could be a form of delayed replication, because under favorable conditions such as in nutrient rich medium or in the presence of amoebae, filaments could fragment and give rise to numerous short *L. pneumophila* cells (Piao et al., 2006; C.C. Sze, Nanyang Technological University, Singapore).

L. pneumophila survives but does not replicate in the absence of live amoebae in complex multi-species biofilms formed in model freshwater-based systems (Murga et al., 2001; Kuiper et al., 2004; Declerck et al., 2009). In these biofilms, despite an abundance of non-legionellae bacteria that could potentially stimulate *L. pneumophila*'s growth, the numbers of *L. pneumophila* did not increase unless live amoebae species were introduced (Declerck et al., 2009). Thus, planktonic *L. pneumophila* can colonize existing biofilms and find shelter. Nonetheless, *L. pneumophila* must rely on intracellular replication to maximize abundance.

L. pneumophila can replicate extracellularly within biofilms (Rogers and Keevil, 1992). However, incorporation of planktonic cells into biofilms was only suggested by later studies. Regarding microcolonies, neither amoebae-grown *L. pneumophila* MIFs nor MIFs differentiated in *Tetrahymena* food vacuoles integrate into existing biofilms as clusters or microcolonies (Bigot et al., 2013). When *L. pneumophila* expressing red or green fluorescent proteins were added as planktonic cells, they were incorporated into biofilms as clusters of green and red cells in the same proportion which they were added. This suggests that the microcolonies had not formed by clonal replication of *L. pneumophila* cells within the biofilm (Bigot et al., 2013). Integration of *L. pneumophila* cells into biofilms can be mediated with quorum sensing. *L. pneumophila* has a quorum sensing mechanism, but *L. pneumophila* can also respond to autoinducers produced by other bacterial species and change its gene expression patterns (Tiaden and Hilbi, 2012; Zeigler-Ballerstein and Barbaree, 2006). This could explain why *L. pneumophila* can find pre-existing biofilms produced by non-legionellae bacteria, or why in some experimental conditions *L. pneumophila* cannot form biofilms in the absence of a complex bacterial community (Fields and Lucas, 2006).

Contrary to the most common forms of biofilms which grow on a solid-water interface, floating biofilms are formed at the water-air interface (Declerck, 2010). *L. pneumophila* is found in floating biofilms either from man-made water systems such as cooling tower basins or natural freshwater environments such as ponds despite that predominant bacteria in these biofilms were *Pseudomonas* spp. and *Aeromonas* spp. Importantly, amoeba species known to support the intracellular growth of *L. pneumophila* were also found, and some of these amoebae were infected with *L. pneumophila* (Declerck et al., 2007b). Therefore, in floating biofilms the growth of *L. pneumophila* also happens intracellularly. In floating biofilms formed under controlled laboratory conditions, using a mixture of *Aeromonas hydrophila*, *Escherichia coli*, *Flavobacterium breve* and *Pseudomonas aeruginosa*, *L. pneumophila* colonized from a planktonic inoculum (Declerck et al., 2007c). The concentration of *L. pneumophila* in floating biofilms remained stable unless *A. castellanii* was introduced. At 48 hours after the introduction of amoebae, 90% of *A. castellanii* cells were infected, and the *L. pneumophila* numbers had increased more than 100-fold (Declerck et al., 2007c). It is more likely to find free-living amoebae in floating biofilms than in conventional biofilms sampled from the same spring water environments (Hsu et al., 2011). Moreover, *Legionella* spp., including *L. pneumophila*, were found in approximately 27% of floating biofilm samples but only in 3% of fixed biofilm samples. Collectively, the data suggests that in relation to conventional biofilms, floating biofilms might be a more suitable environment for *L. pneumophila* from which generation of *L. pneumophila*-laden aerosols would be more efficient.

A non-virulent *L. pneumophila* strain which cannot grow intracellularly can grow extracellularly in a biofilm formed from a natural inoculum growing in a local mains water supply (Surman et al., 2002). The mutant strain obtains the nutrients required for growth from the biofilm, but non-virulent mutants of *L. pneumophila* seem to have different nutritional requirements than their virulent counterparts. Non-virulent *L. pneumophila* forms seem to be naturally counter selected in freshwater environments, as non-virulent *L. pneumophila* are severely predated, ingested and digested by protozoa. When cycloheximide, a drug that inhibits eukaryotic protein synthesis to reduce amoebal populations in biofilms, the growth of the non-virulent mutant increases suggesting reduced predation (Surman et al., 2002). *L. pneumophila* in freshwater environments are under a strong selective pressure to differentiate into MIFs. Strains that cannot differentiate into SPFs or MIFs are digested by amoeba and ciliates while they appear to grow well in mammalian host cells (Faulkner et al., 2008; Robertson et al., 2014). Since differentiation into MIFs does not happen extracellularly it is likely that in nature, *L. pneumophila* is under selective pressure to grow intracellularly to complete its developmental cycle (Garduño et al., 2002b; Garduño et al., 2008). In total, the evidence is clear that virulent *L. pneumophila* does not meaningfully multiply extracellularly in the environment.

Legionella longbeachae and *L. pneumophila* in the soil environment

The species *L. longbeachae* is the leading cause of Legionnaires' disease in Australia and New Zealand (Whiley and Bentham, 2011; Graham et al., 2012). The primary mode of transmission of *L. longbeachae* to humans is also unique, as *L. longbeachae* is transmitted from soil rather than from water (Currie and Beattie, 2015). Both *L. longbeachae* and *L. pneumophila* can survive and potentially thrive in both water and soil (Potočnjak et al., 2016; Thornley et al., 2017). Unfortunately, there is a scarcity of studies on this topic (e.g. Amemura-Maekawa et al., 2012; van Heijnsbergen et al., 2014, 2016). It remains unknown which *L. pneumophila* strains are prevalent in soil and how soil characteristics such as pH, composition, temperature and humidity relate to *L. pneumophila* growth and whether *L. pneumophila* can move from soil into water envionrments (Sakamoto et al., 2009; van Heijnsbergen et al., 2014).

Legionella spp. interacts with soil nematodes, and it was first demonstrated that nematodes ingest *L. pneumophila* and *L. longbeachae* added to sterile soil. The ingested *Legionella* cells then colonize their intestinal tract lumen (Brassinga et al., 2010). Pulse-chase experiments with fluorescent *L. pneumophila* cells expressing either mCherry or GFP suggested that *L. pneumophila* multiplies extracellularly in the intestinal tract of nematodes. Both *Legionella* species grossly accumulated in the nematodes' intestinal tract and produced differentiated forms that were morphologically like MIFs and assumed to be expelled by the nematodes into the soil (Brassinga et al., 2010). *L. pneumophila* is persistent in the intestinal tract of nematodes and colonization of nematodes of elderly age specifically results in severe pathology. In addition, bifidobacteria protect nematodes from the negative effects of *L. pneumophila* colonization (Komura et al., 2010). Although most *L. pneumophila* cells ingested by nematodes remain in the lumen of the intestinal tract, few *L. pneumophila* cells interact with the intestinal microvilli and even penetrate the lining of epithelial cells (Hellinga et al., 2015). Intracellular replication was only equivocally demonstrated while formation of LCVs in extraintestinal locations has been strongly suggested and differentiated *L. pneumophila* cells of unique morphologies have also been identified (Hellinga et al., 2015). Finally, nematodes can ingest *in vitro* grown *L. pneumophila* cells as well as MIFs that emerg from infected amoebae which then colonize the intestinal lumen. Interestingly, the infected amoebae do not survive past the nematodes' pharynx where they are mechanically disrupted (Rasch et al., 2016). The surface *L. pneumophila* protein Mip also plays a role in nematode colonization, and this provides an insight into the mechanisms underlying *L. pneumophila* nematode interaction. Although these studies were done in floating aquatic biofilms, it is possible that similar microbial interactions occur in soil. Collectively, nematodes may be an important reservoir of *L. pneumophila* in soil.

Ecological aspects of developmental *L. pneumophila* forms in freshwater

L. pneumophila is a pleomorphic organism that differentiates into several forms within a developmental network. In total,14 morphological or physiological *L. pneumophila* forms have been indentified (Robertson et al. 2014). Nine of the identified forms are relevant to the freshwater ecology of *L. pneumophila*, namely, MIFs, MIF-VBNCCs, MIF pellets, MIF-VBNCCs pellets, MIF-RPF intermediate, replicative phase forms (RPFs), RPF-MIF intermediate, filamentous forms (FFs) and MIFs contained in vesicles which are counted as a new *L. pneumophila* form (Figure 2). These forms all interact differently with the abiotic and biotic components of the freshwater environment.

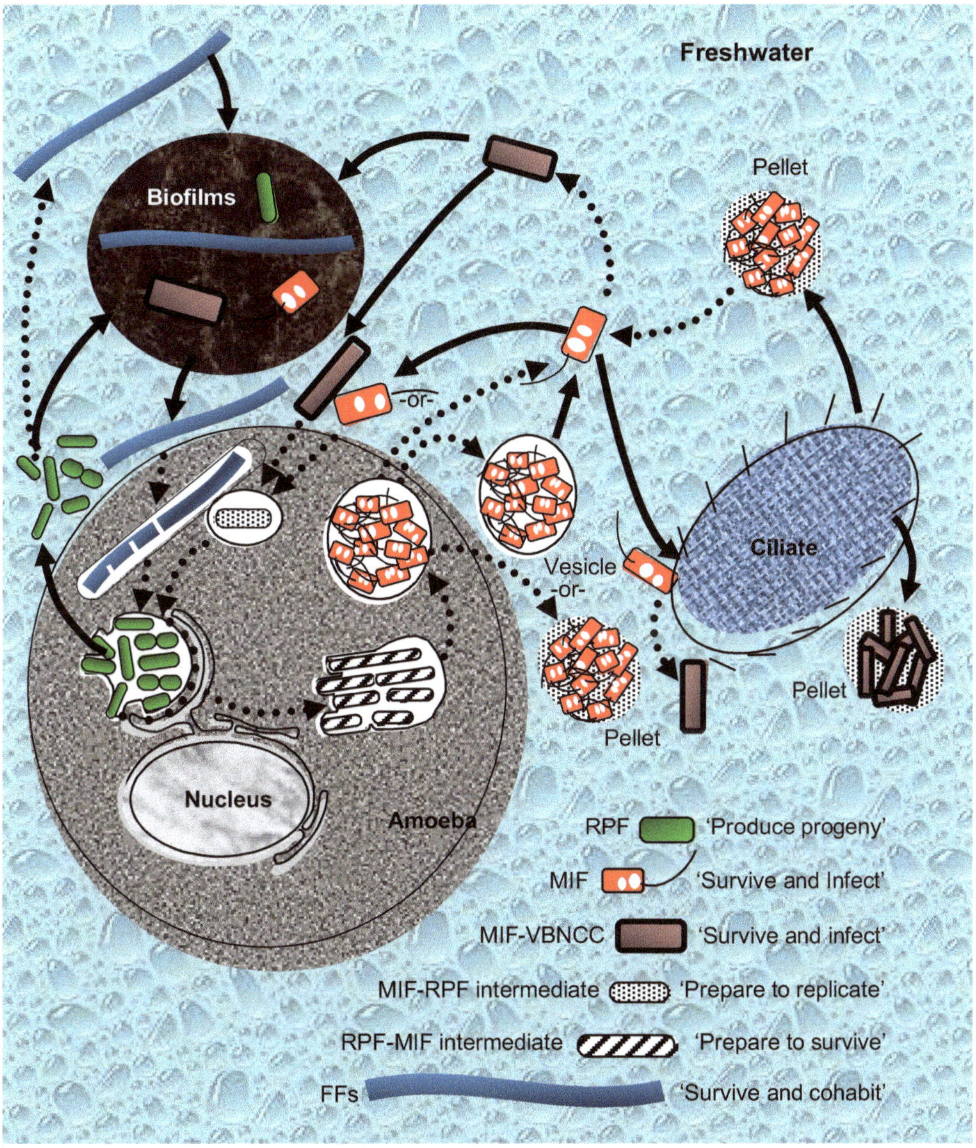

Figure 2. Schematic representation of *L. pneumophila*'s developmental network showing developmental and morphological forms relevant for the ecology of *L. pneumophila* in the freshwater environment (background blue square). The interactions (black solid arrows) or developmental links (black dotted arrows) are different for each of the forms. The MIFs are the most abundant form in this environment and have the most interactions and developmental links. MIFs could be present as free planktonic forms, sessile forms in biofilms or packaged into pellets or vesicles. MIFs can produce VBNCCs in response to environmental stress. VBNCCs can also be packaged into pellets by ciliates, become sessile in biofilms or infect amoebae. RPFs are likely released from infected and lysed amoebae. These RPFs could integrate into biofilms and give rise to FFs.

MIFs and the various MIF presentations

The progeny of *L. pneumophila* produced in freshwater are the result of intracellular replication in amoebae. Consequently, since intracellular replication of *L. pneumophila* in amoeba is linked to differentiation into MIFs, the bulk of new *L. pneumophila* cells incorporated into the environment must be MIFs. Therefore, MIFs are the developmental *L. pneumophila* forms released from wasted amoebae. These developmental forms were first noticed as short and transiently motile *L. pneumophila* cells appearing upon lysis of infected host amoebae (Rowbotham, 1983). Since authors have determined that MIFs from amoebae are resistant to biocides, have altered fatty acid and protein profiles, show bright red staining with Giménez stain, have increased infectivity to mammalian cells and amoebae, have increased virulence towards mice, are more resistant to antibiotics, have increased environmental fitness, show a unique cellular ultrastructure and produce a high yield of VBNCCs. (Barker et al., 1992; Barker et al., 1993; Cirillo et al., 1994; Abdelhady and Garduño, 2013; Brieland et al., 1997; Cirillo et al. 1999; Barker et al., 1995; Abu Kwaik et al., 1997 and 1998a; Greub and Raoult, 2003; C. Lima and R. Garduño, unpublished results; Al-Bana et al., 2014. It is clear from these studies that MIFs are well equipped to withstand environmental stress and infect new hosts.

MIFs do not utilize nutrients nor replicate, and thus MIF interactions with the abiotic component of freshwater could be characterized as quiescent but efficient at survival against stress. In interactions with the biotic component of freshwater, planktonic MIFs can incorporate into biofilms (Mampel et al., 2006; Declerck et al., 2007b and 2007c; Bigot et al., 2013). As highly infectious forms, MIFs would infect new amoebal hosts either as motile planktonic cells or as attached cells within biofilms. MIFs in biofilms do not grow or acquire nutrients. However, when phagocytosed by amoeba they initiate a growth cycle. If planktonic MIFs are ingested by ciliates, they survive digestion to be packaged and re-integrated into the freshwater environment as MIF pellets (McNeally et al., 2002; Berk et al., 2008). MIF pellets can be recycled by ciliates or phagocytosed whole by amoebae (Hojo et al., 2012). When MIFs are unable to find a new host or face unfavourable environmental conditions such as high temperature, starvation or presence of biocides, they may enter the VBNC state. MIF-VBNCCs can persist for extended periods in the freshwater environment and be packaged by ciliates into pellets (Al-Bana et al., 2014). MIF-VBNCCs can also be ingested by amoeba after which they initiate a growth cycle. In fact, infection of amoebae was the first reported method to resuscitate *L. pneumophila* VBNCCs (Steinert et al., 1997). Finally, similarly to MIF pellets, MIF vesicles have increased fitness and withstand desiccation better than free unpackaged MIFs or MIF-VBNCCs. The freshwater ecology of MIFs and their related forms could therefore be simply defined as 'survive and infect'.

The MIF-RPF intermediate

When MIFs, MIF-VBNCCs, MIF pellets or MIF-VBNCC pellets are phagocytosed by amoeba, they start a growth cycle. However, they must avoid digestion inside phagosomes, delay phagosome-lysosome fusion and initiate differentiation into replicative forms through an MIF-RPF intermediate. Although the physiological activation of MIFs could happen within complex biofilms in freshwater, it is possible that MIF-RPF intermediates only form intracellularly in amoebae assumming that no *L. pneumophila* form present in freshwater can replicate or differentiate into a replicative form extracellularly.

Morphological characterization of MIF-RPF intermediates has not been accomplished, as their differentiation process is fast and difficult to capture by electron microscopy (Faulkner and Garduño, 2002). However, intermediates called exponential phase forms (EPFs) produced by the differentiation of MIFs into replicative forms in culture broth *in vitro* are characterized by a mixed Giménez staining. The presence of large intraperiplasmic vesicles and a unique envelope ultrastructure suggest that they are a distinct developmental form of *L. pneumophila* (Garduño et al., 2002a; Faulkner and Garduño, 2002). Similar morphological changes may characterize the differentiation of MIFs into MIF-RPF intermediates. The trigger for differentiation is the presence of nutrients in particular, amino acids (Sauer et al., 2005). Differentiation results in the breakdown of ppGpp by SpoT, downregulation of LetS activity and increased activity of CsrA collectively resulting in low expression of virulence traits and increased expression of genes involved in metabolism (Molofsky and Swanson, 2004; and Robertson et al., 2014). After MIFs delay phagosome-lysosome fusion, the MIF-RPF intermediates equipped to establish the early LCV attract ER-derived vesicles as well as mitochondria to the LCV membrane and set the influx of nutrients into the LCV lumen. The MIF-RPF intermediate ecology is spatially restricted to the interaction with amoeba phagosomes and could be summarized as 'prepare to replicate'.

RPFs

The primary mission of RPFs is to produce *L. pneumophila* progeny. The functions of active biosynthetic metabolism, protein synthesis, genome replication and cell division are restored after shutting down in the MIFs as the MIF-RPF intermediates differentiate into RPFs. The RPF deploys factors and mechanisms which *L. pneumophila* uses to get nutrients from the infected host cell. The RPF's ecology is spatially restricted to the established LCV within amoebae and is defined as 'produce progeny'.

RPF-MIF intermediate

As intracellular nutrients are exhausted and *L. pneumophila* host cells become wasted, SpoT and RelA activate and the concentration of ppGpp increases inside *L. pneumophila*. DksA and ppGpp act in concert to trigger the signaling cascade regulating virulence trait expression and gradually shutting down active metabolism, protein synthesis and cell division

(Molofsky and Swanson, 2004; Robertson et al., 2014). This change in gene expression is accompanied by a metabolic shift into usage of carbohydrates, synthesis of PHB and morphological differentiation of *L. pneumophila* cells characterized by ultrastructural changes in their cell envelope (Eylert et al., 2010). RPF-MIF intermediates typically are long rods which stain purple after the Giménez stain and have a dense cytoplasm. They also begin to display inclusions of PHB, invaginations of the inner membrane and outer membrane either appearing smooth and straight or with sharp ripples in ultra-thin sections in transmission electron microscopy (Faulkner and Garduño, 2002; Abdelhady and Garduño, 2013).

RPF-MIF intermediates are *L. pneumophila* progeny prematurely released from amoebae that die before *L. pneumophila* can complete differentiation into MIFs, and they are also initially contained in MIF vesicles (Bouyer et al., 2007). RPF-MIF intermediates plausibly complete differentiation within released vesicles or as planktonic cells in freshwater. Once RPF-MIF intermediates differentiate into MIFs, the developmental cycle of *L. pneumophila* is completed and a new one can begin. Therefore, the ecology of the RPF-MIF intermediates is spatially restricted to interactions with late LCVs and could be summarized as 'prepare to survive'.

FFs

Although FFs are formed by *L. pneumophila* under stressful conditions, the mechanisms controlling their production are still an enigma. It is common to see FFs or long rods in direct microscopy preparations of concentrated samples of tap water labeled with an *L. pneumophila*-OmpS-specific antibody (unpublished results). Though, it is unkown where these FFs come from. *In vitro*, FFs are formed from SPFs, but also replicating cultures of EPFs in defined media can be filamentous (Pine et al., 1979; Warren and Miller, 1979; Ristroph et al., 1981). Since RPFs are not expected to exist in significant numbers in the freshwater environments,

Table 1. Rationale supporting the notion that *L. pneumophila* does not grow extracellularly in the freshwater environment.

	Rationale	**References / subsection**
Facts	*L. pneumophila* is overgrown by other freshwater bacteria in biofilms.	Gião et al., 2009
	There are many microbes capable of inhibiting the extracellular growth of *L. pneumophila*.	Flesher et al., 1980; Verdon et al., 2008
	Under certain conditions, bacteria that support nutritional symbioses with *L. pneumophila* can be inhibitory.	Toze et al. 1990
	Sessile *L. pneumophila* cells in monospecies biofilms do not replicate.	Mampel et al., 2006
	At high temperatures, multinucleated *L. pneumophila* filaments form biofilms but do not replicate unless placed in fresh nutrient-rich media.	Piao et al., 2006
	In complex, multi-species biofilms *L. pneumophila* does not replicate in the absence of live amoebae.	Murga et al., 2001; Kuiper et al., 2004; Declerck et al., 2009
	Amoeba-grown MIFs incorporate into biofilms without replication.	Bigot et al., 2013
	The growth of *L. pneumophila* in floating biofilms depends on the presence of amoebae.	Declerck et al, 2007b; Declerck et al, 2007c
Arguments	*L. pneumophila* is well equipped to establish an intracellular niche for growth.	*Nutrients*
	In all studies with complex biofilms, *L. pneumophila* replication in amoebae is clearly evident.	*Biofilms*
	No necrotrophic growth or nutritional symbiosis studies have been done in freshwater environments to compare the inputs.	*Freshwater bacteria and algae*
	In nature, *L. pneumophila* is under selective pressure to grow intracellularly.	*Biofilms*

the FFs found in freshwater may originate from RPFs prematurely released from amoebae that have lysed or from non-replicating but metabolically active *L. pneumophila* cells in biofilms that can still acquire nutrients and increase their size without dividing.

As *in vitro* FFs form under stressful conditions, the overexpression of heat shock protein 60, chaperonin HtpB, which is upregulated by stress and prevents protein aggregation also induces filamentation (Garduño et al., 2011). In addition, reduced signaling through quorum sensing induces filamentation and overexpression of LqsA which mediates the synthesis of the *L. pneumophila* autoinducer and reduces filamentation (Tiaden et al., 2010). This suggests that *L. pneumophila* might be prone to form FFs when it is present at low densities. However, this does not correspond with the observed formation of highly dense masses of *L. pneumophila* FFs unless FFs downregulate quorum sensing (Piao et al., 2006).

Since filamentation of *L. pneumophila* in the freshwater environment is poorly understood, filamentation could be used by *L. pneumophila* as an antipredation and(or) delayed replication strategy regardless of the mechanisms involved or which cells give rise to FFs. Ciliates do not support the intracellular replication of *L. pneumophila* and cannot ingest bacterial filaments, but *A. castellani* trophozoites support *L. pneumophila*'s replication and phagocytose *L. pneumophila* FFs (unpublished results). Also, once FFs are in contact with amoebae, they rapidly fragment giving rise to many *L. pneumophila* rods (C.C. Sze, Nanyang Technological University, Singapore). The rapid fragmentation of FFs has been seen in broth cultures and intracellularly after FFs have been taken up by macrophages (Piao et al., 2006; Prashar et al., 2013). Therefore, the ecology of FFs seems to focus on survival and co-existence with other organisms in complex biofilms and can be simplified as 'survive and cohabit.' The key facts and arguments of this chapter are laid out in Table 1.

References

Abdelhady, H., and Garduño, R.A. (2013). The progeny of *Legionella pneumophila* in human macrophages shows unique developmental traits. FEMS Microbiol. Lett. 349, 99-107. doi:10.1111/1574-6968.12300.

Abdel-Nour, M., Duncan, C., Low, D.E., and Guyard, C. (2013). Biofilms: The stronghold of *Legionella pneumophila*. Int. J. Mol. Sci. 14, 21660-21675. doi:10.3390/ijms141121660.

Abu Kwaik, Y., Gao, L.-Y., Harb, O.S., and Stone, B.J. (1997). Transcriptional regulation of the macrophage-induced gene (*gspA*) of *Legionella pneumophila* and phenotypic characterization of a null mutant. Mol. Microbiol. 24, 629-642.

Abu Kwaik, Y., Gao, L.-Y., Stone, B.J., Venkataraman, C., and Harb, O.S. (1998a). Invasion of protozoa by *Legionella pneumophila* and its role in bacterial ecology and pathogenesis. Appl. Environ. Microbiol. 64, 3127-3133.

Abu Kwaik, Y., Venkataraman, C., Harb, O.S., and Gao, L.-Y. (1998b). Signal transduction in the protozoan host *Hartmanella vermiformis* upon attachment and invasion by *Legionella micdadei*. Appl. Environ. Microbiol. 64, 3134-3139.

Al-Bana, B.H., Haddad, M.T., and Garduño, R.A. (2014). Stationary phase- and mature infectious-forms of *Legionella pneumophila* produce distinct viable but non-culturable cells. Environ. Microbiol. 16, 382-395. doi: 10.1111/1462-2920.12219.

Albert-Weissenberger, C., Cazalet, C., and Buchrieser, C. (2007). *Legionella pneumophila* - a human pathogen that co-evolved with fresh water protozoa. Cell. Mol. Life Sci. 64, 432-448. DOI 10.1007/ s00018-006-6391-1.

Amaro, F., Wang, W., Gilbert, J.A., Anderson, O.R., and Shuman, H.A. (2015). Diverse protist grazers select for virulence-related traits in Legionella ISME J. 9, 1607-1618. doi:10.1038/ismej.2014.248

Amemura-Maekawa, J., Kikukawa, K., Helbig, J.H., Kaneko, S., Suzuki-Hashimoto, A., Furuhata, K., Chang, B., Murai, M., Ichinose, M.,

Ohnishi, M., Kura, F., and the Working Group for *Legionella* in Japan. (2012). Distribution of monoclonal antibody subgroups and sequence-based types among *Legionella pneumophila* serogroup 1 isolates derived from cooling tower water, bathwater, and soil in Japan. Appl. Environ. Microbiol. 78, 4263-4270. doi: 10.1128/AEM.06869-11

Andreozzi, E., Di Cesare, A., Sabatini, L., Chessa, E., Sisti, D., Rocchi, M., and Citterio, B. (2014). Role of biofilm in protection of the replicative form of *Legionella pneumophila*. Curr. Microbiol. 69, 769-774. DOI 10.1007/s00284-014-0648-y

Barbeau, J., Gauthier, C., and Payment, P. (1998). Biofilms, infectious agents and dental unit waterlines: a review. Can. J. Microbiol. 44, 1019-1028.

Barker, J., Brown, M.R.W., Collier, P.J., Farrell, I., and Gilbert, P. (1992). Relationship between *Legionella pneumophila* and *Acanthamoeba polyphaga*: physiological status and susceptibility to chemical inactivation. Appl. Environ. Microbiol. 58, 2420-2425.

Barker, J., Lambert, P.A., and Brown, M.R. (1993). Influence of intra-amoebic and other growth conditions on the surface properties of *Legionella pneumophila*. Infect. Immun. 61, 3503-3510.

Barker, J., Scaife, H., and Brown, M.R. (1995). Intraphagocytic growth induces an antibiotic-resistant phenotype of *Legionella pneumophila*. Antimicrob. Agents Chemother. 39, 2684-2688. doi: 10.1128/AAC. 39.12.2684

Baumgartner, M., Neu, T.R., Blom, J.F., and Pernthaler, J. (2016) Protistan predation interferes with bacterial long-term adaptation to substrate restriction by selecting for defence morphotypes. J. Evol. Biol. 29, 2297-2310. DOI: 10.11116/jeb.12957

Berendt, R.F. (1981). Influence of blue-green algae (cyanobacteria) on survival of *Legionella pneumophila* in aerosols. Infect. Immun. 32, 690-692.

Berk, S.G., Ting, R.S., Turner, G.W., and Ashburn, R.J. (1998). Production of respirable vesicles containing live *Legionella pneumophila* cells by two *Acanthamoeba* spp. Appl. Environ. Microbiol. 64, 279-286.

Berk, S.G., Gunderson, J.H., Newsome, A.L., Farone, A.L., Hayes, B.J., Redding, K.S., Uddin, N., Williams, E.L., Johnson, R.A., Farsian, M., Reid, A., Skimmyhorn, J., and Farone, M.B. (2006). Occurrence of infected amoebae in cooling towers compared with natural aquatic environments: Implications for emerging pathogens. Environ. Sci. Technol. 40, 7440-7444.

Berk, S.G., Faulkner, G., Garduño, E., Joy, M.C., Ortiz-Jimenez, M.A., and Garduño, R.A. (2008). Packaging of live *Legionella pneumophila* into pellets expelled by *Tetrahymena* spp. does not require bacterial replication and depends on a Dot/Icm-mediated survival mechanism. Appl. Environ. Microbiol. 74, 2187-2199. doi:10.1128/AEM.01214-07

Bigot, R., Bertaux, J., Frere, J., Berjeaud, J.-M. (2013). Intra-amoeba multiplication induces chemotaxis and biofilm colonization and formation for *Legionella*. PLoS ONE 8(10): e77875. doi:10.1371/journal.pone.0077875

Bohach, G.A., and Snyder, I.S. (1983a). Cyanobacterial stimulation of growth and oxygen uptake by *Legionella pneumophila*. Appl. Environ. Microbiol. 46, 528-531.

Bohach, G.A., and Snyder, I.S. (1983b). Characterization of surfaces involved in adherence of *Legionella pneumophila* to *Fischerella* species. Infect. Immun. 42, 318-325.

Bopp, C.A., Sumner, J.W., Morris, G.K., and Wells, J.G. (1981) Isolation of *Legionella* spp. from environmental water samples by low-pH treatment and use of a selective medium. J. Clin. Microbiol. 13, 714-719.

Borella, P., Guerrieri, E., Marchesi, I., Bondi, M., and Messi, P. (2005). Water ecology of *Legionella* and protozoa: environmental and public health perspectives. Biotechnol. Annu. Rev. 11, 355-380. DOI: 10.1016/S1387-2656(05)11011-4

Bouyer, S., Imbert, C., Rodier, M.-H., and Héchard, Y. (2007). Long-term survival of *Legionella pneumophila* associated with *Acanthamoeba castellanii* vesicles. Environ. Microbiol. 9, 1341-1344. doi:10.1111/j.1462-2920.2006.01229.x

Brassinga, A.K.C., Kinchen, J.M., Cupp, M.E., Day, S.R., Hoffman, P.S., and Sifri, C.D. (2010). *Caenorhabditis* is a metazoan host for *Legionella*. Cell. Microbiol. 12, 343-361. doi:10.1111/j.1462-5822.2009.01398.x

Brieland, J.K., Fantone, J.C., Remick, D.G., LeGendre, M., McClain, M., and Engleberg, N.C. (1997). The role of *Legionella pneumophila*-infected *Hartmanella vermiformis* as an infectious particle in a murine model of Legionnaires' disease. Infect. Immun. 65, 5330-5333.

Brissaud, F., Blin, E., Hemous, S., and Garrelly, L. (2008). Water reuse for urban landscape irrigation: aspersion and health related regulations. Water Sci. Technol. 57, 781-787. DOI: 10.2166/wst.2008.162

Brooks, T., Osicki, R.A., Springthorpe, V.S., Sattar, S.A., Filion, L., Abrial, D., and Riffard, S. (2004). Detection and identification of *Legionella* species from groundwaters. J. Toxicol. Environ. Health, Part A 67, 1845-1859. http://dx.doi.org/10.1080/15287390490492449

Brügemann, H., Hagman, A., Jules, M., Sismeiro, O., Dillies, M.-A., Gouyette, C., Kunst, F., Steinert, M., Heuner, K., Coppée, J.-Y., and Buchrieser, C. (2006). Virulence strategies for infecting phagocytes deduced from the *in vivo* transcriptional program of *Legionella pneumophila*. Cell. Microbiol. 8, 1228-1240. doi:10.1111/j.1462-5822.2006.00703.x

Burnside, D.M., Wu, Y., Shafaie, S., and Cianciotto, N.P. (2015). The *Legionella pneumophila* siderophore legiobactin is a polycarboxylate that is identical in structure to rhizoferrin. Infect. Immun. 83, 3937-3945. doi: 10.1128/IAI.00808-15

Buse, H.Y., and Ashbolt, N.J. (2011). Differential growth of *Legionella pneumophila* strains within a range of amoebae at various temperatures associated with in-premise plumbing. Lett. Appl. Microbiol. 53, 217-224. doi:10.1111/j.1472-765X.2011.03094.x

Byrne, B., and Swanson, M.S. (1998). Expression of *Legionella pneumophila* virulence traits in response to growth conditions. Infect. Immun. 66, 3029-3034.

Catrenich, C.E., and Johnson, W. (1989). Characterization of the selective inhibition of growth of virulent *Legionella pneumophila* by supplemented Mueller-Hinton medium. Infect. Immun. 57, 1862-1864.

Cazalet, C., Rusniok, C., Brüggemann, H., Zidane, N., Magnier, A., Ma, L., Tichit, M., Jarraud, S., Bouchier, C., Vandenesch, F., Kunst, F., Etienne, J., Glaser, P., and Buchrieser, C. (2004). Evidence in the *Legionella pneumophila* genome for exploitation of host cell functions and high genome plasticity. Nat. Genet.36, 1165-1173. DOI: 10.1038/ng1447

Cianciotto, N.P. (2007). Iron acquisition by *Legionella pneumophila*. Biometals 20, 323-331. DOI: 10.1007/s10534-006-9057-4

Cianciotto, N.P. (2015). An update on iron acquisition by *Legionella pneumophila*: new pathways for siderophore uptake and ferric iron reduction. Future Microbiol. 10, 841-851. doi:10.2217/fmb.15.21

Cirillo, J.D., Falkow, S., and Tompkins, L.S. (1994). Growth of *Legionella pneumophila* in *Acanthamoeba castellani* enhances invasion. Infect Immun 62, 3254-3261.

Cirillo, J.D., Cirillo, S.L.G., Yan, L., Bermudez, L.E., Falkow, S., and Tompkins, L.S. (1999). Intracellular growth in *Acanthamoeba castellani* affects monocyte entry mechanisms and enhances virulence of *Legionella pneumophila*. Infect. Immun. 67, 4427-4434.

Cirillo, S.L.G., Yang, L., Littman, M., Samrakandi, M.M., and Cirillo, J.D. (2002). Role of the *Legionella pneumophila rtxA* gene in amoebae. Microbiology 148, 1667-1677. doi: 10.1099/00221287-148-6-1667

Costa, J., Tiago, I., da Costa, M.S., and Veríssimo, A. (2005). Presence and persistence of *Legionella* spp. in groundwater. Appl. Environ. Microbiol. 71, 663-671. doi:10.1128/AEM.71.2.663-671.2005

Currie, S.L., and Beattie, T.K. (2015). Compost and *Legionella longbeachae*: an emerging infection? Perspect. Public Health 135, 309-315. doi: 10.1177/1757913915611162

Declerck, P. (2010). Biofilms: the environmental playground of *Legionella pneumophila*. Environ. Microbiol. 12, 557-566. doi:10.1111/j. 1462-2920.2009.02025.x

Declerck, P., Behets, J., Delaedt, Y., Margineanu, A., Lammertyn, E., and Ollevier, F. (2005). Impact of non-*Legionella* bacteria on the uptake and intracellular replication of *Legionella pneumophila* in *Acanthamoeba castellanii* and *Naegleria lovaniensis*. Microb. Ecol. 50, 536-549. DOI: 10.1007/s00248-005-0258-0

Declerck, P., Behets, J., De Keersmaecker, B., and Ollevier, F. (2007a). Receptor-mediated uptake of *Legionella pneumophila* by *Acanthamoeba castellanii* and *Naegleria lovaniensis*. J. Appl. Microbiol. 103, 2697-2703. doi:10.1111/j.1365-2672.2007.03530.x

Declerck, P., Behets, J., vanHoef, V., and Ollevier, F. (2007b). Detection of *Legionella pneumophila* and some of its amoeba hosts in floating biofilms from anthropogenic and natural aquatic environments. Water Res. 41, 3159-3167. doi: 10.1016/j.watres.2007.04.011

Declerck, P., Behets, J., vanHoef, V., and Ollevier, F. (2007c). Replication of *Legionella pneumophila* in floating biofilms. Curr. Microbiol. 55, 435-440. DOI 10.1007/s00284-007-9006-7

Declerck, P., Behets, J., Margineanu, A., van Hoef, V., De Keersmaecker, B., and Ollevier, F. (2009). Replication of *Legionella pneumophila* in biofilms of water distribution pipes. Microbiol. Res. 164, 593-603. doi: 10.1016/j.micres.2007.06.001

Denoncourt, A.M., Paquet, V.E. and Charette, S.J. (2014). Potential role of bacteria packaging by protozoa in the persistence and transmission of pathogenic bacteria. Front. Microbiol. 5, 240. doi:10.3389/fmicb. 2014.00240

Dey, R., Bodennec, J., Mameri, M.O., and Pernin, P. (2008). Free-living freshwater amoebae differ in their susceptibility to the pathogenic bacterium *Legionella pneumophila*. FEMS Microbiol. Lett. 290, 10-17. DOI:10.1111/j.1574-6968.2008.01387.x

Ducret, A., Chabalier, M., and Dukan, S. (2014). Characterization and resuscitation of 'non-culturable' cells of *Legionella pneumophila*. BMC Microbiol. 14, 3, http://www.biomedcentral.com/1471-2180/14/3. doi: 10.1186/1471-2180-14-3

Dupuy, M., Binet, M., Bouteleux, C., Herbelin, P., Soreau, S., and Héchard, Y. (2016). Permissiveness of freshly isolated environmental strains of amoebae for growth of *Legionella pneumophila*. FEMS Microbiol. Lett. 363, 1-6. doi: 10.1093/femsle/fnw022

Edelstein, P.H., Hu, B., Shinzato, T., Edelstein, M.A.C., Xu, W., and Bessman, M.J. (2005). *Legionella pneumophila* NudA is a Nudix hydrolase and virulence factor. Infect. Immun. 73, 6567-6576. doi: 10.1128/IAI.73.10.6567-6576.2005

Eisenreich, W., and Hauner, K. (2016). The life stage-specific pathometabolism of *Legionella pneumophila*. FEBS Lett. 590, 3868-3886. doi:10.1002/1873-3468.12326

Escoll, P., Rolando, M., Gomez-Valero, L., and Buchrieser, C. (2013). From amoeba to macrophages: exploring the molecular mechanisms of *Legionella pneumophila* infection in both hosts. Curr. Top. Microbiol. Immunol. 376, 1-34. doi: 10.1007/82_2013_351

Escoll, P., Song, O-R, Viana, F., Steiner, B., Lagache, T., Olivo-Marin, J.-C., Impens, F., Brodin, P., Hilbi, H., and Buchrieser, C. (2017). *Legionella pneumophila* modulates mitochondrial dynamics to trigger metabolic repurposing of infected macrophages. Cell Host and Microbe *in press*. http://dx.doi.org/10.1016/j.chom.2017.07.020

Ewann, F., and Hoffman, P.S. (2006). Cysteine metabolism in *Legionella pneumophila*: Characterization of an L-cystine-utilizing mutant. Appl. Environ. Microbiol. 72, 3993-4000. doi:10.1128/AEM.00684-06

Eylert, E., Herrmann, V., Jules, M., Gillmaier, N., Lautner, M., Buchrieser, C., Eisenreich, W., and Heuner, K. (2010). Isotopologue profiling of *Legionella pneumophila*. Role of serine and glucose as carbon substrates. J. Biol. Chem. 285, 22232-22243. DOI 10.1074/jbc.M110.128678

Faulkner, G., and Garduño, R.A. (2002). Ultrastructural analysis of differentiation in *Legionella pneumophila*. J. Bacteriol. 184, 7025-7041. doi: 10.1128/JB.184.24.7025-7041.2002

Faulkner, G., Berk, S.G., Garduño, E., Ortiz-Jiménez, M.A., and Garduño. R.A. (2008). Passage through *Tetrahymena tropicalis* triggers a rapid morphological differentiation in *Legionella pneumophila*. J. Bacteriol. 190, 7728-7738. doi: 10.1128/JB.00751-08

Feeley, J.C., Gibson, R.J., Gorman, G.W., Langford, N.C., Rasheed, J.K., Mackel, D.C., and Baine, W.B. (1979). Charcoal-yeast extract agar: Primary isolation medium for *Legionella pneumophila*. J. Clin. Microbiol. 10, 437-441.

Fernandez, R.C., Logan, S.M., Lee, S.H.S., and Hoffman, P.S. (1996). Elevated levels of *Legionella pneumophila* stress protein Hsp60 early in infection of human monocytes and L929 cells correlate with virulence. Infect. Immun. 64, 1968-1976.

Fields, B.S. (1997). Legionellae and Legionnaires' disease. In Manual of Environmental Microbiology, Vol. 1., C.J. Hurst, G.R. Knudsen, M.J. McInerney, L.D. Stetzenbach and M.V. Walter, eds. (Washington, USA: American Society for Microbiology), pp. 666-675.

Fields, B.S., and Lucas, C.E. (1997). Characterization of sessile and planktonic *Legionella pneumophila* in model biofilms. In *Legionella*: State of the Art 30 Years after its Recognition, N.P. Cianciotto, Y. Abu Kwaik, P.H. Edelstein, B.S. Fields, D.F. Geary, T.G. Harrison, C.A. Joseph, R.M. Ratcliff, J.E. Stout, and M.S. Swanson, eds. (Washington, USA: American Society for Microbiology), pp. 383-389.

Fisher-Hoch, S.P., Smith, M.G., and Colbourne, J.S. (1982). *Legionella pneumophila* in hospital hot water cylinders. Lancet 1, 1073.

Flesher, A.R., Kasper, D.L., Modern, P.A., and Mason, E.O. Jr. (1980). *Legionella pneumophila*: Growth inhibition by human pharyngeal flora. J. Infect. Dis. 142, 313-317.

Fliermans, C.B. (1996). Ecology of *Legionella*: From data to knowledge with a little wisdom. Microb. Ecol. 32, 203-228.

Fliermans, C.B., and Tyndall, R.L. (1992). Decade of monitoring *of Legionella pneumophila* with natural ecosystems. In Abstracts book of the 4th International Symposium on *Legionella,* January 26-29, 1992, Orlando, Florida (Washington, USA: American Society of Microbiology Conference), p. 30.

Fliermans, C.B., Cherry, W.B., Orrison, L.H., and Thacker, L. (1979). Isolation of *Legionella pneumophila* from nonepidemic-related aquatic habitats. Appl. Environ. Microbiol. 37, 1239-1242.

Fliermans, C.B., Cherry, W.B., Orrison, L.H., Smith, S.J., Tison, D.L., and Pope, D.H. (1981a). Ecological distribution of *Legionella pneumophila.* Appl. Environ. Microbiol. 41: 9-16.

Fliermans, C.B., Soracco, R.J., and Pope, D.H. (1981b). Measure of *Legionella pneumophila* activity in situ. Curr. Microbiol. 6, 89-94.

Flores-Díaz, M., Monturiol-Gross, L., Naylor, C., Alape-Girón, A., and Flieger, A. (2016). Bacterial sphingomyelinases and phospholipases as virulence factors. Microbiol. Mol. Biol. Rev. 80, 597- 628. doi:10.1128/MMBR.00082-15

Fonseca, M.V., and Swanson, M.S. (2014). Nutrient salvaging and metabolism by the intracellular pathogen *Legionella pneumophila.* Front. Cell. Infect. Microbiol. 4, 12. doi:10.3389/fcimb.2014.00012

Gao, L.-Y., and Abu Kwaik, Y. (2000). The mechanism of killing and exiting the protozoan host *Acanthamoeba polyphaga* by *Legionella pneumophila.* Environ. Microbiol. 2, 79-90. doi: 10.1046/j.1462-2920.2000.00076.x

Garduño, R.A. (2008). Life cycle, growth cycles and developmental cycle of *Legionella pneumophila.* In *Legionella pneumophila*: Pathogenesis and Immunity, P. Hoffman, H. Friedman, and M. Bendinelli, eds. A volume in the series Infectious Diseases and Pathogenesis, H. Friedman, and M. Bendinelli, eds. (New York, USA, Springer), pp. 65-84.

Garduño, R.A., Quinn, F., and Hoffman, P.S. (1998). HeLa cells as a model to study the invasiveness and biology of *Legionella pneumophila.* Can. J. Microbiol. 44, 430-440.

Garduño, R.A., Garduño, E., Hiltz, M., Allan, D., and Hoffman, P.S. (2002a). Morphological and physiological evidence for a developmental cycle in *Legionella pneumophila*. In *Legionella*, R. Marre, Y. Abu Kwaik, C. Bartlett, N.P. Cianciotto, B.S. Fields, M. Frosch, J. Hacker, and P.C. Lück, eds. (Washington, USA: ASM Press), pp. 82-85.

Garduño, R.A., Garduño, E., Hiltz, M., and Hoffman, P.S. (2002b). Intracellular growth of *Legionella pneumophila* gives rise to a differentiated form dissimilar to stationary-phase forms. Infect. Immun. 70, 6273-6283. DOI: 10.1128/IAI.70.11.6273-6283.2002

Garduño, R.A., Chong A., and Faulkner G. (2008). Developmental cycle - Differentiation of *Legionella pneumophila*. In *Legionella*: Molecular Microbiology, K. Heuner and M. Swanson, eds. (Norfolk, UK: Caister Academic Press), pp. 55-73.

Garduño, R.A., Chong, A., Nasrallah, G.K., and Allan, D.S. (2011). The *Legionella pneumophila* chaperonin - an unusual multifunctional protein in unusual locations. Front. Microbiol. 2, 122. doi:10.3389/fmicb.2011.00122

Gast, R.J., Moran, D.M., Dennett, M.R., Wurtsbaugh, W.A., and Amaral-Zettler, L.A. (2011). Amoebae and *Legionella pneumophila* in saline environments. J. Water Health 9, 37-52. doi:10.2166/wh.2010.103

George, J.R., Pine, L., Reeves, M.W., and Harrell, W.K. (1980). Amino acid requirements of *Legionella pneumophila*. J. Clin. Microbiol. 11, 286-291.

Gião, M.S., Wilks, S.A., Azevedo, N.F., Vieira, M.J., and Keevil, C.W. (2009). Comparison between standard culture and peptide nucleic acid 16S rRNA hybridization quantification to study the influence of physico-chemical parameters on *Legionella pneumophila* survival in drinking water biofilms. Biofouling 25, 335-343. DOI: 10.1080/08927010902802232

Graham, F.F., White, P.S., Harte, D.J., and Kingham, S.P. (2012). Changing epidemiological trends of legionellosis in New Zealand,

1979-2009. Epidemiol. Infect. 140, 1481-1496. DOI: 10.1017/S0950268811000975

Greub, G., and Raoult, D. (2003). Morphology of *Legionella pneumophila* according to their location within *Hartmanella vermiformis*. Res. Microbiol. 154, 619-621. DOI: 10.1016/j.resmic.2003.08.003

Greub, G., and Raoult, D. (2004). Microorganisms resistant to free-living amoebae. Clin. Microbiol. Rev. 17, 413-433. DOI: 10.1128/CMR.17.2.413-433.2004

Groothuis, D.G., and Veenendaal, H.R. (1983). Heat treatment as an aid for the isolation of *Legionella pneumophila* from clinical and environmental samples. Zentralbl. Bakteriol. Mikrobiol. Hyg. A 255, 39-43.

Harb, O.S., Gao, L-Y, and Abu Kwaik, Y. (2000). From protozoa to mammalian cells: a new paradigm in the life cycle of intracellular bacterial pathogens. Environ. Microbiol. 2, 251-265.

Heller, R., Höller, C., Süßmuth, R., and Gundermann, K.-O. (1998). Effect of salt concentration and temperature on survival of *Legionella pneumophila*. Lett. Appl. Microbiol. 26, 64-68

Hellinga, J.R., Garduño, R.A., Kormish, J.D., Tanner, J.R., Kahn, D., Buchko, K., Jimenez, C., Pinette M.M., and Brassinga, A.K. (2015). Identification of vacuoles containing extraintestinal differentiated forms of *Legionella pneumophila* in colonized *Caenorhabditis elegans* soil nematodes. MicrobiologyOpen 4, 660-681. doi: 10.1002/mbo3.271

Herrmann, V., Eidner, A., Rydzewski, K., Blädel, I., Jules, M., Buchrieser, C., Eisenreich, W., and Heuner, K. (2011). GamA is a eukaryotic-like glucoamylase responsible for glycogen- and starch-degrading activity of *Legionella pneumophila*. Int. J. Med. Microbiol. 301, 133-139. doi: 10.1016/j.ijmm.2010.08.016

Heuner, K., and Eisenreich, W. (2013). Chapter 8. The intracellular metabolism of *Legionella* by isotopologue profiling. In *Legionella*: Methods and protocols, Methods in Molecular Biology Vol. 954, C.

Buchrieser, H. Hilbi, eds. (New York, USA, Springer Science+Business), pp. 163-181. DOI 10.1007/978-1-62703-161-5_8

Hindré, T., Brüggemann, H., Buchrieser, C., and Héchard, Y. (2008). Transcriptional profiling of *Legionella pneumophila* biofilm cells and the influence of iron on biofilm formation. Microbiology 154, 30-41. DOI 10.1099/mic.0.2007/008698-0

Hoffman, P.S. (2008). Microbial physiology. In *Legionella pneumophila*: Pathogenesis and Immunity, P. Hoffman, H. Friedman, and M. Bendinelli, eds. A volume in the series Infectious Diseases and Pathogenesis, H. Friedman, and M. Bendinelli, eds. (New York, USA, Springer), pp. 113-131.

Hoffman, P.S., Butler, C.A., and Quinn, F.D. (1989). Cloning and temperature-dependent expression in *Escherichia coli* of a *Legionella pneumophila* gene coding for a genus-common 60-kilodalton antigen. Infect. Immun. 57, 1731-1739.

Hojo, F., Sato, D., Matsuo, J., Miyake, M., Nakamura, S., Kunichika, M., Hayashi, Y., Yoshida, M., Takahashi, K., Takemura, H., Kamiya, S., and Yamaguchi, H. (2012). Ciliates expel environmental *Legionella*-laden pellets to stockpile food. Appl. Environ. Microbiol. 78, 5247-5257. doi: 10.1128/AEM.00421-12

Hovel-Miner, G., Pampou, S., Faucher, S.P., Clarke, M., Morozova, I., Morozov, P., Russo, J.J., Shuman, H.A., and Kalachikov, S. (2009). σS controls multiple pathways associated with intracellular multiplication of *Legionella pneumophila*. J. Bacteriol. 191, 2461-2473. doi: 10.1128/JB.01578-08

Hovel-Miner, G., Faucher, S.P., Charpentier, X., and Shuman, H.A. (2010). ArgR-regulated genes are derepressed in the *Legionella*-containing vacuole. J. Bacteriol. 192, 4504-4516. doi:10.1128/JB.00465-10

Hsu, B.-M., Huang, C.-C., Chen, J.-S., Chen, N.-H., and Huang, J.-T. (2011). Comparison of potentially pathogenic free-living amoeba hosts by *Legionella* spp. in substrate-associated biofilms and floating biofilms from spring environments. Water Res. 45, 5171-5183. doi:10.1016/j.watres.2011.07.019

Hsu, T.-K., Wu, S.-F., Hsu, B.-M., Kao, P.-O., Tao, C.-W., Shen, S.-M., Ji, W.-T., Huang, W.-C., and Fan, C.-W. (2015). Surveillance of parasitic *Legionella* in surface waters by using immunomagnetic separation and amoebae enrichment. Pathogens and Global Health 109, 328-335. DOI 10.1179/2047773215Y.0000000034

Hume, R.D., and Hann, W.D. (1984a). Growth relationships of *Legionella pneumophila* with green algae (Chlorophyta). In *Legionella*: Proceedings of the 2nd International Conference, C. Thornsberry, A. Balows, J.C. Feeley, and W. Jakubowski, eds. (Washington, USA: American Society of Microbiology), pp. 323-324.

Hume, R.D., Hann, W.D. (1984b). Growth of *Legionella, Escherichia* and *Pseudomonas* among cells of the green algae *Chlorella, Gloeocystis*, and *Scenedesmus* as seen by scanning electron microscopy. Micron Microsc. Acta 15, 107-108.

ISO (1998). Water Quality - Detection and enumeration of *Legionella*. International Standard ISO 11731:1998 (Geneva, Switzerland: International Organization for Standardization).

ISO (2017). Water Quality - Enumeration of *Legionella*. International Standard ISO 11731:2017 (Geneva, Switzerland: International Organization for Standardization).

Isaac, D.T., Laguna, R.K., Valtz, N., and Isberg, R.R. (2015). MavN is a *Legionella pneumophila* vacuole-associated protein required for efficient iron acquisition during intracellular growth. Proc. Natl. Acad. Sci. USA, 112, E5208-E5217. www.pnas.org/cgi/doi/10.1073/pnas.1511389112

Jacquier, N., Aeby, S., Lienard, J., and Greub, G. (2013). Discovery of new intracellular pathogens by amoebal coculture and amoebal enrichment approaches. J. Vis. Exp. 80, e51055. http://www.jove.com/video/51055. doi:10.3791/51055

Jeon, K.W. (2004). Genetic and physiological interactions in the amoeba-bacteria symbiosis. J. Euk. Microbiol. 51, 502-508. DOI: 10.1111/j.1550-7408.2004.tb00277.x

Joly, J.R., Dewailly, E., Bernard, L., Ramsay, D., and Brisson, J. (1985) *Legionella* and domestic water heaters in the Quebec City area. Can. Med. Assoc. J. 132, 160.

Jousset, A. (2012). Ecological and evolutive implications of bacterial defences against predators. Environ. Microbiol. 14, 1830-1843. doi: 10.1111/j.1462-2920.2011.02627.x

Katz, S.M., and Hammel, J.M. (1987). The effect of drying, heat, and pH on the survival of *Legionella pneumophila*. Ann. Clin. Lab. Sci. 17, 150-156.

Komura, T., Yasui, C., Miyamoto, H., and Nishikawa, Y. (2010). *Caenorhabditis elegans* as an alternative model host for *Legionella pneumophila*, and protective effects of *Bifidobacterium infantis*. Appl. Environ. Microbiol. 76, 4105-4108. doi:10.1128/AEM.03021-09

Konishi, T., Yamashiro, T., Koide, M., and Nishizono, A. (2006). Influence of temperature on growth of *Legionella pneumophila* biofilm determined by precise temperature gradient incubator. J. Biosc. Bioeng. 101, 478-484. DOI: 10.1263/jbb.101.478

Koubar, M. Rodier, M.-H., Garduño, R.A., and Frère, J. (2011). Passage through *Tetrahymena tropicalis* enhances the resistance to stress and the infectivity of *Legionella pneumophila*. FEMS Microbiol. Lett. 325, 10-15. doi: 10.1111/j.1574-6968.2011.02402.x

Koubar, M., Rodier, M.-H., and Frère, J. (2013). Involvement of minerals in adherence of *Legionella pneumophila* to surfaces. Curr. Microbiol. 66, 437-442. DOI 10.1007/s00284-012-0295-0

Kuiper, M.W., Wullings, B.A., Akkermans, A.D.L., Beumer, R.R., and van der Kooij, D. (2004). Intracellular proliferation of *Legionella pneumophila* in *Hartmanella vermiformis* in aquatic biofilms grown on plasticized polyvinylchloride. Appl. Environ. Microbiol. 70, 6826-6833. doi: 10.1128/AEM.70.11.6826-6833.2004

Kusnetsov, J.M., Jousimies-Somer, H.R., Nevalainen, A.I., and Martikainen, P.J. (1994). Isolation of *Legionella* from water samples using various culture methods. J. Appl. Bacteriol. 76, 155-162.

Kusnetsov, J.M., Ottoila, E., and Martikainen, P.J. (1996). Growth, respiration and survival of *Legionella pneumophila* at high temperatures. J. Appl. Bacteriol. 81, 341-347.

La Scola, B., Birtles, R.J., Greub, G., Harrison, T.J., Ratcliff, R.M., and Raoult, D. (2004). *Legionella drancourtii* sp. nov., a strictly intracellular amoebal pathogen. Int. J. Syst. Evol. Microbiol. 54, 699-703. DOI: 10.1099/ijs.0.02455-0

Lau, H.Y., and Ashbolt, N.J. (2009). The role of biofilms and protozoa in *Legionella* pathogenesis: implications for drinking water. J. Appl. Microbiol. 107, 368-378. doi:10.1111/j.1365-2672.2009.04208.x

Lee, J.V., and West, A.A. (1991). Survival and growth of *Legionella* species in the environment. J. Appl. Bacteriol. (Symp. Suppl.) 70, 121S-129S.

Lema, M.W., Brown, A., Butler, C.A., and Hoffman, P.S. (1988). Heat-shock response in *Legionella pneumophila*. Can. J. Microbiol. 34, 1148-1153.

Lesnik, R., Brettar, I., and Höfle, M.G. (2016). *Legionella* species diversity and dynamics from surface reservoir to tap water: from cold adaptation to thermophily. ISME J. 10, 1064-1080. doi:10.1038/ismej.2015.199

Li, L. and Faucher, S.P. (2016). The membrane protein LasM promotes the culturability of *Legionella pneumophila* in water. Front. Cell. Infect. Microbiol. 6,113. doi: 10.3389/fcimb.2016.00113

Li, X., Zeng, Y., Gao, Y., Zheng, X., Zhang, Q., Zhou, S., and Lu, Y. (2010). The ClpP protease homologue is required for the transmission traits and cell division of the pathogen *Legionella pneumophila*. BMC Microbiol. 10, 54. http://www.biomedcentral.com/1471-2180/10/54

Li, L., Mendis, N., Trigui, H., Oliver, J.D. and Faucher, S.P. (2014). The importance of the viable but non-culturable state in human bacterial pathogens. Front. Microbiol. 5, 258. doi:10.3389/fmicb.2014.00258

Li, L., Mendis, N., Trigui, H., and Faucher, S.P. (2015a). Transcriptomic changes of *Legionella pneumophila* in water. BMC Genomics 16, 637. DOI 10.1186/s12864-015-1869-6

Li, H., Xin, H., and Li, S.F.Y. (2015b). Multiplex PMA-qPCR assay with internal amplification control for simultaneous detection of viable *Legionella pneumophila, Salmonella typhimurium* and *Staphylococcus aureus* in environmental waters. Environ. Sci. Technol. 49, 14249-14256. DOI: 10.1021/acs.est.5b03583

Litwin, C.M., Asebiomo, B., Wilson, K., Hafez, M., Stevens, V., Fliermans, C.B., Fields B.S., and Fisher, J.F. (2013). Recreational vehicle water tanks as a possible source for *Legionella* infections. Case Rep. Infect. Dis. Vol. 2013, Article ID 286347. http://dx.doi.org/10.1155/2013/286347

Mampel, J., Spirig, T., Weber, S.S., Haagensen, A.J., Molin, S., and Hilbi, H. (2006). Planktonic replication is essential for biofilm formation by *Legionella pneumophila* in a complex medium under static and dynamic flow conditions. Appl. Environ. Microbiol. 72, 2885-2895. doi: 10.1128/AEM.72.4.2885-2895.2006

Manske, C. and Hilbi, H. (2014). Metabolism of the vacuolar pathogen *Legionella* and implications for virulence. Front. Cell. Infect. Microbiol. 4, 125. doi: 10.3389/fcimb.2014.00125

Marchand, A., Verdon, J., Lacomb, C., Crapart, S., Héchard, Y., and Berjeaud, J.M. (2011). Anti-*Legionella* activity of staphylococcal hemolytic peptides. Peptides 32, 845-851. doi:10.1016/j.peptides.2011.01.025

Marchand, A., Augenstreich, J., Loiseau, C., Verdon, J., Lecomte, S., and Berjeaud, J.-M. (2015). Effect of amino acid substitution in the staphylococcal peptides warnericin RK and PSMα on their anti-*Legionella* and hemolytic activities. Mol. Cell. Biochem. 405, 159-167. DOI 10.1007/s11010-015-2407-1

Marra, A., Blander, S.J., Horwitz, M.A., and Shuman, H.A. (1992). Identification of a *Legionella pneumophila* locus required for intracellular multiplication in human macrophages. Proc. Natl. Acad. Sci. USA 89, 9607-9611.

Matz, C., and Kjelleberg, S. (2005). Off the hook - how bacteria survive protozoan grazing. Trends Microbiol. 13, 302-307. doi:10.1016/j.tim.2005.05.009

McNealy, T., Newsome, A.L., Johnson, R.A., and Berk, S.G. (2002). Impact of amoebae, bacteria and *Tetrahymena* on *Legionella pneumophila* multiplication and distribution in an aquatic environment. In *Legionella*, R. Marre, Y. Abu Kwaik, C. Bartlett, N.P. Cianciotto, B.S. Fields, M. Frosch, J. Hacker, and P.C. Lück, eds. (Washington, USA: ASM Press), pp.170-175.

Mendis, N., McBride, P., Faucher, S.P. (2015). Short-term and long-term survival and virulence of *Legionella pneumophila* in the defined freshwater medium Fraquil. PLoS ONE 10(9): e0139277. doi:10.1371/journal.pone.0139277

Mengaud, J.M., and Horwitz, M.A. (1993). The major iron-containing protein of *Legionella pneumophila* is an aconitase homologous with the human iron-responsive element-binding protein. J. Bacteriol. 175, 5666-5676. doi: 10.1128/jb.175.17.5666-5676.1993

Molmeret, M., Horn, M., Wagner, M., Santic, M., and Abu Kwaik, Y. (2005). Amoebae as training grounds for intracellular bacterial pathogens. Appl. Environ. Microbiol. 71, 20-28. doi:10.1128/AEM.71.1.20-28.2005

Molofsky, A.B., and Swanson, M.S. (2004). Differentiate to thrive: lessons from the *Legionella pneumophila* life cycle. Mol. Microbiol. 53, 29-40. doi:10.1111/j.1365-2958.2004.04129.x

Murga, R., Forster, T.S., Brown, E., Pruckler, J.M., Fields, B.S., and Donlan, R.M. (2001). Role of biofilms in the survival of *Legionella pneumophila* in a model potable-water system. *Microbiology* 147, 3121-3126. doi: 10.1099/00221287-147-11-3121

Nasrallah, G.K., Riveroll, A.L., Chong, A., Murray, L.E., Lewis, P.J., and Garduño, R.A. (2011). *Legionella pneumophila* requires polyamines for optimal intracellular growth. J. Bacteriol. 193, 4346-4360. Author correction for this paper: J. Bacteriol. Vol. 194, Issue 11, page 3032, 2012.

Nasrallah, G.K., Abdelhady, H., Tompkins, N.P., Carson, K.R., and Garduño, R.A. (2014). Deletion of *potD*, encoding a putative spermidine-binding protein, results in a complex phenotype in *Legionella pneumophila*. Int. J. Med. Microbiol. 304, 703-716. http://dx.doi.org/10.1016/j.ijmm.2014.05.004

Newsome, A.L., Farone, M.B., Berk, S.G., and Gunderson, J.H. (2001). Free living amoebae as opportunistic hosts for intracellular bacterial parasites. J. Eukaryot. Microbiol. Suppl., 13S-14S.

O'Connor, T.J., Zheng, H., Van Rheenen, S.M., Ghosh, S., Cianciotto, N.P. and Isberg, R.R. (2016). Iron limitation triggers early egress by the intracellular bacterial pathogen *Legionella pneumophila*. Infect. Immun. 84, 2185-2197. doi:10.1128/IAI.01306-15

Ohno, A., Kato, N., Yamada, K., and Yamaguchi, K. (2003). Factors influencing survival of *Legionella pneumophila* serotype 1 in hot spring water and tap water. Appl. Environ. Microbiol. 69, 2540-2547. doi: 10.1128/AEM.69.5.2540-2547.2003

Ortiz-Roque, C.M., and Hazen, T.C. (1987). Abundance and distribution of *Legionellaceae* in Puerto Rican waters. Appl. Environ. Microbiol. 53, 2231-2236.

Parthuisot, N., Wes, N.J., Lebaron, P., and Baudart, J. (2010). High diversity and abundance of *Legionella* spp. in a pristine river and impact of seasonal and anthropogenic effects. Appl. Environ. Microbiol. 76, 8201-8210.

Pécastaings, S., Bergé, M., Dubourg, K.M., and Roques, C. (2010). Sessile *Legionella pneumophila* is able to grow on surfaces and generate structured monospecies biofilms. Biofouling 26, 809-819. DOI: 10.1080/08927014.2010.520159

Pernthaler, J. (2005). Predation on prokaryotes in the water column and its ecological implications. Nat. Rev. Microbiol. 3, 537-546. doi:10.1038/nrmicro1180

Piao, Z., Sze, C.C., Barysheva, O., Iida, K., and Yoshida, S. (2006). Temperature-regulated formation of mycelial mat-like biofilms by *Legionella pneumophila*. Appl. Environ. Microbiol. 72, 1613-1622. doi: 10.1128/AEM.72.2.1613-1622.2006

Pine, L., George, J.R., Reeves, M.W., and Harrell, W.K. (1979). Development of a chemically defined liquid medium for growth of *Legionella pneumophila*. J. Clin. Microbiol. 9, 615-626.

Pine, L., Hoffman, P.S., Malcolm, G.B., Benson, R.F., and Franzus, M.J. (1986). Role of keto acids and reduced-oxygen-scavenging enzymes in the growth of *Legionella* species. J. Clin. Microbiol. 23, 33-42.

Pope, D.H., Soracco, R.J., Gill, H.K., andFliermans, C.B. (1982). Growth of *Legionella pneumophila* in two-membered cultures with green algae and cyanobacteria. Curr. Microbiol. 7, 319-321.

Portier, E., Zheng, H., Sahr, T., Burnside, D.M., Mallama, C., Buchrieser, C., Cianciotto, N.P., and Héchard, Y. (2015). *iroT/mavN*, a new iron-regulated gene involved in *Legionella pneumophila* virulence against amoebae and macrophages. Environ. Microbiol. 17, 1338-1350. doi: 10.1111/1462-2920.12604

Portier, E., Bertaux, J., Labanowski, J., and Héchard, Y. (2016). Iron availability modulates the persistence of *Legionella pneumophila* in complex biofilms. Microbes Environ. 31, 387-394. doi:10.1264/jsme2.ME16010; https://www.jstage.jst.go.jp/browse/jsme2

Potočnjak, M., Magdalenić, Z., Dijan, M., Rebić, D., and Gobin, I. (2016). Environmental factors affecting the survival of soil dwelling *Legionella longbeachae* in water. Ann. Agric. Environ. Med. 23, 452-455. doi: 10.5604/12321966.1219186

Prashar, A., Bhatia, S., Gigliozzi, D., Martin, T., Duncan, C., Guyard, C., and Terebiznik, M.R. (2013). Filamentous morphology of bacteria delays the timing of phagosome morphogenesis in macrophages. J. Cell Biol. 203, 1081-1097. doi: 10.1083/jcb.201304095

Price, C.T.D., Richards, A.M., and Abu Kwaik, Y. (2014a). Nutrient generation and retrieval from the host cell cytosol by intra-vacuolar

Legionella pneumophila. Front. Cell. Infect. Microbiol. 4, 111. doi: 10.3389/fcimb.2014.00111

Price, C.T.D., Richards, A.M., Von Dwingelo, J.E., Samara, H.A., and Abu Kwaik, Y. (2014b). Amoeba host-*Legionella* synchronization of amino acid auxotrophy and its role in bacterial adaptation and pathogenic evolution. Environ. Microbiol. 16, 350-358. doi:10.1111/1462-2920.12290

Rasch, J., Krüger, S., Fontvieille, D., Ünal, C.M., Michel, R., Labrosse, A., and Steinert, M. (2016). *Legionella*-protozoa-nematode interactions in aquatic biofilms and influence of Mip on *Caenorhabditis elegans* colonization. Int. J. Med. Microbiol. 306, 443-451. http://dx.doi.org/10.1016/j.ijmm.2016.05.012

Reeves, M.W., Pine, L., Hutner, S.H., George, J.R., and Harrell, W.K. (1981). Metal requirements of *Legionella pneumophila*. J. Clin. Microbiol. 13, 688-695.

Reeves, M.W., Pine, L., Neilands, J.B. and Balows, A. (1983). Absence of siderophore activity in *Legionella* species grown in iron-deficient media. J. Bacteriol. 154, 324-329.

Richards, A.M., Von Dwingelo, J.E., Price, C.T. and Abu Kwaik, Y. (2013). Cellular microbiology and molecular ecology of *Legionella*-amoeba interaction. Virulence 4, 307-314. DOI: 10.4161/viru.24290

Riffard, S., Douglass, S., Brooks, T., Springthorpe, S., Filion, L.G., and Sattar, S.A. (2001). Occurrence of *Legionella* in groundwater: an ecological study. Water Sci. Technol. 43, 99-102.

Ristroph, J.D., Hedlund, K.W., and Allen, R.G. (1980). Liquid medium for growth of *Legionella pneumophila*. J. Clin. Microbiol. 11, 19-21.

Ristroph, J.D., Hedlund, K.W., and Gowda, S. (1981). Chemically defined medium for *Legionella pneumophila* growth. J. Clin. Microbiol. 13, 115-119.

Robertson, P., Abdelhady, H., and Garduño, R.A. (2014). The many forms of a pleomorphic bacterial pathogen - The developmental network of *Legionella pneumophila*. Front. Microbiol. **5:** 670. doi: 10.3389/fmicb.2014.00670

Rogers, J., and Keevil, C.W. (1992). Immunogold and fluorescein immunolabelling of *Legionella pneumophila* within an aquatic biofilm visualized by using episcopic differential interference contrast microscopy. Appl. Environ. Microbiol. 58, 2326-2330.

Rowbotham, T.J. (1980). Preliminary report on the pathogenicity of *Legionella pneumophila* for freshwater and soil amoebae. J. Clin. Pathol. 33, 1179-1183.

Rowbotham, T.J. (1983). Isolation of *Legionella pneumophila* from clinical speciments via amoebae, and the interaction of those and other isolates with amoebae. J. Clin. Pathol. 36, 978-986.

Sadowsky, A.B., Wilson, J.W., Steinman, H.M., and Shuman, H.A. (1994). The iron superoxide dismutase of *Legionella pneumophila* is essential for viability. J. Bacteriol. 176, 3790-3799. doi: 10.1128/jb.176.12.3790-3799.1994

Sakamoto, R., Ohno, A., Nakahara, T., Satomura, K., Iwanaga, S. Kouyama, Y., Kura, F., Kato, N., Matsubayashi, K., Okumiya, K., and Yamaguchi, K. (2009). *Legionella pneumophila* in rainwater on roads. Emerg. Infect. Dis. 15, 1295-1297. DOI: 10.3201/eid1508.090317

Sauer, J.D., Bacham, M.A., and Swanson, M.S. (2005). The phagosomal transporter A couples threonine acquisition to differentiation and replication of *Legionella pneumophila* in macrophages. Proc. Natl. Acad. Sci. USA 102, 9924-9929.

Schofield, G.M. (1985). A note on the survival of *Legionella pneumophila* in stagnant tap water. J. Appl. Bacteriol. 59, 333-335.

Schuetz, P., Haubitz, S., Christ-Crain, M., Albrich, W.C., Zimmerli, W., Mueller, B., and the ProHOSP study group. (2013). Hyponatremia and anti-diuretic hormone in Legionnaires' disease. BMC Infect. Dis. 13, 585. http://www.biomedcentral.com/1471-2334/13/585

Schunder, E., Gillmaier, N., Kutzner, E., Herrmann, V., Lautner, M., Heuner, K., and Eisenreich, W. (2014). Amino acid uptake and metabolism of *Legionella pneumophila* hosted by *Acanthamoeaba castellanii*. J. Biol. Chem. 289, 21040-21054. DOI 10.1074/jbc.M114.570085

Schuster, F.L. (1979). Small amebas and amoeboflagellates. In Biochemistry and physiology of protozoa Vol. 1, M. Levandowsky, and S. H. Hutner eds. (New York, USA: Academic Press), p. 215-285.

Sharaby, Y., Rodríguez-Martínez, S., Oks, O., Pecellin, M., Mizrahi, H., Peretz, A., Brettar, I., Höfle, M.G., Halpern, M. (2017). Temperature-dependent growth modeling of environmental and clinical *Legionella pneumophila* multilocus variable-number tandem-repeat analysis (MLVA) genotypes. Appl. Environ. Microbiol. 83: e03295-16, pages 1-13. doi.org/10.1128/AEM.03295-16.

Shen, S.M., Chou, M.Y., Hsu, B.M., Ji, W.T., Hsu, T.K., Tsai, H.F., Huang, Y.L., Chiu, Y.C., Kao, E.S., Kao, P.M., and Fan, C.W. (2015). Assessment of *Legionella pneumophila* in recreational spring water with quantitative PCR (Taqman) assay. Path. Global Health 109, 236-241. DOI: 10.1179/2047773215Y.0000000023

Simmons, G., Jury, S., Thornley, C., Harte, D., Mohiuddin, J., and Taylor, M. (2008). A Legionnaires' disease outbreak: A water blaster and roof-collected rainwater systems. Water Res. 42, 1449-1458.

Skaliy, D., and McEachern, H.V. (1979). Survival of the Legionnaires' disease bacterium in water. Ann. Intern. Med. 90, 662-663.

Söderberg, M.A., and Cianciotto, N.P. (2008). A *Legionella pneumophila* peptidyl-prolyl *cis-trans* isomerase present in culture supernatants is necessary for optimal growth at low temperatures. Appl. Environ. Microbiol. 74, 1634-1638. doi:10.1128/AEM.02512-07

Söderberg, M.A., Rossier, O., and Cianciotto, N.P. (2004). The type II protein secretion system of *Legionella pneumophila* promotes growth at low temperatures. J. Bacteriol. 186, 3712-3720.

Spriggs, D.R. (1987). *Legionella*, microbial ecology, and inconspicuous consumption. J. Inf. Dis. 155, 1086-1087.

States, S.J., Conley, L.F., Ceraso, M., Stephenson, T.E., Wolford, R.S., Wadowsky, R.M., McNamara, A.M., and Yee, R.B. (1985). Effects of metals on *Legionella pneumophila* growth in drinking water plumbing systems. Appl. Environ. Microbiol. 50, 1149-1154.

Steinert, M., Ëmödy, L., Amann, R., and Hacker, J. (1997). Resuscitation of viable but nonculturable *Legionella pneumophila* Philadelphia JR32 by *Acanthamoeba castellanii*. Appl. Environ. Microbiol. 63, 2047-2053.

Steudel, C., Helbig, J.H., and Lück, P.C. (2002). Characterization of a 16-kilodalton species-specific protein of *Legionella pneumophila* promoting uptake in amoebae. In *Legionella*, R. Marre, Y. Abu Kwaik, C. Bartlett, N.P. Cianciotto, B.S. Fields, M. Frosch, J. Hacker, and P.C. Lück, eds. (Washington, USA: ASM Press), pp.165-169.

Stout, J.E., Yu, V.L., and Best, M.G. (1985). Ecology of *Legionella pneumophila* within water distribution systems. Appl. Environ. Microbiol. 49, 221-228.

Stout, J.E., Best, M.G., Yu, V.L., and Rihs, J.D. (1986). A note on symbiosis of *Legionella pneumophila* and *Tatlockia micdadei* with human respiratory flora. J. Appl. Bacteriol. 60, 297-299. DOI: 10.1111/j. 1365-2672.1986.tb01736.x

Surman, S., Morton, G., Keevil, C.W., and Fitzgeorge, R. (2002). *Legionella pneumophila* proliferation is not dependent on intracellular replication. In *Legionella*, R. Marre, Y. Abu Kwaik, C. Bartlett, N.P. Cianciotto, B.S. Fields, M. Frosch, J. Hacker, and P.C. Lück, eds. (Washington, USA: ASM Press), pp. 86-89.

Taylor, M., Ross, K., and Bentham, R. (2009). *Legionella*, protozoa, and biofilms: Interactions within complex microbial systems. Microb. Ecol. 58, 538-547. DOI 10.1007/s00248-009-9514-z

Temmerman, R., Vervaeren, H., Noseda, B., Boon, N., and Verstraete, W. (2006). Necrotrophic growth of *Legionella pneumophila*. Appl. Environ. Microbiol. 72, 4323-4328. doi: 10.1128/AEM.00070-06

Tesh, M.J., and Miller, R.D. (1981). Amino acid requirements for *Legionella pneumophila* growth. J. Clin. Microbiol. 13, 865-869.

Tesh, M.J., and Miller, R.D. (1982). Growth of *Legionella pneumophila* in defined media: requirement for magnesium and potassium. Can. J. Microbiol. 28, 1055-1058.

Tesh, M.J., and Miller, R.D. (1983). Arginine biosynthesis in *Legionella pneumophila*: absence of N-acetylglutamate synthetase. Can. J. Microbiol. 29, 1230-1233.

Thornley, C.N., Harte, D.J., Weir, R.P., Allen, L.J., Knightbridge, K.J., and Wood, P.R.T. (2017). *Legionella longbeachae* detected in an industrial cooling tower linked to a legionellosis outbreak, New Zealand, 2015; possible waterborne transmission? Epidemiol. Infect. 145, 2382-2389. DOI: 10.1017/S0950268817001170

Tiaden, A., and Hilbi, H. (2012). α-hydroxyketone synthesis and sensing by *Legionella* and *Vibrio*. Sensors 12, 2899-2919. doi:10.3390/s120302899

Tiaden, A., Spirig, T., Weber, S.S., Brüggemann, H., Bosshard, R., Buchrieser, C., and Hilbi, H. (2007). The *Legionella pneumophila* response regulator LqsR promotes host cell interactions as an element of the virulence regulatory network controlled by RpoS and LetA. Cell. Microbiol. 9, 2903-2920. doi:10.1111/j.1462-5822.2007.01005.x

Tiaden, A., Spirig, T., Sahr, T., Wälti, M.A., Boucke, K., Buchrieser, C. and Hilbi, H. (2010). The autoinducer synthase LqsA and putative sensor kinase LqsS regulate phagocyte interactions, extracellular filaments and a genomic island of *Legionella pneumophila*. Environ. Microbiol. 12, 1243-1259. doi: 10.1111/j.1462-2920.2010.02167.x

Tison, D.L. (1987). Microbial ecology of *Legionella*. J. Infect.Dis. 156, 852.

Tison, D.L., Pope, D.H., Cherry, W.B., and Fliermans, C.B. (1980). Growth of *Legionella pneumophila* in association with blue-green algae (cyanobacteria). Appl. Environ. Microbiol. 39, 456-459.

Toze, S., Sly, L.I., MacRae, I.C., and Fuerst, J.A. (1990) Inhibition of growth of *Legionella* species by heterotrophic plate-count bacteria isolated from chlorinated drinking water. Curr. Microbiol. 21, 139-143. https://doi-org.ezproxy.library.dal.ca/10.1007/BF02091832

Toze, S., Cahill, M., Sly, L.I., and Fuerst, J.A. (1994). The effect of *Aeromonas* strains on the growth of *Legionella*. J. Appl. Microbiol. 77, 169-174. DOI: 10.1111/j.1365-2672.1994.tb03061.x

Trigui, H., Dudyk, P., Oh, J., Hong, J.-I., and Faucher, S.P. (2015). A regulatory feedback loop between RpoS and SpoT supports the survival of *Legionella pneumophila* in water. Appl. Environ. Microbiol. 81, 918 -928. doi:10.1128/AEM.03132-14

van der Kooij, D., Brouwer-Hanzens, A.J., Veenendaal, H.R., and Wullings, B.A. (2016). Multiplication of *Legionella pneumophila* sequence types 1, 47, and 62 in buffered yeast extract broth and biofilms exposed to flowing tap water at temperatures of 38 °C to 42 °C. Appl. Environ. Microbiol. 82, 6691-6700. doi:10.1128/AEM.01107-16

van Heijnsbergen, E., Husman, A.M.D., Lodder, W.J., Bouwknegt, M., van Leeuwen, A.E.D., Bruin, J.P., Euser, S.M., den Boer, J.W., and Schalk, J.A.C. (2014). Viable *Legionella pneumophila* bacteria in natural soil and rainwater puddles. J. Appl. Microbiol. 117, 882-890. DOI: 10.1111/jam.12559

van Heijnsbergen, E., van Deursen, A., Bouwknegt, M., Bruin, J.P., de Roda Husman, A.M., and Schalk, J.A.C. (2016). Presence and persistence of viable, clinically relevant *Legionella pneumophila* bacteria in garden soil in the Netherlands. Appl. Environ. Microbiol. 82, 5125-5131. doi: 10.1128/AEM.00595-16

Veenendaal. H.R., Brouwer-Hanzens, A.J., and van der Kooij, D. (2017). Incubation of premise plumbing water samples on buffered charcoal yeast extract agar at elevated temperature and pH selects for *Legionella pneumophila*. Water Res. 123, 439-447. http://dx.doi.org/10.1016/j.watres.2017.06.077

Venkataraman, C., Haack, B.J., Bondada, S., and Abu Kwaik, Y. (1997). Identification of a Gal/GalNAc lectin in the protozoan *Hartmanella vermiformis* as a potential receptor for attachment and invasion by the Legionnaires' disease bacterium. J. Exp. Med. 186, 537-547. DOI: 10.1084/jem.186.4.537

Verdon, J., Berjeaud, J.-M., Lacombe, C., and Héchard, Y. (2008). Characterization of anti-*Legionella* activity of warnericin RK and delta-lysin I from *Staphylococcus warneri*. Peptides 29, 978-984. doi:10.1016/j.peptides.2008.01.017

Vogel, J.P., Roy, C., and Isberg, R.R. (1996). Use of salt to isolate *Legionella pneumophila* mutants unable to replicate in macrophages. Ann. NY Acad. Sci. 797, 271-272.

Vogel, J.P., Andrews, H.L., Wong, S.K., and Isberg, R.R. (1998). Conjugative transfer by the virulence system of *Legionella pneumophila*. Science 279, 873-876. DOI: 10.1126/science.279.5352.873

Wadowsky, R.M., and Yee, R.B. (1983). Satellite growth of *Legionella pneumophila* with an environmental isolate of *Flavobacterium breve*. Appl. Environ. Microbiol. 46, 1447-1449.

Wadowsky, R.M., and Yee, R.B. (1985). Effect of non-*Legionellaceae* bacteria on the multiplication of *Legionella pneumophila* in potable water. Appl. Environ. Microbiol. 49, 1206-1210.

Wadowsky, R.M., Yee, R.B., Mezmar, L., Wing, E.J., and Dowling, J.N. (1982). Hot water systems as sources of *Legionella pneumophila* in hospital and nonhospital plumbing fixtures. Appl. Environ. Microbiol. 43, 1104-1110.

Wadowsky, R.M., Wolford, R., McNamara, A.M., and Yee, R.B. (1985). Effect of temperature, pH and oxygen level on the multiplication of naturally occurring *Legionella pneumophila* in potable water. Appl. Environ. Microbiol. 49, 1197-1205.

Warren, W.J., and Miller, R.D. (1979). Growth of legionnaires disease bacterium (*Legionella pneumophila*) in chemically defined medium. J. Clin. Microbiol. 10, 50-55.

Whiley, H., and Bentham, R. (2011). *Legionella longbeachae* and legionellosis. Emerg. Infect. Dis. 17, 579-583. DOI: 10.3201/eid1704.100446

White, P.A., Kalff, J., Rasmussen, J.B., and Gasol, J.M. (1991). The effect of temperature and algal biomass on bacterial production and specific growth rate in freshwater and marine habitats. Microb. Ecol. 21, 99-118. doi:10.1007/BF02539147

Wieland, H., Ullrich, S., Lang, F., and Neumeister, B. (2005). Intracellular multiplication of *Legionella pneumophila* depends on host cell amino acid transporter SLC1A5. Mol. Microbiol. 55, 1528-1537. DOI: 10.1111/j.1365-2958.2005.04490.x

Yee, R.B., and Wadowsky, R.M. (1982). Multiplication of *Legionella pneumophila* in unsterilized tap water. Appl. Environ. Microbiol. 43, 1330-1334.

Zamboni, N., Fendt, S.M., Ruhl, M., and Sauer, U. (2009). ^{13}C-based metabolic flux analysis. *Nat. Protoc.* 4, 878-892. doi:10.1038/nprot.2009.58

Zeigler-Ballerstein, S.D., and Barbaree, J.M. (2006). Evaluation of signaling between *Legionella pneumophila* and multiple prokaryotes. In *Legionella*: State of the Art 30 Years after its Recognition, N.P. Cianciotto, Y. Abu Kwaik, P.H. Edelstein, B.S. Fields, D.F. Geary, T.G. Harrison, C.A. Joseph, R.M. Ratcliff, J.E. Stout, and M.S. Swanson, eds. (Washington, USA: American Society for Microbiology), pp. 403-406.

Chapter 3

A "Secreted Army" for the Invasion and Survival of *Legionella pneumophila* Within Host Cells

Elisabeth Kay[1*], Virginie Lelogeais[1], Sophie Jarraud[1,2], Christophe Gilbert[1] and Patricia Doublet[1]

[1]CIRI, Centre International de Recherche en Infectiologie (Team: *Legionella* pathogenesis), Univ Lyon, Inserm, U1111, Université Claude Bernard Lyon 1, CNRS, UMR5308, ENS de Lyon, F-69007, Lyon, France
[2]Centre de Biologie et de Pathologie Est, Centre National de Référence des Légionelles, 69677 Bron Cedex, France

*elisabeth.kay@univ-lyon1.fr

DOI: https://doi.org/10.21775/9781913652531.03

Abstract
Intracellular multiplication within protozoans is an essential step in the emergence of pathogenic *Legionella pneumophila* strains. Key features of intracellular survival and multiplication are secretion systems propelling virulence factors and other important substances into their surroundings. These systems are highly diverse and likely contribute to the versatility of *Legionella* species which replicate in a wide spectrum of hosts. This chapter discusses secretion systems with emphasis on the highly conserved Dot/Icm T4BSS and its regulators. First, the unique features of each type of secretion system employed by *Legionella* is described. Then, a review of the current knowledge about the Dot/ICM T4BSS highlighting

the cohort of secreted proteins which determine the intracellular fate of *Legionella* is presented. In addition, regulators coordinating responses to environmental cues and determining virulence will be discussed. This shows how *Legionella* is an excellent model for the study of intracellular pathogens, as they survive inside a wide variety of hosts and even manipulate the intracellular environment to increase virulence.

Introduction

L. pneumophila are facultative intracellular bacteria that multiply within phagocytic cells. *L. pneumohila* pathogenic strains emerge from the environment after intracellular multiplication in protozoans, especially amoebae, and become disseminated by contaminated aerosols permitting direct access to human lungs. In the respiratory tract they infect alveolar macrophages thus leading to pneumonia legionellosis. Within environmental phagocytic cells and human macrophages, *L. pneumophila* display a similar infectious cycle that involves four main steps (Figure 1): (1) Bacteria evasion from endocytic degradation. After the bacterial uptake, the pH of the *Legionella*-containing phagosome is around 6.1 compared to less than five within the *Escherichia coli*-containing vacuole (Horwitz and Maxfield, 1984), and the phagosome does not exhibit early or late endosomal and lysosomal markers such as Rab5, Rab7 and LAMP-1 (Clemens *et al.*, 2000a, Clemens *et al.*, 2000b, Roy *et al.*, 1998). (2) Biogenesis of a *Legionella*-containing vacuole (LCV), a rough endoplasmic reticulum-like compartment permissive for its intracellular replication. Within 15 minutes of uptake, the LCV is surrounded and fused with ER-derived smooth vesicles and mitochondria (Horwitz, 1983), and four hours post-contact it is decorated by host-cell ribosomes thus resulting in a replication-permissive vacuole (Roy and Tilney, 2002, Horwitz and Silverstein, 1981). (3) Intensive intracellular replication. In the rough ER-like compartment, *L. pneumophila* proliferates in a replicative form. (4) Genetic re-programming to support the synthesis of virulence traits such as motility and virulence factors. When vacuolar nutrients

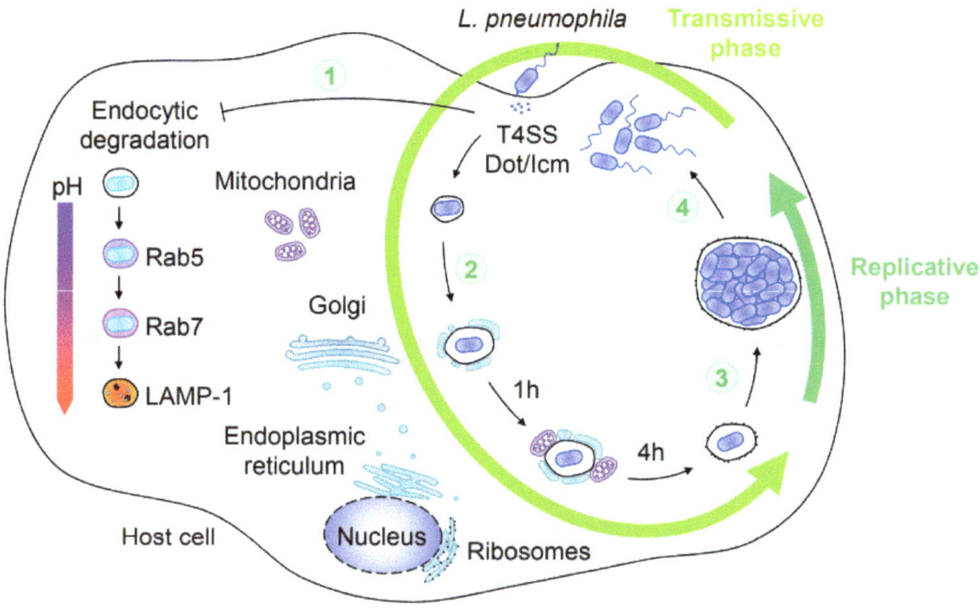

Figure 1. After contact with the host cell, Dot/Icm effectors are secreted and interfere with various host cell pathways. 1) *L. pneumophila* inhibits endocytic degradation, acquisition of endosomal or lysosomal markers and vacuole acidification. 2) *L. pneumophila* induces recruitment of ER vesicles and interacts with mitochondria and ribosomes. 3) After biogenesis of the LCV, *L. pneumophila* replicates efficiently in a replicative phase. 4) Nutrients starvation triggers an extensive genetic re-programming of *L. pneumophila* that allows expression of transmissive traits such as flagella.

become limited, the progeny differentiate into the transmissive phase, repressing multiplication and expressing several traits. The new traits allow *L. pneumophila* to kill and escape from its host cell and survive as planktonic cells until it can re-establish a replicative niche within a new phagocytic cell, environmental or human.

L. pneumophila is a highly adapted intravacuolar pathogen that has set up sophisticated biochemical strategies to hijack host cell processes. Crucial for manipulating the host cell to the benefit of the bacteria is the secretion of bacterial proteins by a type I secretion system (T1SS), a type II secretion system (T2SS) and most of all the type IV secretion system

(T4SS) defective in organella trafficking (Dot) or intracellular multiplication (Icm) proteins. In this chapter, we revisit the role of these macromolecular systems and some of their secreted proteins and review how *Legionella* spp. orchestrate expression in order to survive and evade protozoan or mammalian host cells.

Secretion systems in *Legionella*

A wide range of substrates such as proteins, DNA and small molecules have been identified as substances secreted by bacteria. While small molecule transport systems are coordinated by efflux pumps, DNA and proteins transport is mediated by machineries called secretion systems. In Gram negative bacteria, the diverse protein complex machineries are

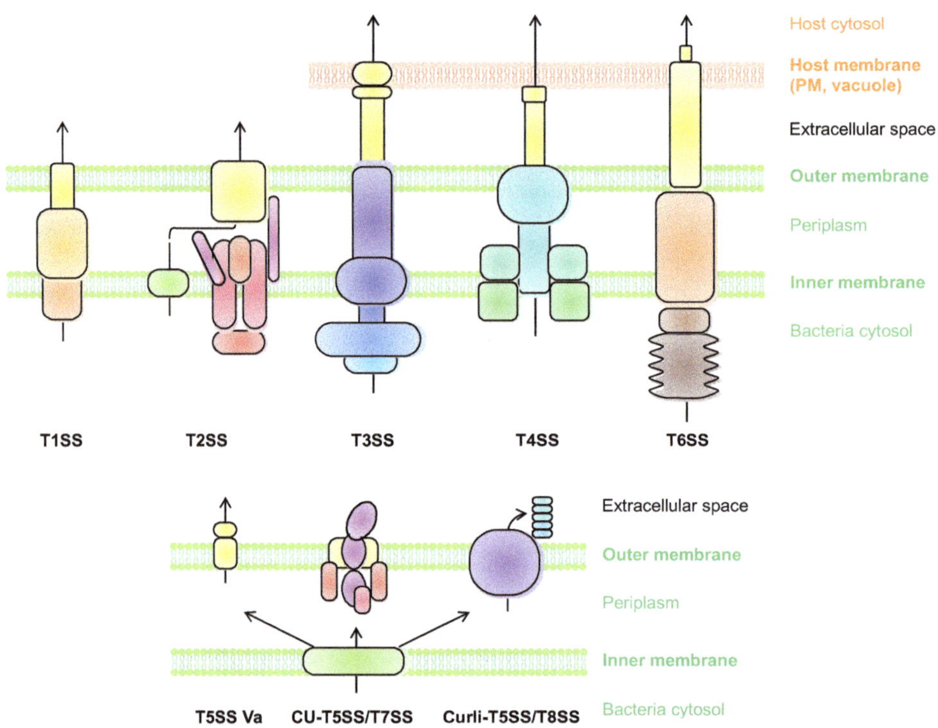

Figure 2. Models of secretion mechanisms (TSS) in Gram negative bacteria. CU: Chaperone Usher.

classified in 6 types of secretion systems (T1SS-T6SS; Figure 2; for review see Costa *et al.*, 2015). Among these systems, T2SS and T5SS are characterized by a two-step secretion mechanism in which substrates are translocated from the cytoplasm into the periplasm via the SecYEG or Tat systems and afterward through the outer membrane to the extracellular space (Collinson *et al.*, 2015). On the contrary, T1SS, T3SS, T4SS and T6SS translocate the substrates in a one-step mechanism from the cytoplasm to the extracellular space (T1SS) or into the target/host cell cytoplasm by crossing a third membrane being either host plasma or vacuole membrane (T3SS, T4SS, T6SS).

The T5SS are divided into five subclasses (a-e) all sharing common features such as a ß-barrel protein domain localized in the outer membrane and a SecYEG dependent pathway through the inner membrane. Briefly, the type Va corresponds to monomeric autotransporters (Gawarzewski *et al.*, 2014), Vb to two-partner systems (Rahman *et al.*, 2014), Vc to trimeric autotransporters (Lyskowski *et al.*, 2011), Vd to patatin-like autotransporters (Salacha *et al.*, 2010) and Ve to the inverse autotransporters (intimin-invasin protein family) (Leo *et al.*, 2015). However, authors have recently proposed modifications renaming some type V secretion systems as type VII (corresponding to the chaperone/usher (CU; Type 1 pilus secretion system) and type VIII (curli secretion) (Figure 2) (Desvaux *et al.*, 2009). However, this new nomenclature does not seem to be adopted by all authors, as the Type VII secretion system nomenclature still refers to a specific system identified in *Mycobacterium* species with no common features of the type 1 pilus system (Groschel *et al.*, 2016). Clarifications will be necessary in the future to avoid any confusion between the T7SS of *Mycobacteria* and T7SS in Gram negative bacteria. Since 2009, a new secretion system discovered in *Porphyromonas gingivalis* has been proposed as type IX (T9SS) (Sato *et al.*, 2010, Sato *et al.*, 2013). This T9SS with no significant homologies with the already known T9SS is involved in the gliding motility and has been identified in other Gram negative bacteria species (Kita *et al.*, 2016). Among *Legionella* species currently four TSS have been found including a putative T5SS and the functionally characterized T1SS, T2SS and T4SS.

T1SS in Legionella

The T1SS has the simplest structure among secretion systems in Gram negative bacteria, and the archetype is the secretion system of hemolysin in *Escherichia coli* composed by the tripartite structure HlyB/HlyD/TolC (Figure 2). It is composed by an inner membrane ATP Binding Cassette (ABC) protein; HlyB in *E. coli* transporter which are homo or heterodimer complex proteins linked to a homotrimer outer membrane protein, TolC in *E. coli* or TolC-like protein, via an adapter protein located in the periplasmic space. This adapter protein, HlyD in *E. coli*, is an homotrimer complex formed by proteins anchored on the periplasmic face of the inner membrane, therefore displaying a large domain in the periplasm. A T1SS composed by a complex named LssB/LssD/TolC has been recently characterized in *L. pneumophila* (Figure 3) (Fuche *et al.*, 2015) along with the unique substrate, RtxA, a huge protein of 700 kDa in strain *L. pneumophila* Paris. Like all T1SS secreted substrates, the secretion signal is in the C-terminus domain of RtxA and is not cleaved during secretion. This is a major characteristic of T1SS. However, no consensus secretion signal sequence can be drawn for T1SS, and the recognition is thought to be structural.

Three families of T1SS can be distinguished on the basis of the inner membrane transporter protein and of the substrate family: (i) T1SS triggering the secretion of bacteriocins, which are maturated during the transport by action of C39-peptidase activity of the inner membrane transporter cleaving an N terminus sequence, (ii) T1SS involved in the secretion of repeat-in ToXins (RTX) based on GGxDxDxxx repeat motifs substrate family including the *E. coli* hemolysin (HlyA), (iii) T1SS with non-

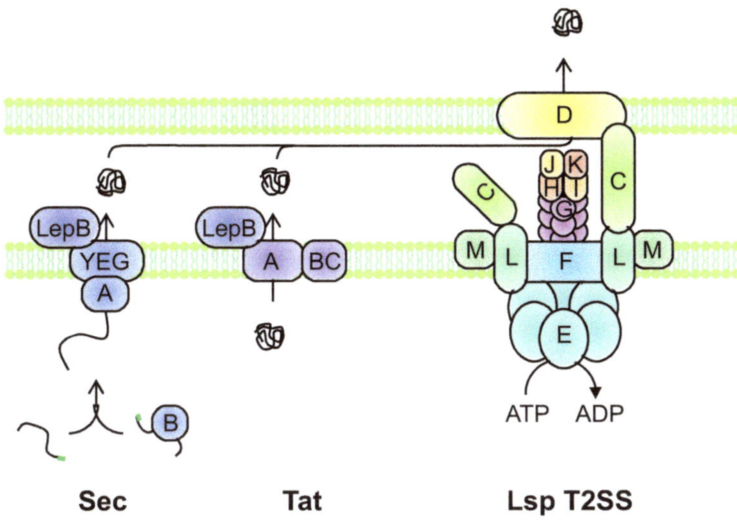

Figure 3. Type II secretion system in *Legionella*. Unfolded or folded proteins are transported from the cytoplasm to the periplasm via the Sec and Tat machineries, respectively. In the second step, the matured (signal sequence cut by LepB) folded substrates are secreted in the extracellular space, being "pushed" through the outer membrane by the pseudopilus "piston" action of Lsp T2SS.

RTX and non-pro-protein substrates (Delepelaire, 2004, Kanonenberg *et al.*, 2013). Based on this classification, the *Legionella* T1SS is clearly a member of the second family involved in a RTX substrate. As other members of this family, the inner membrane transporter LssB possesses a C39-like domain that lost the C39 activity resulting in no cleavage of RtxA protein by its own. However, recent work on *Pseudomonas fluorescens* T1SS showed that RTX substrates such as LapA in *P. fluorescens* can be found embedded in the outer membrane or released in the extracellular space depending on the N-terminus cleavage by a periplasmic protease LapG. Moreover, this cleavage is controlled by concentration of cyclic di-GMP via an inner membrane protein LapD interacting with LapG (Boyd *et al.*, 2014). Interestingly, homologous LapG/LapD proteins have been described in *L. pneumophila,* and LapG from *L. pneumophila* can cleave *P. fluorescens* LapA (Chatterjee *et al.*, 2012). LapA is described as an adhesin involved in biofilm formation in *P. fluorescens.* Recently, Smith *et al.* proposed to classify theses giant adhesins in a new subfamily of T1SS

(Smith *et al.*, 2018). Indeed, their work suggests that uncleaved LapA is retained inside LapE (TolC-like) outer membrane pore. Therefore, it is displayed at the bacterial surface untill cleavage and release by LapG. Currently, the mechanism of RtxA is still unclear in *Legionella*, but its role during the primary steps of amoeba or macrophage infection has been characterized by the fact that T1SS or RtxA deleted strains are less virulent and deficient in early stage of infection (Ferhat *et al.*, 2009, Fuche *et al.*, 2015). However, a role of RtxA during later stages of infection such as intracellular replication or host cell evasion cannot be ruled out, as TolC has been shown to be involved into cellular trafficking during infection of *Paramecium tetraurelia* (Nishida *et al.*, 2018).

Among *Legionella* species, all *L. pneumophila* genomes sequenced to date harbour the T1SS LssB/LssD/TolC which is also present in other species such as *L. moravica*, *L. quateiensis*, *L. shakespearei*, *L. israelensis* and more. In this case, T1SS is always associated with the presence of a putative RTX protein. RTX proteins, however, vary both in length and amino-acid sequences in the repeat region, which is the middle part of the protein, even between strains within the same species. Although, neither T1SS nor RTX protein can be found in the known *L. longbeachae* genomes, and this may suggest different processes involved in early stages of infection for this species.

T2SS in Legionella

As in all known Gram negative bacterial T2SSs, the substrate proteins are first translocated from cytoplasm to periplasm via the SecYEG or the Tat pathways (Figure 3). Currently, the *sec* and *tat* family genes have been identified in all *Legionella* characterized genomes. The SecYEG system is composed by an inner membrane pore formed by the three proteins SecY, SecE and SecG. Two accessory proteins, SecD and SecF, have been identified that interact with SecYEG complex in conjunction with proteins YajC and YidC to form the holo-translocon involved in translocation and insertion of inner membrane proteins (Du Plessis *et al.*, 2011, Collinson *et al.*, 2015). In the T2SS associated SecYEG pathway, the substrate protein harbouring an N-terminus hydrophobic signal sequence is recognized by the cytoplasmic protein SecB (chaperon) and is delivered unfolded to SecA (ATPase). The Substrate-SecA complex binds to SecYEG and successive hydrolysis of ATP provides the energy to translocate the protein through the inner membrane (Figure 3). This secretion process is stimulated by the proton motive force by a SecDF and SecY dependent manner. At the end of the translocation process, the periplasmic peptidase, LepB, plays a crucial role by cleaving the signal sequence which is initially 20 to 25 amino acids in length after the recognition motif A-x-A. On the contrary, the Twin arginine translocation (Tat) system enables the secretion of folded proteins (Arambula *et al.*, 2013). Similar to that of other Gram negative bacteria, the Tat system in *Legionella* seems to be composed of three essential inner membrane proteins TatA, TatB and TatC (Cleon *et al.*, 2015, De Buck *et al.*, 2004). Studies showed that components of this Tat systems were essential for *Legionella* to translocate different substrates (Rossier and Cianciotto, 2005, De Buck *et al.*, 2008), and that Tat secretion susbtrates are involved in *Legionella* biofilm formation and virulence (De Buck *et al.*, 2005). The substrate proteins of Tat system possess a signal sequence with highly conserve motifs which is S-R-R-x-F-L-K close to the N terminus and A-x-A and the cleavage site of LepB. The Tat signal sequences are also less hydrophobic than the SecYEG signal sequences. The two arginines are characteristic of the Tat signal sequence. However, the motif would still functionally translocate proteins through the inner membrane though with less efficiency if a serine replaced the first aginine.

Following the translocation across the inner membrane and leader peptide cleavage, the protein susbtrates are recognized by a multiprotein complex specifically dedicated to T2SS, the Lsp apparatus of *Legionella*. This machinery is composed by 12 proteins (Figure 3). There are three inner membrane proteins establishing a platform, LspF, LspL and LspM. The periplasmic pseudopilus is formed by the major pseudopilin LspG and four minor pseudopilins LspH, LspI, Lsp J and LspK that are integrated in the structure after being processed by the inner membrane prepilin peptidase PilD (Liles *et al.*, 1998, Soderberg *et al.*, 2004). This pilus-like structure spans the periplasm. The last part of this T2SS structure corresponds to the outer membrane pore, an oligomer of the secretin protein LspD. The link between the inner membrane platform and the secretion outer membrane pore is maintained by the periplasmic protein LspC (Cianciotto, 2009). In this model, the protein substrates are somehow recognized by the Lsp machinery and pushed through the outer membrane secretin pore by the pseudopilus LspG/H/I/J/K acting like a piston using the energy generated from ATP hydrolysis by the ATPase LspE. Defective *Legionella* Lsp mutants reveals that at least 25 protein substrates were dependant of this T2SS secretion including lipases (PlaA, PlcA, LipA, LipB...), aminopeptidases and proteases (LapA, LapB, MspA, LegP) phosphatases and one ribonuclease (SrnA) (DebRoy *et al.*, 2006). The role of all the T2SS substrates still warrants investigation, but it appears clearly that many of them are essential for virulence of *Legionella* and for its persistence in the environment or within the lung (Soderberg *et al.*, 2004, Tyson *et al.*, 2014, Cianciotto, 2014, DebRoy *et al.*, 2006). For example, a mutant of *L. pneumophila* strain AA200 deficient in expression of zinc metalloprotease MspA showed impaired growth within *Hartmannella vermiformis* host and *Naegleria lovaniensis*, but still could infect *Acanthamoeba castellanii* and macrophages (Rossier *et al.*, 2008, Tyson *et al.*, 2013). This pointed out both the role of some T2SS substrates in virulence and the differential role of effectors in each host. Recently T2SS was also shown to be implicated in the LCV formation by two ways. The first involves the recruitment of Rab1B, a host GTPase that facilitates the tethering of endoplasmic reticulum (ER)-derived vesicles to LCVs, and the second is independent of Rab1B (White and Cianciotto, 2016). Although the case of IcmX substrate is more intriguing. This protein member of

T4BSS is translocated into the periplasm via a Sec-dependent pathway and has been identified in the extracellular medium in a truncated form during the growth in liquid medium, and therefore it is potentially a substrate of T2SS (Matthews and Roy, 2000). However, no secretion in the host cell cytoplasm was detected during the infection cycle. This indicates that this T2SS translocation may not occurred in all life cycles and suggests the possible links between T2SS and the functionality of T4BSS in *Legionella*.

In conclusion, T2SS is mainly studied in *L. pneumophila* species. However, genome sequence analysis reports the presence of highly conserved Lsp machinery and of many substrate proteins in other *Legionella* species which confirms the importance of this *Legionella* T2SS for survival in the environment. The function of all substrate proteins is still under investigation, and *L. pneumophila* genome analysis prediction revealed 60 proteins harbouring a potential Sec or Tat signal sequence which is one of the largest T2SS substrate libraries in the realm of bacteria (DebRoy *et al.*, 2006). Both the number of T2SS effectors and the impact of Lsp system in virulence suggest the essential role of T2SS in *Legionella* pathogenicity and may be even as essential as the T4BSS, especially during infection of amoeba host cells.

T5SS in Legionella

The potential autotransporter protein corresponding to T5aSS, Lpp0779, was first describes in the *L. pneumonila* paris genome (Cazalet *et al.*, 2004). It contains an N-terminal leader peptide for secretion across inner membrane and a C-terminal domain forming a pore in the outer membrane. The autotransporter passenger domain composed of hemagglutinin repeats can pass to the cell surface via this pore and presents high homologies with *E. coli* autotransporters AIDA-I and Ag43 which are two proteins implicated in virulence mediating adherence to mammalian cell and cell-cell aggregation. Astonishingly, the presence of this autotransporter seems to be restricted to a minimal panel of strains in *L. pneumophila* species such as Paris, Leg01/20 and *Legionella* strain sp. 39-23. However, the role of this T5aSS in *Legionella* virulence has yet to be demonstrated.

T4SS in Legionella

Various type IV secretion systems have been described in *Legionella*. T4SS are classified as T4ASS and T4BSS based on homologies with the VirB/VirD4 T4SS of *Agrobacterium tumefaciens* or with the conjugative T4SS of IncI plasmid family respectively (Voth *et al.*, 2012).

Within the T4ASS, three subclasses can be proposed in *Legionella*. The first one named Tra is homologous to the Tra system characterized in *E. coli* conjugative F-plasmid. The second one, Trb system, corresponds to a *Pseudomonas aeruginosa* conjugative P-type homologous system. These two systems seem to be present in many *Legionella* species such as *pneumophila*, *hackeliae*, *falloni* or l*ongbeachae* and are encoded within integrative conjugative elements (ICE) (Khodr *et al.*, 2016, Gomez-Valero *et al.*, 2011). Nevertheless, not all strains in each species contain these Tra/Trb T4SS. This notion adheres to the hypothesis of horizontal genes transfer origin rather than vertical transmission from a common ancestor. Moreover, some strains have more than one system as three are present in *L. longbeachae* NSW150, two Tra and one Trb. These Tra and Trb systems correspond to conjugative systems that enable direct transfer of DNA from a donor to a recipient bacterium and are not associated with virulence. Recently, a class of genomic islands (GI-like) lacking in T4SS machinery but capable of mobility and substrate of conjugation, potentially mobilized via Tra or Trb, have been described in *L. pneumophila* (Schroeder *et al.*, 2010, Wee *et al.*, 2013). Up to four GI-like elements have been identified in *L. drancourti* LLAP12 genome and many other species harbours either none, one or two of them depending on strain, and this is a mark of high genome plasticity in *Legionella*. The third T4ASS identified is the Lvh system encoded by a plasmid-like element that can be integrated into a *L. pneumophila* chromosome or can exist as a multi copy plasmid depending on growth phase (Cazalet *et al.*, 2004, Doleans-Jordheim *et al.*, 2006). Few studies demonstrated its importance in the infection process at low temperatures (30°C) (Ridenour *et al.*, 2003). Moreover, its role in infection appeared link to stress activation as a *L. pneumophila* Δ*dotA* mutant strain, which is therefore impaired for T4BSS secretion and virulence as described below and submitted to a water stress for restoration of virulent capacity towards amoeba (Bandyopadhyay *et al.*, 2007, Bandyopadhyay *et al.*, 2013). It was shown that the VirD4 component of Lvh is responsible of this complementation by interacting with the T4BSS secretion system which implicates Lvh in some effectors translocation process. Interestingly, the contribution of Lvh in conjunction with T4BSS to DNA conjugation process in *Legionella* was described earlier (Vogel *et al.*, 1998, Segal *et al.*, 1999). Therefore, Lvh system identified in *L. pneumophila,* which is also present in few other species such as *micdadei* and *longbeachae,* seems to be an important partner of T4BSS in the translocation process of DNA and effector proteins.

The T4BSS known as Dot/Icm T4SS is the most studied secretion system of *Legionella* spp. due to its essential role in virulence. Indeed, it was first identified by isolation of *L. pneumophila* mutant strains unable to replicate within the macrophage and therefore displaying an avirulent phenotype (Berger *et al.*, 1994, Brand *et al.*, 1994, Segal *et al.*, 1998, Vogel *et al.*, 1998). Moreover, currently, chromosomal Dot/Icm T4SS genes have been identified and highly conserved in all *Legionella* genome across species and strains (Gomez-Valero *et al.*, 2014, Burstein *et al.*, 2016). This supports the hypothesis that a functional T4BSS is necessary for *Legionella* to survive within the amoeba in its natural habitat. From the late 90s, few research groups are working to elucidate the structure and the functioning of this Dot/Icm T4SS (Kuroda *et al.*, 2015, Coers *et al.*, 2000,

Nakano *et al.*, 2010, Kubori *et al.*, 2014, Watarai *et al.*, 2001, VanRheenen *et al.*, 2004, Buscher *et al.*, 2005, Kubori and Nagai, 2016). Among the proposed 26 proteins involved in Dot/Icm T4SS, the role of many components are still not clear, though a working model has been proposed (Farelli *et al.*, 2013). New insights in Dot/Icm molecular architecture using electron cryotomography have been recently published providing more insight into the structure and function (Chetrit *et al.*, 2018, Ghosal *et al.*, 2017, Ghosal *et al.*, 2018). Based on these studies, the proteins are well organized and building a central stalk channel in the inner membrane that opens in a periplasmic secretion chamber (Figure 4).

Taking a closer look at the proteins involved in this complex, DotC and DotD are two outer membrane lipoproteins associated with DotH for which the location in the outer membrane is dependent of DotC/DotD presence. The lipoprotein DotK is also located on the internal face of the outer membrane in association with the DotC/DotD/DotH complex and seems to be important during infection of certain hosts (Qin *et al.*, 2012, Yerushalmi *et al.*, 2005). Astonishingly, an unusual number of 13-fold symmetry, compared to 14-fold symmetry seen in most T4ASS, has been identified in the ring formed by this complex meaning 13 of each protein association. DotF and DotG are two inner membrane interacting proteins with extensions into the periplasmic space. This complex also forms a 13-fold symmetry ring which could be stabilized by DotU and IcmF proteins that are proposed to be located inside the ring (Isberg *et al.*, 2009, Ghosal *et al.*, 2018).

By analogies to Vir T4SS from *E. coli* R388 conjugative plasmid, the characterized *Legionella* DotI/DotJ complex potentially 12-hetero-complexes to form a ring and may be associated with the DotO ATPase (Low *et al.*, 2014). Moreover, the hexameric ring organisation of DotB/DotO ATPases complex on the inner face of the inner membrane has been recently clarified (Prevost and Waksman, 2018, Chetrit *et al.*, 2018).

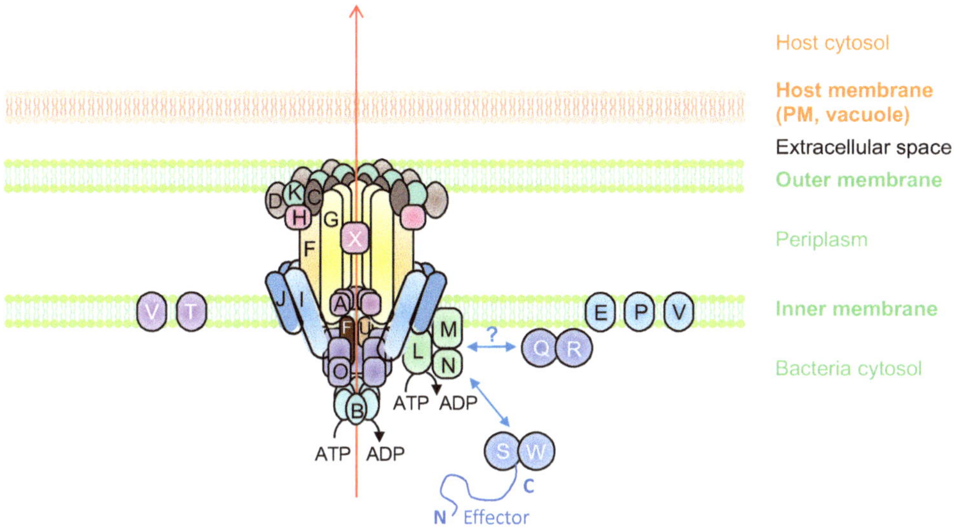

Figure 4. Model of Dot/Icm T4BSS in *L pneumophila*. Dot proteins are in black letters and Icm proteins are in white. Dot nomenclature has been privileged to draw the figure but depending on the authors, many proteins of Dot/Icm system have been annotated with two names as follow: DotE=IcmC, DotF=IcmG, DotG=IcmE, DotH=IcmK, DotI=IcmL, DotJ=IcmM, DotK=IcmN, DotL=IcmO, DotM=IcmP, DotN=IcmJ, DotO=IcmB, DotP=IcmD, DotU=IcmH, DotV=IcmC, IcmX=IcmY.The blue arrows correspond to protein-protein interactions.

The cytoplasmic ATPase DotB is crucial to Dot/Icm function (Sexton *et al.*, 2005, Sexton *et al.*, 2004). It is also suggested that a complex of the three proteins able to bind ATP, DotL, DotM and DotN, can have two functions: (i) to act as a gatekeeper towards effectors to translocate through the T4SS pore (ii) to interact with IcmS/IcmW which are two cytosolic acidic proteins able to bind the effectors by recognition of a carboxy-terminal signal (substrates of Dot/Icm T4SS; Figure 4) (Vincent *et al.*, 2012, Meir *et al.*, 2018). Interestingly, structural interactions have been proposed between DotL and DotB/DotO complex suggesting a hexameric ring structure for DotL (Ghosal *et al.*, 2018). The acidic phospholipids membrane binding by IcmQ/IcmR complex has been shown and seems to be dependent of the NAD+/IcmQ binding in *Legionella* (Farelli *et al.*, 2013). Therefore, a possible NAD-dependent role of interaction with other Dot/Icm components is possible.

IcmX is a periplasmic protein that plays a role in establishing the LCV during the infection cycle (Matthews and Roy, 2000). IcmX is also newly proposed to be located inside the Dot/Icm secretion chamber (Figure 4). Finally, DotA, an essential protein of the T4BSS, contains eight transmembrane hydrophobic domains and is often reported as an inner membrane component. Recent studies have contemplated a particular location of the periplasmic DotA domains ring at the top of the stalk channel where linking with the periplasmic secretion chamber occurs (Ghosal *et al.*, 2018). This location is compatible with the observation of Nagai and Roy in 2001 suggesting that DotA could also be a substrate of the Dot/Icm machinery and may even be released into culture media as oligomeric ring like structures of 10 nm diameter (Nagai and Roy, 2001).

To conclude, the precise role of each Dot/Icm component is still under investigation and few cannot yet be placed in the model. However, its essential, and perhaps universal, role in *Legionella* pathogenicity is enabling the translocation of a huge number of effectors in host cell. This is relevant to the high number of studies on T4BSS effectors, as discussed in the next paragraphs.

Control of host cell pathways by the Dot/Icm effectors

Crucial for intracellular fate and intracellular multiplication of *L. pneumophila,* is the T4SS Dot/Icm and its effector proteins. Although the specific host cell pathways targeted by the 300 Dot/Icm effectors have not been fully identified, these bacterial proteins have been demonstrated to interfere with various host cell processes, such as (i) vesicular trafficking to inhibit phagosome maturation and to trigger LCV biogenesis, (ii) protein ubiquitination to contribute to the nutrient acquisition necessary for efficient intravacuolar multiplication, (iii) the host apoptotic program and innate immune defence-associated pathways such as NF-κB pathway and autophagy process (Figure 5).

Hijack vesicular trafficking for escape from endocytic degradation

Immediately after a bacteria is internalized by a phagocytic cell, the phagosome maturates by sequentially fusing with early endosomes, late endosomes that strongly acidify the vacuole and finally with lysosomes that discharge degradative enzymes, lysis peptides and reactive oxygen species to degrade the pathogen by endocytosis. L. *pneumophila* secretes two Dot/Icm effectors, VipA and LegK2, which interfere with actin cytoskeleton and contribute to bacterial evasion from endocytic degradation (Figure 5). While VipA localizes to the early endosomes and activates actin polymerisation (Franco *et al.*, 2012), LegK2 is a protein kinase that localizes at the surface of the LCV where phosphorylates the actin nucleator ARP2/3. Phosphorylation of ARP2/3 results in the inhibition of actin polymerisation on the LCV and subsequently in the inhibition of late endosome and lysosome trafficking to the LCV (Michard *et al.*, 2015, Hervet *et al.*, 2011). In particular, two other effectors, Ceg14 and RavK, target host actin cytoskeleton despite that the biological effect during infectious cycle has not yet been established (Guo *et al.*, 2014, Liu *et al.*, 2017). The Dot/Icm substrate VipD is a Rab5-activated phospholipase that targets phosphatidylinositol 3-phosphate [PI(3)P] on the LCV and protects the bacteria from endosomal fusion (Gaspar and Machner, 2014). Further manipulating endosomal trafficking, *L. pneumophila* secretes SidK that inhibits the vacuolar ATPase A-subunit that inhibits protons transport into the LCV and subsequently LCV acidification (Xu *et al.*, 2010).

Hijack vesicular trafficking for LCV biogenesis

A particularly interesting aspect of the intracellular fate of *L. pneumophila* is the biogenesis of a replication-permissive vacuole. The LCV membrane is characterized by (i) phosphatidylinositol 4-phosphate [PI(4)P]

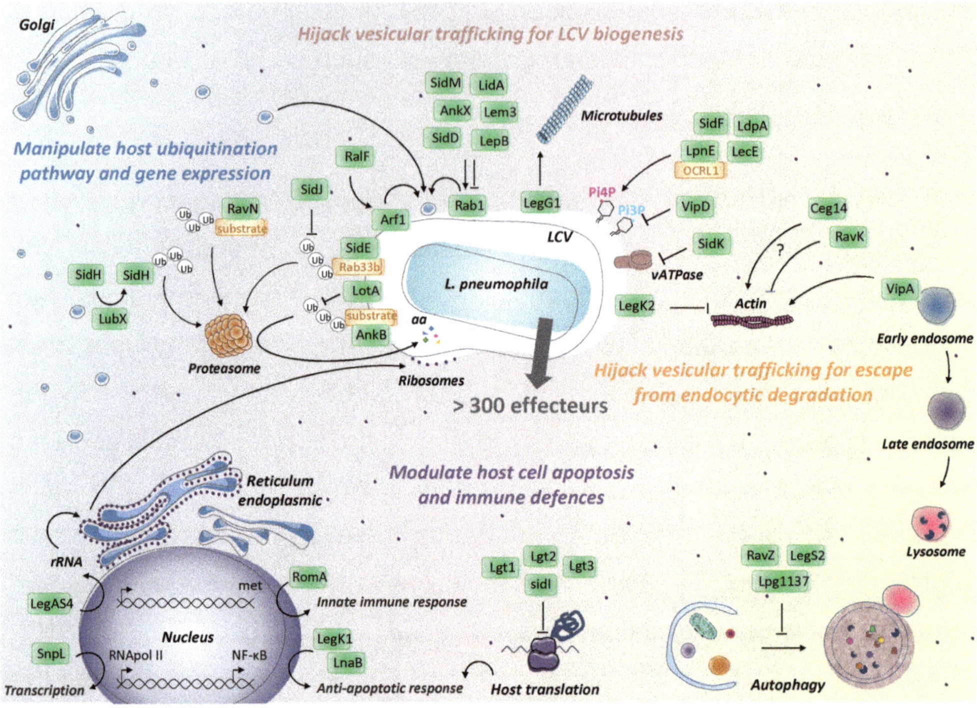

Figure 5. *L. pneumophila* effectors interfere with various host cell pathways. 300 Dot/Icm effectors are secreted and interfere with various host cell pathways. Some effectors hijack vesicular trafficking: LecE, SidF, LpdA, LpnE effectors accumulate PI4P at the surface of the LCV; RalF, SidM, LidA, AnkX, Lem3, SidD and LepB effectors recruit and control the small GTPases Rab1 and Arf1 recruiting ER vesicles at the surface of the LCV. Meanwhile, RidL inhibits retrograde trafficking between Golgi and ER. *L. pneumophila* also interacts with cytoskeleton, notably microtubules through LegG1 effector and controls actin remodelling with LegK2, VipA, Ceg14 and RavK, in order to escape from endocytic degradation. *L. pneumophila* control host cell apoptosis and immune defenses by controlling both host cell genes transcription (RomA, LegK1 and LnaB) and host translation (Lgt1, Lgt2, Lgt3 and SidI). *L. pneumophila* hijacks autonomous cell immunity, named autophagy, through RavZ, LegS2 and Lpg1137 effectors. LegAS4 is an effector that localizes within host cell nucleolus and triggers an epigenetic control of rRNA transcription, possibly for ribosomal recruitment at the LCV. SnpL is a nuclear effector that regulates RNA-pol II dependent mRNA processing and transcription elongation. Finally, *L. pneumophila* uses ubiquitination and proteasome pathway for addressing cytosolic host proteins to the proteasome providing amino acids to bacteria for replication (AnkB) and regulating the effect of bacterial effectors (meta-effector LubX on SidH). LotA localizes on the LCV and its deubiquitinase activity is crucial for unbiquitin removal from vacuoles. The SidE family of effectors (SdeA, SdeB, SdeC a,d SidE) target ER-associated Rabs, such as Rab 33b, by a phosphoribosyl-ubiquitination mechanism, completely independent of the host ubiquitination machinery. This reaction is reversed by SidJ, a ubiquitin-deconjugating enzyme that functions to impose temporal regulation on the activity of SidE effectors family.

enrichment and (ii) recruitment/activation of host proteins such as GTPases, thus promoting fusion with smooth ER vesicles. Phosphoinositides (PIs) are critical hallmarks of different cellular compartments. The [PI(4)P] enrichment of the LCV membrane, which is originally derived from plasma membrane (PM), gives the advantage of both hiding the phagosomal origin of the LCV and mimicking the Golgi compartment. This enrichment consequently also favours fusion with ER-vesicles. LCV [PI(4)P] enrichment results from the effects of three Dot/Icm substrates, LecE, LpdA and SidF (Figure 5). LecE and LpdA manipulate host phospholipides biosynthesis and could be indirectly involved in the host cell PI4KIIIβ activation that adds a phosphate at position 4 of PI to yield PI(4)P (Viner *et al.*, 2012). In particular, SidF harbours a phosphoinositide phosphatase activity which dephosphorylates [PI(3,4,5)P3] and [PI(4,5)P2] to produce [PI(4,5)P2] and [PI(4)P] (Hsu *et al.*, 2012). SidF co-exists on the LCV surface with the host phosphoinositide phosphatase OCRL1. The OCRL1 binds to the LCV in association with LpnE, another Dot/Icm substrate, and is specialized for dephosphorylating [PI(3,4,5)P3] and [PI(4,5)P2] (Weber *et al.*, 2009). Thus, SidF would act synergistically to give only [PI(4)P] as a final product.

While *L. pneumophila* remodels the lipid composition of its LCV, the bacteria hijacks vesicles trafficking between the ER and the Golgi apparatus to intercept smooth ER vesicles. This allows for both the conversion of the plasma membrane of the LCV into membranes with ER-characteristics and for provisions of membrane materials to expand the LCV for intracellular proliferation. The manipulation of host cell vesicular trafficking by *L. pneumophila* is dependent on many Dot/Icm substrates that target two small GTPases of the host cell, namely Arf1 and Rab1. This is essential for the regulation of vesicles trafficking between the ER and the Golgi apparatus (Figure 5). The Dot/Icm substrate, RalF, displays guanine nucleotide exchange factor (GEF) activity that promotes the exchange of GDP to GTP on Arf1. The GEF thus mediates the recruitment and activation of Arf1 on the LCV (Nagai *et al.*, 2002). Another Dot/Icm effector, SidM, also known as DrrA, recruits and activates the small GTPases Rab1 on the LCV. Despite any homology with eukaryotic Rab-GEF, the central domain of SidM displays both GEF activity and guanine nucleotide dissociation inhibitor (GDI) displacement factor (GDF) activity. This results in Rab1 release from GDI and in LCV membrane associated GTP-coupled Rab1, respectively (Ingmundson *et al.*, 2007, Suh *et al.*, 2010). The SidM-mediated activation of Rab1 promotes the fusion of the ER vesicle membrane with the PM-derived membrane of the LCV (Arasaki *et al.*, 2012). Moreover, the N-terminal domain of SidM catalyzes a post-translational modification of Rab1, namely AMPylation or adenylylation, which locks Rab1 in the GTP-bound active state on the LCV (Muller *et al.*, 2010). The activation of Rab1 by SidM is counteracted by two other Dot/Icm effectors, SidD and LepB. SidD removes AMP from Rab1 (Neunuebel *et al.*, 2011, Tan *et al.*, 2011) which makes Rab1 accessible for GTPase-activating protein (GAP) activity such as that exhibited by LepB. Despite any similarity with eukaryotic Rab-GAPs, LepB harbours a Rab1-specific GAP activity that promotes GTP hydrolysis and subsequent removal of Rab1 from the LCV (Gazdag *et al.*, 2013).

Three other Dot/Icm effectors target the Rab1 GTPase and participate in the temporal control of its activation during *Legionella* infection cycle. AnkX harbours a novel post-translational modification activity, namely phosphocholination, that results in the same biochemical consequence as the SidM-mediated AMPylation, i.e. locking Rab1 in the active form (Mukherjee *et al.*, 2011). Similar to AMPylation, phosphocholination is also reversible. The Dot/Icm effector Lem3 has been shown to possess antagonistic activity to that of AnkX by removing the phosphocholine from Rab1 (Tan *et al.*, 2011, Goody *et al.*, 2012). Additionally, LidA protects Rab1 from inactivation by GAP activity and nucleotide extraction. Finally, LidA also interferes with Rab1-AMPylation and phosphocholination and blocks deAMPylation by SidD and dephosphocholination by Lem3 (Neunuebel *et al.*, 2011). Thus, Rab1 is directly targeted, and its activity is controlled by at least six different Dot/Icm effectors, namely SidM, SidD, LepB, AnkX, Lem3 and LidA.

L. pneumophila also translocates RidL which inhibits the retrograde vesicle trafficking pathway. The pathway guides transport from early, late and recycling endosomes to the *trans* Golgi network and through the Golgi apparatus to ER, thus promoting the intracellular replication (Finsel *et al.*, 2013) (Barlocher *et al.*, 2017). Finally, another Dot/Icm substrate, LegG1, displays a GEF activity specific to the RAN GTPase on the surface of the LCV which triggers microtubules polymerisation. Thus, LegG1 would contribute to the LCV motion from plasma membrane to ER (Rothmeier *et al.*, 2013).

Manipulation of host ubiquitination pathways and gene expression to promote bacterial intracellular replication

One of the host cell pathways especially targeted by *L. pneumophila* during its infectious cycle is protein ubiquitination. Nearly ten Dot/Icm substrates exhibit the U-box and F-box domains typical of eukaryotic E3 ubiquitin ligases. The Dot/Icm effector AnkB harbours two ankyrin (ANK) protein-protein interaction domains and an F-box domain which together could function as a platform for the docking of polyubiquitinated proteins to the LCV membrane. This action promotes proteasome-mediated generation of free amino acids that are essential as energy and carbon sources for *L. pneumophila* intracellular proliferation (Price *et al.*, 2011) (Figure 5). AnkB is not directly involved in LCV biogenesis. In addition, there are controversial results regarding its genetic requirement and its biological function during infection (Price *et al.*, 2009, Ensminger and Isberg, 2010, Lomma *et al.*, 2010). However, it directly affects the LCV cytosolic face by recruiting ubiquitinated proteins which become enriched on LCV during infection. Other Dot/Icm substrates with E3 ligase domains promote ubiquitination of other Dot/Icm effectors and thus temporally control the effects of targeted effectors. This is the case of LubX, considered as the first "meta-effector" described which triggers the ubiquitination of SidH and subsequently addresses SidH for degradation by host cell proteasome (Kubori *et al.*, 2010). In particular, structural studies revealed a new group of *Legionella* E3 ubiquitin ligases family including RavN that displays an only remotely similar U-box domain (Lin *et al.*, 2018). Markedly, the SidE family of effectors, SdeA, SdeB, SdeC and SidE, were shown to catalyze a new kind of ubitiquitination of host proteins on seryl residues by a different mechanism and without engaging any conventional ubiquitination machinery (Qiu *et al.*, 2016, Akturk *et al.*, 2018, Kalayil *et al.*, 2018, Wang *et al.*, 2018). Recently, SidJ was shown to reverse ubiquitination of SidEs-modified substrates (Qiu *et al.*, 2017). Finally, MavC and MvcA were revealed as ubiquitin deamidases (Valleau *et al.*, 2018) and LotA a deubiquitinase (Kubori *et al.*, 2018). Together, this data highlights that *L. pneumophila* dedicates a considerable fraction effector arsenal to the manipulation of the host ubiquitination pathway.

Moreover, *L. pneumophila* is able to control host gene transcription to its benefit (Figure 5). The Dot/Icm effector LegAS4 from the Philadelphia strain was recently shown to localize into the nucleolus of the host cell. LegAS4 displays H3 methylase activity that modifies the H3 histone and thus resulting in an epigenetic upregulated transcription of ribosomic RNAs. It has been hypothesized that LegAS4 could promote ribosomal recruitment at the surface of LCV and subsequently could contribute to the successful intravacuolar replication of the bacteria (Li *et al.*, 2013). A second example of nuclear-localized *Legionella* effector is SnpL. The SnpL regulates RNA Polymerase II dependent mRNA processing and transcription elongation. SnpL's ectopic expression leads to massive upregulaton of host gene expression and macrophage cell death, and despite of the fact that its precise functional role during infection remains to be elucidated (Schuelein *et al.*, 2018).

Modulation of host cell apoptosis and immune defences

To sustain efficient intracellular replication, *L. pneumophila* modulates the apoptosis of its host cell. *Legionella* infection triggers host cell apoptosis via the activation of caspase 3, and this is most likely a part of autonomous cellular immune defences. However, *L. pneumophila* can counteract this deleterious induction of apoptosis by inducing the NF-KB pathway (Figure 5). The Dot/Icm substrate LegK1 is a protein kinase that specifically phosphorylates the NF-κB inhibitor IκB. This results in IκB ubiquitination and degradation, subsequent nuclear translocation of NF-κB and induction of anti-apoptotic genes transcription (Ge *et al.*, 2009). *L. pneumophila* also indirectly controls the NF-KB pathway by secreting three glucosyltransferases, Lgt1, Lgt2 and Lgt3, that all target elongation factor eEF1A and subsequently inhibit overall host cell translation which includes translation of IκB encoding genes (Fontana *et al.*, 2011). Notably, it has been proposed that the Lgt effector family activates mTORC1 through inhibition of host translation while the SidE effector family acts as mTORC1 inhibitors. The two families of effectors thus work in concert to liberate host amino acids for consumption by *Legionella* (De Leon *et al.*, 2017). SidI is another Dot/Icm substrate that targets eEF1A leading to inhibition of protein synthesis and induction of host stress response (Shen *et al.*, 2009). LnaB is a Dot/Icm substrate also shown to activate the NF-KB pathway despite that a mechanism has not yet been described (Losick *et al.*, 2010). It remains noteworthy that the induction of NF-KB pathway by these effectors induces the host inflammatory response in addition to controlling the host cell death.

L. pneumophila infection of macrophages induces extracellular and intracellular immune recognition. The immune recognition subsequently triggers host immune defenses including production of type I interferon, activation of various inflammasomes and induction of lysosomal cell death (Zhu and Luo, 2016). The Dot/Icm substrate RomA from the Paris strain displays an H3 methyl transferase that trimethylates the H3 histone and down-regulates the transcription of some cellular genes including genes involved in innate immune defences (Rolando *et al.*, 2013). Moreover, two Dot/Icm effectors, RavZ and LegS2, can inhibit the host cell's autonomous immunity process, namely autophagy (Choy *et al.*, 2012, Rolando *et al.*, 2016). Finally, Lpg1137 is a serine protease that cleaves syntaxin 17 at the surface of mitochondria and thus inhibiting not only autophagy but also staurosporine-induced apoptosis (Arasaki *et al.*, 2017).

Control of virulence in *Legionella pneumophila*

As a pathogenic bacteria, *L. pneumophila* must cope with numerous specific challenges such as distinguishing between intra- or extracellular niches, counteracting host cell defenses, optimizing intracellular replication inside host and subsequent release and survival in low nutrient environments until infection of a new host. Consequently, the production of virulence factors must be tightly regulated to ensure their expression at the right time and appropriate level. *Legionella,* thus, possesses multiple regulatory elements including RNA-binding proteins, CsrA and Hfq, four distinct two-component systems, LetA/S, CpxR/A, PmrA/B and the LqsR/ST, the sigma factors RpoS, non-coding RNAs and nucleoid-associated proteins (NAPs) coordinating at the transcriptional and post-transcriptional levels. These numerous virulence factors of *L. pneumophila* together promote successful infection.

RNA-binding proteins and the regulation of the biphasic life cycle

As *Legionella* spp. are facultative intracellular bacteria that alternate between intra- and extracellular niches, they have developed a biphasic developmental cycle consisting of a replicative non-virulent form and a transmissive virulent form. (Manske and Hilbi, 2014, Molofsky and Swanson, 2004). Each phase of this biphasic cycle features characteristic phenotypic traits and distinct gene expression profiles (Bruggemann *et al.*, 2006, Faucher *et al.*, 2011). In replicative phase, *Legionella* up-regulates genes involved various metabolic pathways such as energy transfer, replication, transcription, translation and cell division allowing extensive bacterial multiplication. When bacterial growth starts to slow down because of nutrient-limiting conditions, *Legionella* operates strong modifications in genetic expression leading to the down regulation of replicative traits and upregulation of transmissive ones such as motility, cytotoxicity, cell shape shortening, pigmentation, sodium sensitivity and stress resistance (Molofsky & Swanson, 2003). This genetic and phenotypic switch promotes bacterial evasion from the eukaryotic cell until new infection or survival under planktonic or biofilm conditions. The transition between replicative and transmissive form is controlled by multiple regulators working in concert to sense the physiological and/or environmental stimuli and modifying gene expression accordingly.

One of the most important regulators of this biphasic development program is the RNA-binding protein, carbon storage segulator (CsrA) (Fettes *et al.*, 2001, Forsbach-Birk *et al.*, 2004, Molofsky and Swanson, 2003). In many bacterial species, CsrA is known to control carbon metabolism, motility and virulence (Vakulskas *et al.*, 2015). Additionally, CsrA acts as a global post-transcriptional regulator that directly binds messenger RNAs at an ANGGA motif and thus interfers with translation initiation, elongation, mRNA stability and protection against RNaseE. In *Legionella*, CsrA activates bacterial replication in broth culture or inside eukaryotic cells, and it represses transmissive traits (Fettes *et al.*, 2001, Forsbach-Birk *et al.*, 2004, Molofsky and Swanson, 2003). A bioinformatic approach based on the search of the CsrA binding motif ANGGA on mRNA led to identification of 26 Dot/Icm effectors directly repressed by CsrA (Rasis and Segal, 2009, Nevo *et al.*, 2014). In a recent study using genome-wide transcriptomic, proteomic and RNA-Immunoprecipitation followed by deep sequencing (RIPSeq) approaches, Sahr et al. (2017) identified 479 targets of CsrA and further described various modes of action at the transcriptional, translational and protein stability levels. Aside from CsrA's enhancer role on all major metabolic pathways and energy transfer, CsrA affects virulence by acting simultaneously on four different pathways. First, CsrA inhibits the motility of bacteria, the hallmark of virulence in *L. pneumophila* through a base-pairing mechanism preventing the translation of FleQ and FleR, the two major regulators of the regulatory cascade for flagellar biosynthesis. Secondly, CsrA directly interacts with mRNAs of 41 Dot/Icm effectors, of which 14 have been identified as CsrA-targets (Rasis and Segal, 2009, Nevo *et al.*, 2014). Thirdly, CsrA participates in iron homeostasis, which is essential for virulence, by binding and stabilizing the *fur* transcript encoding the main regulator of iron uptake. Finally, CsrA regulates several other major regulators such as sigma factor RpoS, response regulator of the TCS PmrA/B (PmrA), Quorum Sensing response regulator (LqsR) and NAPs. These are all involved in the regulation of the expression and secretion of Dot/Icm effectors or others virulence-related factors (Figure 6).

Another RNA-binding protein, host factor essential for Qβ RNA phage replication (Hfq), may also be involved in the regulation of *Legionella* virulence traits in a life cycle-dependent manner (McNealy *et al.*, 2005, Trigui *et al.*, 2013, Oliva *et al.*, 2017). In *E. coli, and* as in many other bacteria, Hfq impacts many cellular processes including virulence, biofilm formation and quorum sensing (Feliciano *et al.*, 2016, Updegrove *et al.*, 2016). This RNA-chaperone is a key posttranscriptional regulator that facilitates or stabilizes base-pairing between non-coding RNAs and target mRNA and thus inhibits or enhances translation initiation. In *L. pneumophila*, a strain deleted for the *hfq* gene is able to replicate withinin macrophages, but replicationis is impaired intracellularly in amoeba at 20°C (McNealy *et al.*, 2005, Oliva *et al.*, 2017). Additionally, the expression of the flagellin protein FlaA is strongly reduced as shown in the *L. pneumophila* Paris strain (Oliva *et al.*, 2017). The transcriptome analysis of an *hfq* mutant strain reveals few modifications in gene expression. However, it is worth mentioning that the key repressor of the transmissive phase CsrA, the Fur repressor implicated in iron homeostasis as well as the global regulators fis1 and Fis2 (see below) are all differentially expressed compared to the wild type strain (McNealy *et al.*, 2005, Trigui *et al.*, 2013, Oliva *et al.*, 2017). Additionally, genes coding for at least four Dot/Icm effectors, enhance entry proteins involved in host cell infection, EnhA and EnhB, and flagellar assembly genes, *flgG* and *flgH,* are upregulated (Oliva *et al.*, 2017). Hfq is regulated in a growth phase-dependent manner by a cis-encoded sRNA named anti-Hfq. This sRNA expressed in exponential phase binds to the complementary region of *hfq* mRNA leading to low translation of Hfq mRNA. In transmissive phase, as the expression of anti-hfq decreases that of Hfq increases (Oliva *et al.*, 2017). Thus, considering the growth phase-dependent regulation of this RNA-chaperone, the impact on motility and connection with CsrA, it is believed that Hfq plays a role in the biphasic developmental cycle of *L. pneumophila*.

Regulation between replicative and transmissive virulence phases by sigma factor RpoS and LetA/S two-component system

Multiple environmental and metabolic signals regulate the transition from replicative to transmissive or virulent phase (Molofsky and Swanson, 2004). One example is how nutrient-limiting conditions inside host cell stop bacterial growth and trigger the stringent response (Hammer and Swanson, 1999). Amino acid starvation and perturbations in fatty acids biosynthesis lead to the synthesis of the alarmone (p)ppGpp (guanosine 3'-diphosphate 5-' diphosphate) by the means of two ppGpp synthase RelA and SpoT (Hammer and Swanson, 1999, Zusman *et al.*, 2002, Dalebroux *et al.*, 2009). The current model for *Legionella* biphasic life cycle comprises the accumulation of (p)ppGpp in the bacterial cell leading to expression of *rpoS* coding for the alternative sigma factor RpoS (Dalebroux *et al.*, 2010) and activation of the TCS LetA/LetS (Dalebroux *et al.*, 2009). As described below, both RpoS and LetA/LetS promote important transcriptional changes leading to the expression of the transmissive traits and repression of replicative ones.

The LetA/LetS TCS subsides in the LetS sensor histidine kinase which activates the LetA response regulator by a four-step phosphorelay (Edwards *et al.*, 2010, Hammer *et al.*, 2002). How (p)ppGpp or other unidentified molecules bind to and trigger LetS phosphorylation is not currently understood. However, LetA/S activation occurrs when bacterial growth rate decreases at the end of the replicative phase (Lynch *et al.*, 2003, Gal-Mor and Segal, 2003b, Hammer *et al.*, 2002). Activation of LetA then promotes the transcription of non-coding RNAs, RsmX, RsmY and RsmZ by direct binding of LetA to a conserved regulatory element in their promoter regions (Sahr *et al.*, 2009, Rasis and Segal, 2009, Hovel-Miner *et al.*, 2009). These non-coding RNAs contain multiple stem-loop structures exhibiting several GGA motifs that bind and sequester multiple molecules of CsrA (Sahr *et al.*, 2009). Finally, when CsrA is bound to the small RNAs can no longer interact with its target mRN and thus relieving the expression of transmissive traits (Sahr *et al.*, 2009, Rasis and Segal, 2009). Consistent with this model, the *letA* or *rsmYZ* mutants are less cytotoxic and less pigmented, but they are also more stress sensitive, sodium resistant (which is relates to pathogenicity), non-motile (at least for *letA* mutants) and have a reduced capacity to infect amoebae (Sahr *et al.*, 2009, Gal-Mor and Segal, 2003b, Lynch *et al.*, 2003, Hammer *et al.*, 2002).

The sigma factor RpoS also plays a key role in this regulatory cascade. The increase of RpoS, correlated to the accumulation (p)ppGpp at the end of the replicative phase, favors its recruitment by the RNA polymerase which causes drastic changes in gene expression profile. At first, RpoS, as the transcriptional regulator LetA, strongly induces the expression of RNAs RsmX, RsmY and RsmZ and thus participates in the transition from replicative to transmissive phase in an RsmXYZ/CsrA-dependent manner.

Secondly, RpoS enhances expression of numerous Dot/Icm effector-genes (Hovel-Miner *et al.*, 2009). Finally, RpoS indirectly impacts the expression of virulence trait by controlling other regulators such as the chaperone-RNA Hfq, PmrA/B and CpxR/A two-component systems as well as the quorum sensing regulator LqsR (see below) (Hovel-Miner *et al.*, 2009, Tiaden *et al.*, 2007, Oliva *et al.*, 2017).

In conclusion, the regulators RpoS and LetA/S are the top regulators of a hierarchical cascade governing the biphasic life cycle and the expression of many virulence traits in a RsmXYZ/CsrA-Hfq-dependent manner (Figure 6).

Quorum sensing system of L. pneumphila

Quorum sensing allows bacteria to coordinate gene expression and behavior at the level of bacterial population by the mean of diffusible molecules. The Leigonella quorum sensing (Lqs) system relies on the synthesis, release and detection of the small signaling molecule, 3-hydroxy-pentadecan-4-one, also termed *Legionella* autoinducer 1 (LAI-1) (Spirig *et al.*, 2008). As *Legionella* density increases, the signal concentration reaches a threshold and triggers pathways that ultimately alter gene expression. The Lqs system comprises the LAI-1 inducer synthase LqsA (Spirig *et al.*, 2008), two sensor histidine kinases LqsS and LqsT, lqsT being localized elsewhere in the genome (Tiaden *et al.*, 2008, Kessler *et al.*, 2013, Tiaden *et al.*, 2010) and LqsR, their cognate response regulator (Tiaden *et al.*, 2007, Schell *et al.*, 2014). The gene *hdeD* is also present in the gene cluster *lqsA-lqsR-hdeD-lqsS* encoding a putative membrane protein whose function is unknown (Tiaden *et al.*, 2008, Mates *et al.*, 2007). The Lqs sytem plays a key role in *Legionella* pathogenicity. Indeed, strains lacking LqsR (Tiaden *et al.*, 2007), LqsS (Tiaden *et al.*, 2010, Kessler *et al.*, 2013) or the entire *lqs* cluster (*lqsA-lqsR-hdeD-lqsS*) (Tiaden *et al.*, 2008) are impaired in phagocytosis and intracellular growth inside both amoebae and macrophages. However, the *lqsA* and *lqsT* mutant strains are only mildly defective in intracellular replication (Tiaden *et al.*, 2010, Kessler *et al.*, 2013). Nevertheless, a wildtype strain surpassed all *lqs* mutant strains in competition assays with multiple rounds of infection in amoebae (Kessler *et al.*, 2013) and demonstrated reduced motility (Schell *et al.*, 2016). Congruently, the transcriptome and proteome analyses of strains lacking all four genes of the *lqs* cluster showed downregulation of structural and regulatory components of the flagellum and other transmissive traits (Tiaden *et al.*, 2008). In addition, a set of Dot/Icm effectors such as RalF, LidA, SidC, sidD, sidG, sidH, sidM and paralogues of the sidE family are downregulated as well as Dot/Icm-independent factors. Nevertheless, whether the response regulator LqsR affects virulence gene expression directly or indirectly is still unknown as no consensus binding motif has been found. Interestingly, transcriptomic analyses of strains lacking either lqsS, lqsT or both revealed that a large number of genes are inversely regulated by the two sensors suggesting antagonistic functions (Kessler *et al.*, 2013). Among genes differentially expressed in absence of both sensors, the majority were upregulated including components of the Icm/Dot machinery, 26 different effectors, type IV pili or flagellum components and other factors important in *Legionella*/host interaction such as Macrophage infectivity potentiator (Mip) and Enh proteins (Kessler *et al.*, 2013). Furthermore, both sensor kinases, LqsS and LqsT, differ binding affinity towards the response regulator LqsR (Schell *et al.*, 2014) and seem to be differentially producing during growth phase (Kessler *et al.*, 2013).

In the current model proposed by Hilbi's group, both LqsS and LqsT sensor kinases recognize LAI-1 signal molecule (Schell *et al.*, 2016). At low cell density and low concentration of LAI-1, the LqsS and LqsT are autophosphorylated and transfer the phosphoryl group to the response regulator LqsR which then dimerizes (Schell *et al.*, 2014). These conditions promote bacterial replication while transmissive virulence traits are repressed. Inversely, at higher cell densities increased concentration of LAI-1 inhibits autophosphorylation of LqsS and LqsT leading to dephosphorylation of lqsR by presumed phosphatase activity leading to bacterial growth repression, positive regulation of motility and other transmissive traits (Schell *et al.*, 2016). In addition, it has been shown that the Lqs system is connected to other regulators since the production of LqsR depends on the sigma factor RpoS and to a lesser extent on the response regulator LetA (Tiaden *et al.*, 2007) (Figure 6). Moreover, LqsR is also directly regulated at a post-transcriptional level by CsrA (Sahr *et al.*, 2012, Sahr *et al.*, 2017). Collectively, these studies thus suggest that the signaling molecule LAI-1 and the Lqs system are important elements of the biphasic life cycle regulatory network.

Regulations of virulence trait expression by PmrA/B two-component system either directly or in a CsrA-dependent manner

The polymyxin resistant A (PmrA/B) TCS is known to mediate chemical modifications in the lipopolysaccharide (LPS) of numerous bacterial species in response to various stimuli which can alter resistance against antimicrobial coumpounds such as Polymixin (Chen and Groisman, 2013). As in other pathogens, the specific role of TCS PmrA/B in the regulation of Legionella virulence traits has been investigated. The biotic or abiotic signals that activate PmrB sensor histidine kinase are unknown. However, it has been shown that a *Legionella pmrB* mutant is sensitive to citric acid, which suggests that low pH may trigger PmrA/B TCS (Al-Khodor *et al.*, 2009). A *pmrB* mutant strain also has a strong intracellular growth defect in U937 human macrophages, *A. castelanii* and even stronger within ciliate *Tetrahymena pyriformis* (Al-Khodor *et al.*, 2009). In contrast, the role of the response regulator PmrA in intracellular replication is more controversial. This is due to a *pmrA* mutant derived from the JR32 *L. pneumophila* which was partially defective in HL-60 impact (Zusman *et al.*, 2007) while a *pmrA* mutant derived from AA100 strain had no intracellular growth defect in both hMDMs- and U937-dervived macrophages (Al-

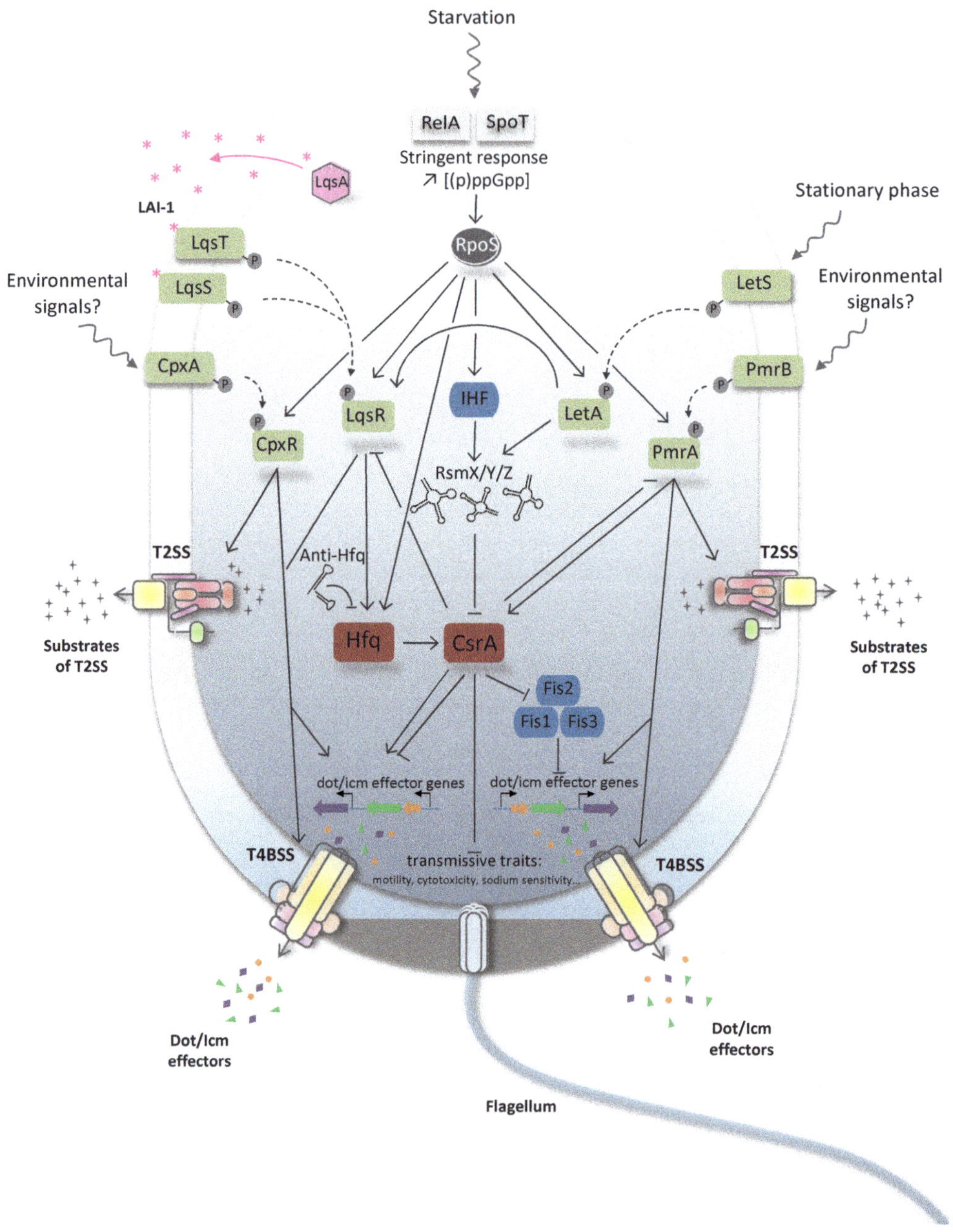

Figure 6. Model of regulatory circuitry that controls the expression of virulent traits in *L. pneumophila*. *L. pneumophila* senses environmental and metabolic signals by the means four signaling transduction pathways each comprising histidine sensor kinases and response regulator proteins (letA/S, CpxR/A, PmrA/B and LqsR/ST, in green) as well as two ppGpp synthases (RelA, SpoT, in grey). Dashed lines indicate phosphorelay mechanism between histidine sensor kinases and their cognate response regulators. The alternative sigma factor RpoS (in black), whose

expression rises with increasing concentration of (p)ppGpp, is at the top of a hierarchical cascade as it is required for the expression of most of its regulatory components. The response regulators PmrA, CpxR and LqsR activate or repress the expression of their target-genes coding for the secretion system components and their substrates while LetA promote the expression of non-coding RNAs (RsmXYZ) that sequester the RNA-binding protein RsmA (in red) thus relieving the CsrA-posttranscriptional repression. The nucleoid-associated proteins (IHF or Fis proteins, in blue) are also connected to the regulatory circuitry as they participate either directly or indirectly to the expression of Icm/Dot components and effectors.

Khodor *et al.*, 2009). To identify target genes of PmrA, a bioinformatic search based on the identification of the PmrA-regulatory element in the promoter regions of several effector encoding genes was performed. This approach led to the identification of 13 genes encoding Dot/Icm substrates that are directly activated by PmrA (Zusman *et al.*, 2007). A transcriptomic approach confirmed these findings and further extended the PmrAB regulon to include genes encoding type IV pilus, flagellum, substrates of the type II secretion system, Dot/Icm apparatus and effectors (Al-Khodor *et al.*, 2009). PmrA also induced the expression of *csrA,* and thus connect the TCS PmrAB to the LetAS/RsmYZ/CsrA regulatory cascade (Rasis and Segal, 2009)(figure 6). According to the model, increased levels of CsrA leads to the repression of effector-encoding genes that are under the control of CsrA. Therefore, activation by PmrA at transcription level and repression by CsrA at post-transcription level would result in opposition on effector gene expression. One explanation for this contradictory regulation may be that these two regulators modulate distinct groups of effector encoding genes (Segal, 2013). In addition, PmrA/B and LetA/S both positively regulate CsrA and likely respond to different environmental stimuli. Consequently, Dot/Icm substrates would be differentially expressed depending on the signal. Hence, PmrA/B and LetA/S could act together to fine-tune expression of Dot/Icm substrates as infection progresses.

Dual role of CpxRA two-component system on transmissive gene expression and connections with other regulators

Originally, the CpxR/A TCS was described to sense and respond to damages or perturbations in the cell envelope of Gram-negative bacteria. However, more recent studies suggest that its regulatory role could be more broad (Vogt and Raivio, 2012, Hunke *et al.*, 2012). The contribution of CpxR/A in *L. pneumophila* virulence was first described in three independent genetic screens looking for either direct regulators of the Dot/Icm components (Gal-Mor and Segal, 2003a), suppressors of the *ΔicmO/dotL* lethality phenotype (Vincent *et al.*, 2012) or mutants with defects in recruitment of the host GTPase Rab1 to the LCV (Murata *et al.*, 2006). Later, subsequent construction of the response regulator *cpxR* mutant demonstrated its positive regulatory effect on the Dot/Icm components IcmR (Gal-Mor and Segal, 2003a) and IcmW, IcmV and DotA (Vincent et al. 2006). Afterwards, a bioinformatic based approach searching for the consensus binding sequence of CpxR, GTAAA-N6-GWAAA (with W=T or A) identified additional CpxR-regulated genes. In addition to the genes cited above, this approach led to the identification of new CpxR-target genes: *lvgA* gene which codes an accessory protein of the Dot/Icm complex, 11 effector-encoding genes, five genes under positive regulation and six other genes under negative regulation (Altman and Segal, 2008).

Two recent studies have considerably expanded the CpxRA regulon of *L. pneumophila* (Feldheim *et al.*, 2016, Tanner *et al.*, 2016). First, a similar bioinformatic approach to that described above was carried out, this time including an *L. pneumophila* genome-wide search as well as three mismatches tolerance in the CpxR-consensus binding sequence. This screen found the 15 genes that were previously shown to be regulated by CpxR (Altman and Segal, 2008) as well as 23 additional genes including 16 effectors-encoding genes, two regulators-encoding genes (*letE* and *oxyR*) and five genes of unknown functions (Feldheim *et al.*, 2016). CpxR activates the expression of LetE regulatory protein which represses the two sRNA RsmY and RsmZ leading to a CsrA-mediated repression of Dot/Icm effector genes. Thus, CpxR/A is linked to the LetA/S-rsmYZ/CsrA cascade and regulates the expression of effector-encoding genes both directly and indirectly (figure 6). The second approach to identify direct and/or indirect target genes of the CpxRA system is based on a whole transcriptome analysis (Tanner *et al.*, 2016). The regulon members include type II secreted substrates and at least 29 Dot/Icm effectors and some of which were identified in a previous screen (Altman and Segal, 2008, Feldheim *et al.*, 2016). Among these genes, some were found to be positively controlled by CpxR while the others were negatively regulated. CpxR thus has a dual regulatory function, acting as both an activator and repressor (Feldheim *et al.*, 2016). Interestingly, the positioning of the CpxR-binding sequence relative to the promoter element as well as its self-interaction are two important requirements for positive regulation of target genes (Feldheim *et al.*, 2016).

The interplay of nucleoid-associated proteins in the control of virulence-related genes

NAPs are involved in the control of virulence-related gene expression in *L. pneumophila*. NAPs such as FIS, IHF, HU and H-NS are DNA-binding proteins that both play a key role in the chromatin structure and influence transcription initiation by binding at specific promoter (Dame, 2005, Nasser *et al.*, 2001, Dillon and Dorman, 2010, Browning *et al.*, 2010). Of these NAPs, factor for inversion stimulation (FIS) was highlighted as a crucial coordinator of temporal virulence gene expression in many human or plant pathogens (Duprey *et al.*, 2014, Goldberg *et al.*, 2001, Falconi *et al.*, 2001, Lenz and Bassler, 2007). Surprisingly, all organisms belonging to the *Legionella* genus have three distinct copies of *fis* gene while other bacteria have only one copy. This presumably would result from two duplication events that occurred before the divergence of the *Legionella* genus (Zusman *et al.*, 2014) and suggest unusual regulatory functions for these three proteins. Interestingly, only Fis1 and Fis3, and not Fis2, are needed to optimize intracellular growth in both amoeba and macrophages. In addition, Fis1 and Fis3 directly repress the expression level of 18 Dot/Icm effector-encoding genes by binding a specific 17-bp regulatory element in the promoter region of target genes. Fis1 and Fis3 appear to have common and distinct regulon since some effector genes are mainly repressed by either Fis1 or Fis3 while other sets of genes are repressed by both regulators in a similar manner. On the contrary, Fis2 did not modulate the expression of any effector genes (Zusman *et al.*, 2014). These results suggest that after gene duplication, the three copies have diverged and have acquired at least in part, distinct regulatory functions.

Another NAP, named integration host factor (IHF), has been shown to be involved in the virulence control in *Legionella* (Pitre *et al.*, 2013, Morash *et al.*, 2009) as well as in other pathogens (Sieira *et al.*, 2004, Fass and Groisman, 2009, Lee and Zhao, 2016). IHF is a heterodimeric protein encoded by *himA* and *HimD* that recognizes a 13-bp consensus regulatory element and interferes with regulatory processes such as transcription by bending the DNA. Deletion of either *himA*, *himD* or both genes in *L. pneumophila* resulted in a strong attenuated growth phenotype in amoeba *A. castelanii* but not in HeLa cells. Whether IHF directly regulates Dot/Icm genes and effectors is not known. However, it has been shown that IHF binds promoter regions of rsmY and RsmZ regulatory RNAs and is essential, along with LetA, for their full expression (Pitre *et al.*, 2013). In addition, the sigma factor RpoS upregulates IHF level which in turn also upregulates its own expression through a positive feedback loop on *himA* and *himD* gene expression (Pitre *et al.*, 2013). Collectively, these results indicate that IHF is an additional component of the regulatory cascade mediated by RpoS/LetAS/rsmYZ/CsrA which controls expression of virulence related genes (figure 6).

Legionella has evolved sophisticated and interconnected regulatory networks allowing a precise control of virulence trait expression all along the infection process. Undoubtedly, only a small fraction of these control processes has been described so far. The characterisation of novel regulatory elements, the identification of target genes and the elucidation of the nature of signaling molecules of transducing pathways are some of the missing pieces to this elaborate puzzle.

Conclusion

L. pneumophila is an adept intracellular pathogen that has acquired an arsenal of virulence factors. The virulence factors are mainly secreted by different secretion systems allowing it to replicate not only in unicellular protozoa but also in human alveolar macrophages. Although multiple studies have pointed out the major role of T1SS, T2SS and T4ASS in pathogenesis and environmental persistence, the Dot/Icm T4BSS has exceedingly generated the most intrigue. Indeed, T4BSS with its cohort of secreted proteins is of outmost importance for the intracellular fate of *Legionella*. The translocation of effectors through Dot/Icm machinery allow *Legionella* to subvert many host cell processes for its own benefit. This makes *Legionella* an attractive and relevant model for understanding the strategies developed by intracellular pathogens to survive inside hosts. Interestingly, the Dot/Icm machinery is perfectly conserved throughout the *Legionella* genus whereas Dot/Icm effector repertoire is highly variable at both the genetic and functional levels (Burstein *et al.*, 2016). This extensive diversity of effectors between species could explain their ability to replicate in such a wide spectrum of hosts. Of equal importance for successful infection of eukaryotic cells is the regulatory network orchestrating the expression and delivery of these secreted proteins. In *Legionella,* as in other pathogens, virulence is closely related to environmental cues and physiological states of the bacteria at various stages of the infection. Several transduction signaling pathways, global or specific regulators coordinate the expression of different subsets of virulence genes in a growth-phase-dependent manner. Although few links have been identified in this regulatory network (figure 6), other regulatory elements likely participate in the spatiotemporal expression of virulence factors and must be further characterized. For example, the c-diGMP signaling proteins that govern pivotal lifestyle changes in many bacteria, such as switching between planktonic and sessile life, are largely represented in the *Legionella* genus. These GGDEF/EAL proteins are mainly upregulated *in vivo* during the transmissive phase (Bruggemann *et al.*, 2006). Consequently, they could contribute to the control of the virulence program or to the increased pathogenic potential of some strains. Indeed, it has been demonstrated that deregulation of Shiga-toxin production by one particular GGDEF/EAL protein conferred increased virulence to *E. coli* O104:H4 during German outbreak in 2011 (Richter *et al.*, 2014).

Currently it is unclear whether *Legionella*'s pathogenicity towards humans is due to a specific combination of virulence factors, Dot/Icm-dependent or not, combined with tight and optimized control of their expression. Additional factors could explain both the increased virulence in specific species and the varying degrees of severity in legionellosis cases such as: i) inadequate immune response triggered by specific bacterial determinants, ii) human genetic factors that increased the susceptibility of the patient, iii) acquisition of antibiotic resistance during treatment, iv) adaptation of some pathogenic strains to a particular niche that favor human transmission or v) any combination of these different hypotheses. Certainly, future studies should continue exploration of these domains while considering the interplay between these different factors to get a comprehensive and integrated view of *Legionella* pathogenesis.

References

Akturk, A., D.J. Wasilko, X. Wu, Y. Liu, Y. Zhang, J. Qiu, Z.Q. Luo, K.H. Reiter, P.S. Brzovic, R.E. Klevit and Y. Mao, (2018) Mechanism of phosphoribosyl-ubiquitination mediated by a single Legionella effector. *Nature* 557: 729-733.

Al-Khodor, S., S. Kalachikov, I. Morozova, C.T. Price and Y. Abu Kwaik, (2009) The PmrA/PmrB two-component system of Legionella pneumophila is a global regulator required for intracellular replication within macrophages and protozoa. *Infect Immun* 77: 374-386.

Altman, E. and G. Segal, (2008) The response regulator CpxR directly regulates expression of several Legionella pneumophila icm/dot components as well as new translocated substrates. *Journal of bacteriology* 190: 1985-1996.

Arambula, D., W.E. Wong, B.A. Medhekar, H.T. Guo, M. Gingery, E. Czornyj, M.H. Liu, S. Dey, P. Ghosh and J.F. Miller, (2013) Surface display of a massively variable lipoprotein by a Legionella diversity-generating retroelement. *Proceedings of the National Academy of Sciences of the United States of America* 110: 8212-8217.

Arasaki, K., Y. Mikami, S.R. Shames, H. Inoue, Y. Wakana and M. Tagaya, (2017) Legionella effector Lpg1137 shuts down ER-mitochondria communication through cleavage of syntaxin 17. *Nature communications* 8: 15406.

Arasaki, K., D.K. Toomre and C.R. Roy, (2012) The Legionella pneumophila Effector DrrA Is Sufficient to Stimulate SNARE-Dependent Membrane Fusion. *Cell Host andamp; Microbe* 11: 46-57.

Bandyopadhyay, P., E.A.S. Lang, K.S. Rasaputra and H.M. Steinman, (2013) Implication of the VirD4 Coupling Protein of the Lvh Type 4 Secretion System in Virulence Phenotypes of Legionella pneumophila. *Journal of bacteriology* 195: 3468-3475.

Bandyopadhyay, P., S.Q. Liu, C.B. Gabbai, Z. Venitelli and H.M. Steinman, (2007) Environmental mimics and the Lvh type IVA secretion system contribute to virulence-related phenotypes of Legionella pneumophila. *Infection and Immunity* 75: 723-735.

Barlocher, K., C.A.J. Hutter, A.L. Swart, B. Steiner, A. Welin, M. Hohl, F. Letourneur, M.A. Seeger and H. Hilbi, (2017) Structural insights into Legionella RidL-Vps29 retromer subunit interaction reveal displacement of the regulator TBC1D5. *Nature communications* 8: 1543.

Berger, K.H., J.J. Merriam and R.R. Isberg, (1994) Altered intracellular targeting properties associated with mutations in the *Legionella pneumophila dotA* gene. *Molecular microbiology* 14: 809-822.

Boyd, C.D., T.J. Smith, S. El-Kirat-Chatel, P.D. Newell, Y.F. Dufrene and G.A. O'Toole, (2014) Structural Features of the Pseudomonas fluorescens Biofilm Adhesin LapA Required for LapG-Dependent Cleavage, Biofilm Formation, and Cell Surface Localization. *Journal of bacteriology* 196: 2775-2788.

Brand, B.C., A.B. Sadosky and H.A. Shuman, (1994) The *Legionella pneumophila* ICM locus - a set of genes required for intracellular multiplication in human macrophages. *Molecular microbiology* 14: 797-808.

Browning, D.F., D.C. Grainger and S.J. Busby, (2010) Effects of nucleoid-associated proteins on bacterial chromosome structure and gene expression. *Current opinion in microbiology* 13: 773-780.

Bruggemann, H., A. Hagman, M. Jules, O. Sismeiro, M.A. Dillies, C. Gouyette, F. Kunst, M. Steinert, K. Heuner, J.Y. Coppee and C. Buchrieser, (2006) Virulence strategies for infecting phagocytes deduced from the in vivo transcriptional program of *Legionella pneumophila*. *Cell. Microbiol.* 8: 1228-1240.

Burstein, D., F. Amaro, T. Zusman, Z. Lifshitz, O. Cohen, J.A. Gilbert, T. Pupko, H.A. Shuman and G. Segal, (2016) Genomic analysis of 38 Legionella species identifies large and diverse effector repertoires. *Nature genetics*.

Buscher, B.A., G.M. Conover, J.L. Miller, S.A. Vogel, S.N. Meyers, R.R. Isberg and J.P. Vogel, (2005) The DotL protein, a member of the TraG-coupling protein family, is essential for viability of Legionella pneumophila strain Lp02. *Journal of bacteriology* 187: 2927-2938.

Cazalet, C., C. Rusniok, H. Bruggemann, N. Zidane, A. Magnier, L. Ma, M. Tichit, S. Jarraud, C. Bouchier, F. Vandenesch, F. Kunst, J. Etienne, P. Glaser and C. Buchrieser, (2004) Evidence in the Legionella pneumophila genome for exploitation of host cell functions and high genome plasticity. *Nature genetics* 36: 1165-1173.

Chatterjee, D., C.D. Boyd, G.A. O'Toole and H. Sondermann, (2012) Structural Characterization of a Conserved, Calcium-Dependent Periplasmic Protease from Legionella pneumophila. *Journal of bacteriology* 194: 4415-4425.

Chen, H.D. and E.A. Groisman, (2013) The biology of the PmrA/PmrB two-component system: the major regulator of lipopolysaccharide modifications. *Annual review of microbiology* 67: 83-112.

Chetrit, D., B. Hu, P.J. Christie and C.R. Roy, (2018) A unique cytoplasmic ATPase complex defines the Legionella pneumophila type IV secretion channel. *EMBO Rep* 3: 678-686.

Choy, A., J. Dancourt, B. Mugo, T.J. O'Connor, R.R. Isberg, T.J. Melia and C.R. Roy, (2012) The Legionella Effector RavZ Inhibits Host Autophagy Through Irreversible Atg8 Deconjugation. *Science* 338: 1072-1076.

Cianciotto, N.P., (2009) Many substrates and functions of type II secretion: lessons learned from Legionella pneumophila. *Future microbiology* 4: 797-805.

Cianciotto, N.P., (2014) Type II Secretion and Legionella Virulence. In: Molecular Mechanisms in Legionella Pathogenesis. H. Hilbi (ed). Berlin: Springer-Verlag Berlin, pp. 81-102.

Clemens, D.L., B.Y. Lee and M.A. Horwitz, (2000a) Deviant expression of Rab5 on phagosomes containing the intracellular pathogens Mycobacterium tuberculosis and Legionella pneumophila is associated with altered phagosomal fate. *Infection and Immunity* 68: 2671-2684.

Clemens, D.L., B.Y. Lee and M.A. Horwitz, (2000b) Mycobacterium tuberculosis and Legionella pneumophila phagosomes exhibit arrested maturation despite acquisition of Rab7. *Infection and Immunity* 68: 5154-5166.

Cleon, F., J. Habersetzer, F. Alcock, H. Kneuper, P.J. Stansfeld, H. Basit, M.I. Wallace, B.C. Berks and T. Palmer, (2015) The TatC component of the twin-arginine protein translocase functions as an obligate oligomer. *Molecular microbiology* 98: 111-129.

Coers, J., J.C. Kagan, M. Matthews, H. Nagai, D.M. Zuckman and C.R. Roy, (2000) Identification of Icm protein complexes that play distinct roles in the biogenesis of an organelle permissive for Legionella pneumophila intracellular growth. *Molecular microbiology* 38: 719-736.

Collinson, I., R.A. Corey and W.J. Allen, (2015) Channel crossing: how are proteins shipped across the bacterial plasma membrane? *Philos. Trans. R. Soc. B-Biol. Sci.* 370: 13.

Costa, T.R.D., C. Felisberto-Rodrigues, A. Meir, M.S. Prevost, A. Redzej, M. Trokter and G. Waksman, (2015) Secretion systems in Gram-negative bacteria: structural and mechanistic insights. *Nature Reviews Microbiology* 13: 343-359.

Dalebroux, Z.D., R.L. Edwards and M.S. Swanson, (2009) SpoT governs Legionella pneumophila differentiation in host macrophages. *Molecular microbiology* 71: 640-658.

Dalebroux, Z.D., B.F. Yagi, T. Sahr, C. Buchrieser and M.S. Swanson, (2010) Distinct roles of ppGpp and DksA in Legionella pneumophila differentiation. *Molecular microbiology* 76: 200-219.

Dame, R.T., (2005) The role of nucleoid-associated proteins in the organization and compaction of bacterial chromatin. *Molecular microbiology* 56: 858-870.

De Buck, E., D. Hoper, E. Lammertyn, M. Hecker and J. Anne, (2008) Differential 2-D protein gel electrophoresis analysis of Legionella pneumophila wild type and Tat secretion mutants. *International Journal of Medical Microbiology* 298: 449-461.

De Buck, E., I. Lebeau, L. Maes, N. Geukens, E. Meyen, L. Van Mellaert, J. Anne and E. Lammertyn, (2004) A putative twin-arginine translocation pathway in Legionella pneumophila. *Biochem. Biophys. Res. Commun.* 317: 654-661.

De Buck, E., L. Maes, E. Meyen, L. Van Mellaert, N. Geukens, J. Anne and E. Lammertyn, (2005) Legionella pneumophila Philadelphia-1 tatB and tatC affect intracellular replication and biofilm formation. *Biochem. Biophys. Res. Commun.* 331: 1413-1420.

De Leon, J.A., J. Qiu, C.J. Nicolai, J.L. Counihan, K.C. Barry, L. Xu, R.E. Lawrence, B.M. Castellano, R. Zoncu, D.K. Nomura, Z.Q. Luo and R.E. Vance, (2017) Positive and Negative Regulation of the Master Metabolic Regulator mTORC1 by Two Families of Legionella pneumophila Effectors. *Cell reports* 21: 2031-2038.

DebRoy, S., J. Dao, M. Soderberg, O. Rossier and N.P. Cianciotto, (2006) Legionella pneumophila type II secretome reveals unique exoproteins and a chitinase that promotes bacterial persistence in the lung. *Proceedings of the National Academy of Sciences of the United States of America* 103: 19146-19151.

Delepelaire, P., (2004) Type I secretion in gram-negative bacteria. *Biochimica Et Biophysica Acta-Molecular Cell Research* 1694: 149-161.

Desvaux, M., M. Hebraud, R. Talon and I.R. Henderson, (2009) Secretion and subcellular localizations of bacterial proteins: a semantic awareness issue. *Trends Microbiol.* 17: 139-145.

Dillon, S.C. and C.J. Dorman, (2010) Bacterial nucleoid-associated proteins, nucleoid structure and gene expression. *Nature reviews. Microbiology* 8: 185-195.

Doleans-Jordheim, A., M. Akermi, C. Ginevra, C. Cazalet, E. Kay, D. Schneider, C. Buchrieser, D. Atlan, F. Vandenesch, J. Etienne and S. Jarraud, (2006) Growth-phase-dependent mobility of the lvh-encoding region in Legionella pneumophila strain Paris. *Microbiology-Sgm* 152: 3561-3568.

Du Plessis, D.J.F., N. Nouwen and A.J.M. Driessen, (2011) The Sec translocase. *Biochim. Biophys. Acta-Biomembr.* 1808: 851-865.

Duprey, A., S. Reverchon and W. Nasser, (2014) Bacterial virulence and Fis: adapting regulatory networks to the host environment. *Trends Microbiol* 22: 92-99.

Edwards, R.L., M. Jules, T. Sahr, C. Buchrieser and M.S. Swanson, (2010) The Legionella pneumophila LetA/LetS two-component system exhibits rheostat-like behavior. *Infect Immun* 78: 2571-2583.

Ensminger, A.W. and R.R. Isberg, (2010) E3 Ubiquitin Ligase Activity and Targeting of BAT3 by Multiple Legionella pneumophila Translocated Substrates. *Infection and Immunity* 78: 3905-3919.

Falconi, M., G. Prosseda, M. Giangrossi, E. Beghetto and B. Colonna, (2001) Involvement of FIS in the H-NS-mediated regulation of virF gene of Shigella and enteroinvasive Escherichia coli. *Molecular microbiology* 42: 439-452.

Farelli, J.D., J.C. Gumbart, I.V. Akey, A. Hempstead, W. Amyot, J.F. Head, C.J. McKnight, R.R. Isberg and C.W. Akey, (2013) IcmQ in the Type 4b Secretion System Contains an NAD(+) Binding Domain. *Structure* 21: 1361-1373.

Fass, E. and E.A. Groisman, (2009) Control of Salmonella pathogenicity island-2 gene expression. *Current opinion in microbiology* 12: 199-204.

Faucher, S.P., C.A. Mueller and H.A. Shuman, (2011) Legionella Pneumophila Transcriptome during Intracellular Multiplication in Human Macrophages. *Frontiers in microbiology* 2: 60.

Feldheim, Y.S., T. Zusman, Y. Speiser and G. Segal, (2016) The Legionella pneumophila CpxRA two-component regulatory system: new insights into CpxR's function as a dual regulator and its connection to the effectors regulatory network. *Molecular microbiology* 99: 1059-1079.

Feliciano, J.R., A.M. Grilo, S.I. Guerreiro, S.A. Sousa and J.H. Leitao, (2016) Hfq: a multifaceted RNA chaperone involved in virulence. *Future microbiology* 11: 137-151.

Ferhat, M., D. Atlan, A. Vianney, J.C. Lazzaroni, P. Doublet and C. Gilbert, (2009) The TolC Protein of Legionella pneumophila Plays a Major Role in Multi-Drug Resistance and the Early Steps of Host Invasion. *Plos One* 4.

Fettes, P.S., V. Forsbach-Birk, D. Lynch and R. Marre, (2001) Overexpresssion of a Legionella pneumophila homologue of the E. coli regulator csrA affects cell size, flagellation, and pigmentation. *International journal of medical microbiology : IJMM* 291: 353-360.

Finsel, I., C. Ragaz, C. Hoffmann, C.F. Harrison, S. Weber, V.A. van Rahden, L. Johannes and H. Hilbi, (2013) The Legionella Effector RidL Inhibits Retrograde Trafficking to Promote Intracellular Replication. *Cell Host andamp; Microbe* 14: 38-50.

Fontana, M.F., S. Banga, K.C. Barry, X.H. Shen, Y.H. Tan, Z.Q. Luo and R.E. Vance, (2011) Secreted Bacterial Effectors That Inhibit Host Protein Synthesis Are Critical for Induction of the Innate Immune Response to Virulent Legionella pneumophila. *PLoS pathogens* 7: 15.

Forsbach-Birk, V., T. McNealy, C. Shi, D. Lynch and R. Marre, (2004) Reduced expression of the global regulator protein CsrA in Legionella pneumophila affects virulence-associated regulators and growth in Acanthamoeba castellanii. *International journal of medical microbiology : IJMM* 294: 15-25.

Franco, I.S., N. Shohdy and H.A. Shuman, (2012) The Legionella pneumophila Effector VipA Is an Actin Nucleator That Alters Host Cell Organelle Trafficking. *PLoS pathogens* 8: 16.

Fuche, F., A. Vianney, C. Andrea, P. Doublet and C. Gilbert, (2015) Functional Type 1 Secretion System Involved in Legionella pneumophila Virulence. *Journal of bacteriology* 197: 563-571.

Gal-Mor, O. and G. Segal, (2003a) Identification of CpxR as a positive regulator of icm and dot virulence genes of Legionella pneumophila. *Journal of bacteriology* 185: 4908-4919.

Gal-Mor, O. and G. Segal, (2003b) The Legionella pneumophila GacA homolog (LetA) is involved in the regulation of icm virulence genes and is required for intracellular multiplication in Acanthamoeba castellanii. *Microbial pathogenesis* 34: 187-194.

Gaspar, A.H. and M.P. Machner, (2014) VipD is a Rab5-activated phospholipase A(1) that protects Legionella pneumophila from endosomal fusion. *Proceedings of the National Academy of Sciences of the United States of America* 111: 4560-4565.

Gawarzewski, I., F. DiMaio, E. Winterer, B. Tschapek, S.H.J. Smits, J. Jose and L. Schmitt, (2014) Crystal structure of the transport unit of the autotransporter adhesin involved in diffuse adherence from Escherichia coli. *J. Struct. Biol.* 187: 20-29.

Gazdag, E.M., A. Streller, I. Haneburger, H. Hilbi, I.R. Vetter, R.S. Goody and A. Itzen, (2013) Mechanism of Rab1b deactivation by the Legionella pneumophila GAP LepB. *EMBO Rep.* 14: 199-205.

Ge, J.N., H. Xu, T. Li, Y. Zhou, Z.B. Zhang, S. Li, L.P. Liu and F. Shao, (2009) A Legionella type IV effector activates the NF-kappa B pathway by phosphorylating the I kappa B family of inhibitors. *Proceedings of the National Academy of Sciences of the United States of America* 106: 13725-13730.

Ghosal, D., Y.-W. Chang, K.C. Jeong, J.P. Vogel and G. Jensen, (2018) Molecular architecture of the Legionella Dot/Icm type IV secretion system. *bioRxiv.*

Ghosal, D., Y.W. Chang, K.C. Jeong, J.P. Vogel and G.J. Jensen, (2017) In situ structure of the Legionella Dot/Icm type IV secretion system by electron cryotomography. 18: 726-732.

Goldberg, M.D., M. Johnson, J.C. Hinton and P.H. Williams, (2001) Role of the nucleoid-associated protein Fis in the regulation of virulence properties of enteropathogenic Escherichia coli. *Molecular microbiology* 41: 549-559.

Gomez-Valero, L., C. Rusniok, S. Jarraud, B. Vacherie, Z. Rouy, V. Barbe, C. Medigue, J. Etienne and C. Buchrieser, (2011) Extensive recombination events and horizontal gene transfer shaped the Legionella pneumophila genomes. *BMC genomics* 12.

Gomez-Valero, L., C. Rusniok, M. Rolando, M. Neou, D. Dervins-Ravault, J. Demirtas, Z. Rouy, R.J. Moore, H.L. Chen, N.K. Petty, S. Jarraud, J. Etienne, M. Steinert, K. Heuner, S. Gribaldo, C. Medigue, G. Glockner, E.L. Hartland and C. Buchrieser, (2014) Comparative analyses of Legionella species identifies genetic features of strains causing Legionnaires' disease. *Genome biology* 15: 21.

Goody, P.R., K. Heller, L.K. Oesterlin, M.P. Muller, A. Itzen and R.S. Goody, (2012) Reversible phosphocholination of Rab proteins by Legionella pneumophila effector proteins. *Embo J.* 31: 1774-1784.

Groschel, M.I., F. Sayes, R. Simeone, L. Majlessi and R. Brosch, (2016) ESX secretion systems: mycobacterial evolution to counter host immunity. *Nature Reviews Microbiology* 14: 677-691.

Guo, Z.H., R. Stephenson, J.Z. Qiu, S.J. Zheng and Z.Q. Luo, (2014) A Legionella effector modulates host cytoskeletal structure by inhibiting actin polymerization. *Microbes and Infection* 16: 225-236.

Hammer, B.K. and M.S. Swanson, (1999) Co-ordination of legionella pneumophila virulence with entry into stationary phase by ppGpp. *Molecular microbiology* 33: 721-731.

Hammer, B.K., E.S. Tateda and M.S. Swanson, (2002) A two-component regulator induces the transmission phenotype of stationary-phase Legionella pneumophila. *Molecular microbiology* 44: 107-118.

Hervet, E., X. Charpentier, A. Vianney, J.C. Lazzaroni, C. Gilbert, D. Atlan and P. Doublet, (2011) Protein Kinase LegK2 Is a Type IV Secretion System Effector Involved in Endoplasmic Reticulum Recruitment and Intracellular Replication of Legionella pneumophila. *Infection and Immunity* 79: 1936-1950.

Horwitz, M.A., (1983) Formation of a novel phagosome by the legionnaires-disease bacterium (*Legionella pneumophila*) in human-monocytes. *J. Exp. Med.* 158: 1319-1331.

Horwitz, M.A. and F.R. Maxfield, (1984) *Legionella pneumophila* inhibits acidification of its phagosome in human-monocytes. *Journal of Cell Biology* 99: 1936-1943.

Horwitz, M.A. and S.C. Silverstein, (1981) Interaction of the legionnaires-disease bacterium (*Legionella pneumophila*) with human phagocytes .1. *L pneumophila* resists killing by polymorphonuclear leukocytes, antibody, and complement. *J. Exp. Med.* 153: 386-397.

Hovel-Miner, G., S. Pampou, S.P. Faucher, M. Clarke, I. Morozova, P. Morozov, J.J. Russo, H.A. Shuman and S. Kalachikov, (2009) SigmaS controls multiple pathways associated with intracellular multiplication of Legionella pneumophila. *Journal of bacteriology* 191: 2461-2473.

Hsu, F.S., W.H. Zhu, L. Brennan, L.L. Tao, Z.Q. Luo and Y.X. Mao, (2012) Structural basis for substrate recognition by a unique Legionella phosphoinositide phosphatase. *Proceedings of the National Academy of Sciences of the United States of America* 109: 13567-13572.

Hunke, S., R. Keller and V.S. Muller, (2012) Signal integration by the Cpx-envelope stress system. *FEMS Microbiol Lett* 326: 12-22.

Ingmundson, A., A. Delprato, D.G. Lambright and C.R. Roy, (2007) Legionella pneumophila proteins that regulate Rab1 membrane cycling. *Nature* 450: 365-U361.

Isberg, R.R., T.J. O'Connor and M. Heidtman, (2009) The Legionella pneumophila replication vacuole: making a cosy niche inside host cells. *Nature Reviews Microbiology* 7: 12-24.

Kalayil, S., S. Bhogaraju, F. Bonn, D. Shin, Y. Liu, N. Gan, J. Basquin, P. Grumati, Z.Q. Luo and I. Dikic, (2018) Insights into catalysis and function of phosphoribosyl-linked serine ubiquitination. *Nature* 557: 734-738.

Kanonenberg, K., C.K.W. Schwarz and L. Schmitt, (2013) Type I secretion systems - a story of appendices. *Research in Microbiology* 164: 596-604.

Kessler, A., U. Schell, T. Sahr, A. Tiaden, C. Harrison, C. Buchrieser and H. Hilbi, (2013) The Legionella pneumophila orphan sensor kinase LqsT regulates competence and pathogen-host interactions as a component of the LAI-1 circuit. *Environ Microbiol* 15: 646-662.

Khodr, A., E. Kay, L. Gomez-Valero, C. Ginevra, P. Doublet, C. Buchrieser and S. Jarraud, (2016) Molecular epidemiology, phylogeny and evolution of Legionella. *Infect. Genet. Evol.* 43: 108-122.

Kita, D., S. Shibata, Y. Kikuchi, E. Kokubu, K. Nakayama, A. Saito and K. Ishihara, (2016) Involvement of the Type IX Secretion System in Capnocytophaga ochracea Gliding Motility and Biofilm Formation. *Applied and environmental microbiology* 82: 1756-1766.

Kubori, T., T. Kitao, H. Ando and H. Nagai, (2018) LotA, a Legionella deubiquitinase, has dual catalytic activity and contributes to intracellular growth. *Cellular microbiology* 20: e12840.

Kubori, T., M. Koike, X.T. Bui, S. Higaki, S.I. Aizawa and H. Nagai, (2014) Native structure of a type IV secretion system core complex essential for Legionella pathogenesis. *Proceedings of the National Academy of Sciences of the United States of America* 111: 11804-11809.

Kubori, T. and H. Nagai, (2016) The Type IVB secretion system: an enigmatic chimera. *Current opinion in microbiology* 29: 22-29.

Kubori, T., N. Shinzawa, H. Kanuka and H. Nagai, (2010) Legionella Metaeffector Exploits Host Proteasome to Temporally Regulate Cognate Effector. *PLoS pathogens* 6: 8.

Kuroda, T., T. Kubori, X.T. Bui, A. Hyakutake, Y. Uchida, K. Imada and H. Nagai, (2015) Molecular and structural analysis of Legionella DotI gives insights into an inner membrane complex essential for type IV secretion. *Sci Rep* 5: 14.

Lee, J.H. and Y. Zhao, (2016) Integration Host Factor Is Required for RpoN-Dependent hrpL Gene Expression and Controls Motility by Positively Regulating rsmB sRNA in Erwinia amylovora. *Phytopathology* 106: 29-36.

Lenz, D.H. and B.L. Bassler, (2007) The small nucleoid protein Fis is involved in Vibrio cholerae quorum sensing. *Molecular microbiology* 63: 859-871.

Leo, J.C., P. Oberhettinger, M. Schutz and D. Linke, (2015) The inverse autotransporter family: Intimin, invasin and related proteins. *International Journal of Medical Microbiology* 305: 276-282.

Li, T., Q.H. Lu, G.L. Wang, H. Xu, H.W. Huang, T. Cai, B. Kan, J.N. Ge and F. Shao, (2013) SET-domain bacterial effectors target heterochromatin protein 1 to activate host rDNA transcription. *EMBO Rep.* 14: 733-740.

Liles, M.R., V.K. Viswanathan and N.P. Cianciotto, (1998) Identification and temperature regulation of Legionella pneumophila genes involved in type IV pilus biogenesis and type II protein secretion. *Infection and Immunity* 66: 1776-1782.

Lin, Y.H., M. Lucas, T.R. Evans, G. Abascal-Palacios, A.G. Doms, N.A. Beauchene, A.L. Rojas, A. Hierro and M.P. Machner, (2018) RavN is a member of a previously unrecognized group of Legionella pneumophila E3 ubiquitin ligases. *PLoS pathogens* 14: e1006897.

Liu, Y., W. Zhu, Y. Tan, E.S. Nakayasu, C.J. Staiger and Z.Q. Luo, (2017) A Legionella Effector Disrupts Host Cytoskeletal Structure by Cleaving Actin. *PLoS pathogens* 13: e1006186.

Lomma, M., D. Dervins-Ravault, M. Rolando, T. Nora, H.J. Newton, F.M. Sansom, T. Sahr, L. Gomez-Valero, M. Jules, E.L. Hartland and C. Buchrieser, (2010) The Legionella pneumophila F-box protein Lpp2082 (AnkB) modulates ubiquitination of the host protein parvin B and promotes intracellular replication. *Cellular Microbiology* 12: 1272-1291.

Losick, V.P., E. Haenssler, M.Y. Moy and R.R. Isberg, (2010) LnaB: a Legionella pneumophila activator of NF-kappa B. *Cellular microbiology* 12: 1083-1097.

Low, H.H., F. Gubellini, A. Rivera-Calzada, N. Braun, S. Connery, A. Dujeancourt, F. Lu, A. Redzej, R. Fronzes, E.V. Orlova and G. Waksman, (2014) Structure of a type IV secretion system. *Nature* 508: 550-+.

Lynch, D., N. Fieser, K. Gloggler, V. Forsbach-Birk and R. Marre, (2003) The response regulator LetA regulates the stationary-phase stress response in Legionella pneumophila and is required for efficient infection of Acanthamoeba castellanii. *FEMS Microbiol Lett* 219: 241-248.

Lyskowski, A., J.C. Leo and A. Goldman, (2011) Structure and Biology of Trimeric Autotransporter Adhesins. In: Bacterial Adhesion: Chemistry, Biology and Physics. D. Linke and A. Goldman (eds). Berlin: Springer-Verlag Berlin, pp. 143-158.

Manske, C. and H. Hilbi, (2014) Metabolism of the vacuolar pathogen Legionella and implications for virulence. *Frontiers in cellular and infection microbiology* 4: 125.

Mates, A.K., A.K. Sayed and J.W. Foster, (2007) Products of the Escherichia coli acid fitness island attenuate metabolite stress at extremely low pH and mediate a cell density-dependent acid resistance. *Journal of bacteriology* 189: 2759-2768.

Matthews, M. and C.R. Roy, (2000) Identification and subcellular localization of the Legionella pneumophila IcmX protein: a factor essential for establishment of a replicative organelle in eukaryotic host cells. *Infection and Immunity* 68: 3971-3982.

McNealy, T.L., V. Forsbach-Birk, C. Shi and R. Marre, (2005) The Hfq homolog in Legionella pneumophila demonstrates regulation by LetA and RpoS and interacts with the global regulator CsrA. *Journal of bacteriology* 187: 1527-1532.

Meir, A., D. Chetrit, L. Liu, C.R. Roy and G. Waksman, (2018) Legionella DotM structure reveals a role in effector recruiting to the Type 4B secretion system. *Nature microbiology* 9: 507.

Michard, C., D. Sperandio, N. Bailo, J. Pizarro-Cerd, L. LeClaire, E. Chadeau-Argaud, I. Pombo-Gregoire, E. Hervet, A. Vianney, C. Gilbert, M. Faure, P. Cossart and P. Doublet, (2015) The Legionella Kinase LegK2 Targets the ARP2/3 Complex To Inhibit Actin Nucleation on Phagosomes and Allow Bacterial Evasion of the Late Endocytic Pathway. *Mbio* 6.

Molofsky, A.B. and M.S. Swanson, (2003) Legionella pneumophila CsrA is a pivotal repressor of transmission traits and activator of replication. *Molecular microbiology* 50: 445-461.

Molofsky, A.B. and M.S. Swanson, (2004) Differentiate to thrive: lessons from the Legionella pneumophila life cycle. *Molecular microbiology* 53: 29-40.

Morash, M.G., A.K. Brassinga, M. Warthan, P. Gourabathini, R.A. Garduno, S.D. Goodman and P.S. Hoffman, (2009) Reciprocal expression of integration host factor and HU in the developmental cycle and infectivity of Legionella pneumophila. *Applied and environmental microbiology* 75: 1826-1837.

Mukherjee, S., X.Y. Liu, K. Arasaki, J. McDonough, J.E. Galan and C.R. Roy, (2011) Modulation of Rab GTPase function by a protein phosphocholine transferase. *Nature* 477: 103-U122.

Muller, M.P., H. Peters, J. Blumer, W. Blankenfeldt, R.S. Goody and A. Itzen, (2010) The Legionella Effector Protein DrrA AMPylates the Membrane Traffic Regulator Rab1b. *Science* 329: 946-949.

Murata, T., A. Delprato, A. Ingmundson, D.K. Toomre, D.G. Lambright and C.R. Roy, (2006) The *Legionella pneumophila* effector protein DrrA is a Rab1 guanine nucleotide-exchange factor. *Nat. Cell. Biol.* 8: 971-977.

Nagai, H., J.C. Kagan, X.J. Zhu, R.A. Kahn and C.R. Roy, (2002) A bacterial guanine nucleotide exchange factor activates ARF on Legionella phagosomes. *Science* 295: 679-682.

Nagai, H. and C.R. Roy, (2001) The DotA protein from Legionella pneumophila is secreted by a novel process that requires the Dot/Icm transporter. *Embo J.* 20: 5962-5970.

Nakano, N., T. Kubori, M. Kinoshita, K. Imada and H. Nagai, (2010) Crystal Structure of Legionella DotD: Insights into the Relationship between Type IVB and Type II/III Secretion Systems. *PLoS pathogens* 6: 10.

Nasser, W., R. Schneider, A. Travers and G. Muskhelishvili, (2001) CRP modulates fis transcription by alternate formation of activating and repressing nucleoprotein complexes. *The Journal of biological chemistry* 276: 17878-17886.

Neunuebel, M.R., Y. Chen, A.H. Gaspar, P.S. Backlund, A. Yergey and M.P. Machner, (2011) De-AMPylation of the Small GTPase Rab1 by the Pathogen Legionella pneumophila. *Science* 333: 453-456.

Nevo, O., T. Zusman, M. Rasis, Z. Lifshitz and G. Segal, (2014) Identification of Legionella pneumophila effectors regulated by the LetAS-RsmYZ-CsrA regulatory cascade, many of which modulate vesicular trafficking. *Journal of bacteriology* 196: 681-692.

Nishida, T., N. Hara, K. Watanabe, T. Shimizu, M. Fujishima and M. Watarai, (2018) Crucial Role of Legionella pneumophila TolC in the

Inhibition of Cellular Trafficking in the Protistan Host Paramecium tetraurelia. *Journal of bacteriology* 9: 800.

Oliva, G., T. Sahr, M. Rolando, M. Knoth and C. Buchrieser, (2017) A Unique cis-Encoded Small Noncoding RNA Is Regulating Legionella pneumophila Hfq Expression in a Life Cycle-Dependent Manner. *MBio* 8.

Pitre, C.A., J.R. Tanner, P. Patel and A.K. Brassinga, (2013) Regulatory control of temporally expressed integration host factor (IHF) in Legionella pneumophila. *Microbiology (Reading, England)* 159: 475-492.

Prevost, M.S. and G. Waksman, (2018) X-ray crystal structures of the type IVb secretion system DotB ATPases. *Protein science : a publication of the Protein Society.*

Price, C.T., S. Al-Khodor, T. Al-Quadan, M. Santic, F. Habyarimana, A. Kalia and Y. Abu Kwaik, (2009) Molecular Mimicry by an F-Box Effector of Legionella pneumophila Hijacks a Conserved Polyubiquitination Machinery within Macrophages and Protozoa. *PLoS pathogens* 5: 13.

Price, C.T.D., T. Al-Quadan, M. Santic, I. Rosenshine and Y. Abu Kwaik, (2011) Host Proteasomal Degradation Generates Amino Acids Essential for Intracellular Bacterial Growth. *Science* 334: 1553-1557.

Qin, T., K. Iida, Z.Y. Piao, S. Shiota, H.Y. Ren, Z.J. Shao and S. Yoshida, (2012) Importance of the icmN gene in the growth of Legionella pneumophila in amoebic cells at low temperature. *Can. J. Microbiol.* 58: 490-501.

Qiu, J., M.J. Sheedlo, K. Yu, Y. Tan, E.S. Nakayasu, C. Das, X. Liu and Z.Q. Luo, (2016) Ubiquitination independent of E1 and E2 enzymes by bacterial effectors. *Nature* 533: 120-124.

Qiu, J., K. Yu, X. Fei, Y. Liu, E.S. Nakayasu, P.D. Piehowski, J.B. Shaw, K. Puvar, C. Das, X. Liu and Z.Q. Luo, (2017) A unique deubiquitinase that deconjugates phosphoribosyl-linked protein ubiquitination. *Cell research* 27: 865-881.

Rahman, S.U., J. Arenas, H. Ozturk, N. Dekker and P. van Ulsen, (2014) The Polypeptide Transport-associated (POTRA) Domains of TpsB

Transporters Determine the System Specificity of Two-partner Secretion Systems. *J. Biol. Chem.* 289: 19799-19809.

Rasis, M. and G. Segal, (2009) The LetA-RsmYZ-CsrA regulatory cascade, together with RpoS and PmrA, post-transcriptionally regulates stationary phase activation of Legionella pneumophila Icm/Dot effectors. *Molecular microbiology* 72: 995-1010.

Richter, A.M., T.L. Povolotsky, L.H. Wieler and R. Hengge, (2014) Cyclic-di-GMP signalling and biofilm-related properties of the Shiga toxin-producing 2011 German outbreak Escherichia coli O104:H4. *EMBO molecular medicine* 6: 1622-1637.

Ridenour, D.A., S.L.G. Cirillo, S. Feng, M.M. Samrakandi and J.D. Cirillo, (2003) Identification of a gene that affects the efficiency of host cell infection by Legionella pneumophila in a temperature-dependent fashion. *Infection and Immunity* 71: 6256-6263.

Rolando, M., P. Escoll, T. Nora, J. Botti, V. Boitez, C. Bedia, C. Daniels, G. Abraham, P.J. Stogios, T. Skarina, C. Christophe, D. Dervins-Ravault, C. Cazalet, H. Hilbi, T.W.T. Rupasinghe, D. Tull, M.J. McConville, S.Y. Ong, E.L. Hartland, P. Codogno, T. Levade, T. Naderer, A. Savchenko and C. Buchrieser, (2016) Legionella pneumophila S1P-lyase targets host sphingolipid metabolism and restrains autophagy. *Proceedings of the National Academy of Sciences of the United States of America* 113: 1901-1906.

Rolando, M., S. Sanulli, C. Rusniok, L. Gomez-Valero, C. Bertholet, T. Sahr, R. Margueron and C. Buchrieser, (2013) Legionella pneumophila Effector RomA Uniquely Modifies Host Chromatin to Repress Gene Expression and Promote Intracellular Bacterial Replication. *Cell Host andamp; Microbe* 13: 395-405.

Rossier, O. and N.P. Cianciotto, (2005) The Legionella pneumophila tatB gene facilitates secretion of phospholipase C, growth under iron-limiting conditions, and intracellular infection. *Infection and Immunity* 73: 2020-2032.

Rossier, O., J. Dao and N.P. Cianciotto, (2008) The type II secretion system of Legionella pneumophila elaborates two aminopeptidases, as well as a metalloprotease that contributes to differential infection among protozoan hosts. *Applied and environmental microbiology* 74: 753-761.

Rothmeier, E., G. Pfaffinger, C. Hoffmann, C.F. Harrison, H. Grabmayr, U. Repnik, M. Hannemann, S. Wolke, A. Bausch, G. Griffiths, A. Muller-Taubenberger, A. Itzen and H. Hilbi, (2013) Activation of Ran GTPase by a Legionella Effector Promotes Microtubule Polymerization, Pathogen Vacuole Motility and Infection. *PLoS pathogens* 9: 17.

Roy, C.R., K.H. Berger and R.R. Isberg, (1998) Legionella pneumophila DotA protein is required for early phagosome trafficking decisions that occur within minutes of bacterial uptake. *Molecular microbiology* 28: 663-674.

Roy, C.R. and L.G. Tilney, (2002) The road less traveled: transport of Legionella to the endoplasmic reticulum. *Journal of Cell Biology* 158: 415-419.

Sahr, T., H. Bruggemann, M. Jules, M. Lomma, C. Albert-Weissenberger, C. Cazalet and C. Buchrieser, (2009) Two small ncRNAs jointly govern virulence and transmission in Legionella pneumophila. *Molecular microbiology* 72: 741-762.

Sahr, T., C. Rusniok, D. Dervins-Ravault, O. Sismeiro, J.Y. Coppee and C. Buchrieser, (2012) Deep sequencing defines the transcriptional map of L. pneumophila and identifies growth phase-dependent regulated ncRNAs implicated in virulence. *RNA Biol* 9: 503-519.

Sahr, T., C. Rusniok, F. Impens, G. Oliva, O. Sismeiro, J.Y. Coppee and C. Buchrieser, (2017) The Legionella pneumophila genome evolved to accommodate multiple regulatory mechanisms controlled by the CsrA-system. 13: e1006629.

Salacha, R., F. Kovacic, C. Brochier-Armanet, S. Wilhelm, J. Tommassen, A. Filloux, R. Voulhoux and S. Bleves, (2010) The Pseudomonas aeruginosa patatin-like protein PlpD is the archetype of a novel Type V secretion system. *Environ. Microbiol.* 12: 1498-1512.

Sato, K., M. Naito, H. Yukitake, H. Hirakawa, M. Shoji, M.J. McBride, R.G. Rhodes and K. Nakayama, (2010) A protein secretion system linked to bacteroidete gliding motility and pathogenesis. *Proceedings of the National Academy of Sciences of the United States of America* 107: 276-281.

Sato, K., H. Yukitake, Y. Narita, M. Shoji, M. Naito and K. Nakayama, (2013) Identification of Porphyromonas gingivalis proteins secreted by the Por secretion system. *FEMS Microbiol. Lett.* 338: 68-76.

Schell, U., A. Kessler and H. Hilbi, (2014) Phosphorylation signalling through the Legionella quorum sensing histidine kinases LqsS and LqsT converges on the response regulator LqsR. *Molecular microbiology* 92: 1039-1055.

Schell, U., S. Simon, T. Sahr, D. Hager, M.F. Albers, A. Kessler, F. Fahrnbauer, D. Trauner, C. Hedberg, C. Buchrieser and H. Hilbi, (2016) The alpha-hydroxyketone LAI-1 regulates motility, Lqs-dependent phosphorylation signalling and gene expression of Legionella pneumophila. *Molecular microbiology* 99: 778-793.

Schroeder, G.N., N.K. Petty, A. Mousnier, C.R. Harding, A.J. Vogrin, B. Wee, N.K. Fry, T.G. Harrison, H.J. Newton, N.R. Thomson, S.A. Beatson, G. Dougan, E.L. Hartland and G. Frankel, (2010) Legionella pneumophila Strain 130b Possesses a Unique Combination of Type IV Secretion Systems and Novel Dot/Icm Secretion System Effector Proteins. *Journal of bacteriology* 192: 6001-6016.

Schuelein, R., H. Spencer, L.F. Dagley, P.F. Li, L. Luo, J.L. Stow, G. Abraham, T. Naderer, L. Gomez-Valero, C. Buchrieser, C. Sugimoto, J. Yamagishi, A.I. Webb, S. Pasricha and E.L. Hartland, (2018) Targeting of RNA Polymerase II by a nuclear Legionella pneumophila Dot/Icm effector SnpL. *Cellular microbiology.*

Segal, G., (2013) The Legionella pneumophila two-component regulatory systems that participate in the regulation of Icm/Dot effectors. *Current topics in microbiology and immunology* 376: 35-52.

Segal, G., M. Purcell and H.A. Shuman, (1998) Host cell killing and bacterial conjugation require overlapping sets of genes within a 22-kb region of the Legionella pneumophila genome. *Proceedings of the National Academy of Sciences of the United States of America* 95: 1669-1674.

Segal, G., J.J. Russo and H.A. Shuman, (1999) Relationships between a new type IV secretion system and the icm/dot virulence system of Legionella pneumophila. *Molecular microbiology* 34: 799-809.

Sexton, J.A., J.S. Pinkner, R. Roth, J.E. Heuser, S.J. Hultgren and J.P. Vogel, (2004) The Legionella pneumophila PilT homologue DotB exhibits ATPase activity that is critical for intracellular growth. *Journal of bacteriology* 186: 1658-1666.

Sexton, J.A., H.J. Yeo and J.P. Vogel, (2005) Genetic analysis of the Legionella pneumophila DotB ATPase reveals a role in type IV secretion system protein export. *Molecular microbiology* 57: 70-84.

Shen, X.H., S. Banga, Y.C. Liu, L. Xu, P. Gao, I. Shamovsky, E. Nudler and Z.Q. Luo, (2009) Targeting eEF1A by a Legionella pneumophila effector leads to inhibition of protein synthesis and induction of host stress response. *Cellular microbiology* 11: 911-926.

Sieira, R., D.J. Comerci, L.I. Pietrasanta and R.A. Ugalde, (2004) Integration host factor is involved in transcriptional regulation of the Brucella abortus virB operon. *Molecular microbiology* 54: 808-822.

Smith, T.J., M.E. Font, C.M. Kelly and H. Sondermann, (2018) An N-terminal Retention Module Anchors the Giant Adhesin LapA of Pseudomonas fluorescens at the Cell Surface: A Novel Sub-family of Type I Secretion Systems.

Soderberg, M.A., O. Rossier and N.P. Cianciotto, (2004) The type II protein secretion system of Legionella pneumophila promotes growth at low temperatures. *Journal of Bacteriology* 186: 3712-3720.

Spirig, T., A. Tiaden, P. Kiefer, C. Buchrieser, J.A. Vorholt and H. Hilbi, (2008) The Legionella autoinducer synthase LqsA produces an alpha-hydroxyketone signaling molecule. *PLoS genetics* 283: 18113-18123.

Suh, H.Y., D.W. Lee, K.H. Lee, B. Ku, S.J. Choi, J.S. Woo, Y.G. Kim and B.H. Oh, (2010) Structural insights into the dual nucleotide exchange and GDI displacement activity of SidM/DrrA. *Embo J.* 29: 496-504.

Tan, Y.H., R.J. Arnold and Z.Q. Luo, (2011) Legionella pneumophila regulates the small GTPase Rab1 activity by reversible phosphorylcholination. *Proceedings of the National Academy of Sciences of the United States of America* 108: 21212-21217.

Tanner, J.R., L. Li, S.P. Faucher and A.K. Brassinga, (2016) The CpxRA two-component system contributes to Legionella pneumophila virulence. *Molecular microbiology* 100: 1017-1038.

Tiaden, A., T. Spirig, P. Carranza, H. Bruggemann, K. Riedel, L. Eberl, C. Buchrieser and H. Hilbi, (2008) Synergistic contribution of the Legionella pneumophila lqs genes to pathogen-host interactions. *Journal of bacteriology* 190: 7532-7547.

Tiaden, A., T. Spirig, T. Sahr, M.A. Walti, K. Boucke, C. Buchrieser and H. Hilbi, (2010) The autoinducer synthase LqsA and putative sensor kinase LqsS regulate phagocyte interactions, extracellular filaments and a genomic island of Legionella pneumophila. *Environ Microbiol* 12: 1243-1259.

Tiaden, A., T. Spirig, S.S. Weber, H. Bruggemann, R. Bosshard, C. Buchrieser and H. Hilbi, (2007) The Legionella pneumophila response regulator LqsR promotes host cell interactions as an element of the virulence regulatory network controlled by RpoS and LetA. *Cellular microbiology* 9: 2903-2920.

Trigui, H., P. Dudyk, J. Sum, H.A. Shuman and S.P. Faucher, (2013) Analysis of the transcriptome of Legionella pneumophila hfq mutant reveals a new mobile genetic element. *Microbiology (Reading, England)* 159: 1649-1660.

Tyson, J.Y., M.M. Pearce, P. Vargas, S. Bagchi, B.J. Mulhern and N.P. Cianciotto, (2013) Multiple Legionella pneumophila Type II Secretion Substrates, Including a Novel Protein, Contribute to Differential Infection of the Amoebae Acanthamoeba castellanii, Hartmannella vermiformis, and Naegleria lovaniensis. *Infection and Immunity* 81: 1399-1410.

Tyson, J.Y., P. Vargas and N.P. Cianciotto, (2014) The novel Legionella pneumophila type II secretion substrate NttC contributes to infection of amoebae Hartmannella vermiformis and Willaertia magna. *Microbiology-Sgm* 160: 2732-2744.

Updegrove, T.B., A. Zhang and G. Storz, (2016) Hfq: the flexible RNA matchmaker. *Current opinion in microbiology* 30: 133-138.

Vakulskas, C.A., A.H. Potts, P. Babitzke, B.M. Ahmer and T. Romeo, (2015) Regulation of bacterial virulence by Csr (Rsm) systems. *Microbiology and molecular biology reviews : MMBR* 79: 193-224.

Valleau, D., A.T. Quaile, H. Cui, X. Xu, E. Evdokimova, C. Chang, M.E. Cuff, M.L. Urbanus, S. Houliston, C.H. Arrowsmith, A.W. Ensminger and A. Savchenko, (2018) Discovery of Ubiquitin Deamidases in the Pathogenic Arsenal of Legionella pneumophila. *Cell reports* 23: 568-583.

VanRheenen, S.M., G. Dumenil and R.R. Isberg, (2004) IcmF and DotU are required for optimal effector translocation and trafficking of the Legionella pneumophila vacuole. *Infection and Immunity* 72: 5972-5982.

Vincent, C.D., J.R. Friedman, K.C. Jeong, M.C. Sutherland and J.P. Vogel, (2012) Identification of the DotL coupling protein subcomplex of the Legionella Dot/Icm type IV secretion system. *Molecular microbiology* 85: 378-391.

Viner, R., D. Chetrit, M. Ehrlich and G. Segal, (2012) Identification of Two Legionella pneumophila Effectors that Manipulate Host Phospholipids Biosynthesis. *PLoS pathogens* 8: 23.

Vogel, J.P., H.L. Andrews, S.K. Wong and R.R. Isberg, (1998) Conjugative transfer by the virulence system of Legionella pneumophila. *Science* 279: 873-876.

Vogt, S.L. and T.L. Raivio, (2012) Just scratching the surface: an expanding view of the Cpx envelope stress response. *FEMS Microbiol Lett* 326: 2-11.

Voth, D.E., L.J. Broederdorf and J.G. Graham, (2012) Bacterial Type IV secretion systems: versatile virulence machines. *Future microbiology* 7: 241-257.

Wang, Y., M. Shi, H. Feng, Y. Zhu, S. Liu, A. Gao and P. Gao, (2018) Structural Insights into Non-canonical Ubiquitination Catalyzed by SidE. *Cell* 173: 1231-1243.e1216.

Watarai, M., H.L. Andrews and R.R. Isberg, (2001) Formation of a fibrous structure on the surface of Legionella pneumophila associated with exposure of DotH and DotO proteins after intracellular growth. *Molecular microbiology* 39: 313-329.

Weber, S.S., C. Ragaz and H. Hilbi, (2009) The inositol polyphosphate 5-phosphatase OCRL1 restricts intracellular growth of Legionella, localizes to the replicative vacuole and binds to the bacterial effector LpnE. *Cellular microbiology* 11: 442-460.

Wee, B.A., M. Woolfit, S.A. Beatson and N.K. Petty, (2013) A Distinct and Divergent Lineage of Genomic Island-Associated Type IV Secretion Systems in Legionella. *PloS one* 8: 13.

White, R.C. and N.P. Cianciotto, (2016) Type II Secretion Is Necessary for Optimal Association of the Legionella-Containing Vacuole with Macrophage Rab1B but Enhances Intracellular Replication Mainly by Rab1B-Independent Mechanisms. *Infection and Immunity* 84: 3313-3327.

Xu, L., X.H. Shen, A. Bryan, S. Banga, M.S. Swanson and Z.Q. Luo, (2010) Inhibition of Host Vacuolar H+-ATPase Activity by a Legionella pneumophila Effector. *Plos Pathogens* 6: 16.

Yerushalmi, G., T. Zusman and G. Segal, (2005) Additive effect on intracellular growth by Legionella pneumophila Icm/dot proteins containing a lipobox motif. *Infection and Immunity* 73: 7578-7587.

Zhu, W.H. and Z.Q. Luo, (2016) Cell biology and immunology lessons taught by Legionella pneumophila. *Science China-Life Sciences* 59: 3-10.

Zusman, T., G. Aloni, E. Halperin, H. Kotzer, E. Degtyar, M. Feldman and G. Segal, (2007) The response regulator PmrA is a major regulator of the icm/dot type IV secretion system in Legionella pneumophila and Coxiella burnetii. *Molecular microbiology* 63: 1508-1523.

Zusman, T., O. Gal-Mor and G. Segal, (2002) Characterization of a Legionella pneumophila relA insertion mutant and toles of RelA and RpoS in virulence gene expression. *Journal of bacteriology* 184: 67-75.

Zusman, T., Y. Speiser and G. Segal, (2014) Two Fis regulators directly repress the expression of numerous effector-encoding genes in Legionella pneumophila. *Journal of bacteriology* 196: 4172-4183.

Chapter 4

Epidemiology of Legionellosis and a Historical Perspective on *Legionella pneumophila* Strains for the Genomic Era

Natalia A. Kozak-Muiznieks, Jeffrey W. Mercante and Brian H. Raphael*

Respiratory Diseases Branch, Centers for Disease Control and Prevention, Atlanta, Georgia, 30329 USA

*BRaphael@cdc.gov

DOI: https://doi.org/10.21775/9781913652531.04

Abstract

As a relatively newly discovered pathogen, *Legionella*, the cause of Legionnaires' disease and Pontiac fever, emerged at a time of rapidly evolving diagnostic techniques. The sequencing of the first *L. pneumophila* strain in 2004 brought extensive new findings on this unique pathogen regarding virulence, metabolism and diversity of this predominant species. This chapter describes the major *L. pneumophila* reference strains, OLDA, Pontiac-1, Philadelphia, Paris, Lens, Lorraine, Detroit-1 and the strains representing *L. pneumophila* subspecies *pascillei* in the historical contexts in which they were discovered. Each strain contributed significant findings to our current understanding of *Legionella* in the environment and pathologically. The distinct outbreaks leading to the discovery of each strain show the multifarious nature of this organism along with the major public health significance of *Legionella* research.

Facing the era of genomics, much remains to be uncovered including pathogenicity profiles and resolution of the evident diversity within strains.

Introduction

The 1976 outbreak of Legionnaires' disease in Philadelphia, Pennsylvania represented one of the largest outbreak investigations conducted by multiple public health agencies at the time. Within one year, the etiological agent, *Legionella pneumophila*, was isolated from patient samples, which allowed for re-examination of several previous outbreaks subsequently linked to members of the newly identified genus. Over the next four decades, our understanding of the pathogenesis and ecology of this organism markedly increased. The unique life cycle of *Legionella* has been described involving survival and replication within protozoa present in natural, freshwaters sources such as lakes and streams. Although *Legionella* are readily found in freshwater environments, human disease usually results from exposure to man-made water systems where *Legionella* may encounter conditions permitting the organism to grow to larger numbers and then to be spread via aerosols. Interestingly, many of the molecular strategies used to invade and survive within protozoa also contribute to human disease within macrophages in the lungs of susceptible hosts.

Legionnaires' disease and a milder illness called Pontiac fever are collectively termed legionellosis. Much of our understanding regarding the transmission of *Legionella* and the susceptibility of individuals to legionellosis has been learned from outbreak investigations. In addition, many public health efforts aim specifically at the prevention of Legionnaires' disease. In some cases, regulatory systems aimed at reducing *Legionella* colonization in man-made structures, especially in cooling towers, have been implemented. In the health-care setting, water system management and monitoring are important aspects of legionellosis prevention. Recently, the role of water system deficiencies has been analyzed in an effort to inform building and facility owners about reducing known problems that may lead to outbreaks (Garrison et al., 2016).

There are nearly 60 currently recognized species of *Legionella*, and approximately half of which have been associated with human disease (Mercante and Winchell, 2015). However, *L. pneumophila* in particular has been associated with over 90% of community-acquired legionellosis cases (Yu et al., 2002). *Legionella* species associated with much of the remaining cases of legionellosis include *L. longbeachae*, *L. micdadei*, and *L. bozemanii.* Since culture-confirmation of legionellosis is relatively uncommon in the United States, it is probable that the number and distribution of Legionnaires' disease cases due to non-*pneumophila Legionella* spp. is underrepresented. Over 15 serogroups of *L. pneumophila* have been reported (Gomez-Valaero et al, 2009). The highly variable and immunogenic lipopolysaccharide (LPS) serves as the basis of serogroup identification. *L. pneumophila* serogroup 1 is the most commonly isolated serogroup and can be further characterized using panels of monoclonal antibodies (Joly et al., 1986; Lück et al., 1992).

Sequence based typing (SBT) is one of the most commonly used molecular subtyping method for *L. pneumophila* isolates. In this method seven loci, *flaA*, *pilE*, *asd*, *mip*, *mompS*, *proA*, and *neuA*, are amplified and sequenced resulting in an allelic profile that defines a sequence type (ST) (Lück et al., 2013), as described later in the chapter by Fry et al.. Over 2,200 STs have been identified from various countries around the world. Among isolates characterized by the US Centers for Disease Control and Prevention (CDC) (1982-2012), ST1 was the most frequent type among sporadic clinical isolates (25%) and environmental isolates (49%) (Kozak-Muiznieks et al., 2014). Outbreak-associated strains in this study were represented by various STs including ST1, ST35, ST36, ST37, and ST222.

This chapter will focus on the recent trends in legionellosis cases and the historical significance of various reference *L. pneumophila* strains. Whole genome sequencing is currently revolutionizing the way this organism is detected and subtyped. Many strains sequenced in recent years have historical significance for their involvement in notorious outbreaks or for the role that they have played in furthering the understanding of *L. pneumophila* biology.

Epidemiology

Clinical aspects of legionellosis

Legionnaires' disease accounts for 2-9% of community-acquired pneumonia cases, and it is among the most common causes of pneumonia requiring admission to intensive care units (von Baum et al., 2008; Mercante and Winchell., 2015; Vergis et al., 2000). The disease is characterized by non-specific symptoms commonly associated with severe pneumonia such as cough, shortness of breath, fever, muscle aches, and headache. Hospitalizations for Legionnaires' disease are estimated to range from 8,000-18,000 in the United States annually (Marston et al., 1997). Recent active surveillance in the United States revealed that approximately 9% of legionellosis cases were fatal (Dooling et al., 2015).

There is a marked increase in the incidence of Legionnaires' disease with increasing age (Hicks et al., 2011). Interestingly, a majority of cases (>60%) occur in males. Several additional risk factors have been noted including smoking, chronic lung disease, weakened immune system (due to factors such as cancer or chemotherapy), renal failure, and diabetes (Mercante and Winchell, 2015). One study in New York City showed that the age-adjusted incidence of community-acquired Legionnaires' disease was 2.5-fold higher in the highest poverty neighborhoods as compared to the lowest poverty areas (Farnham et al., 2014).

The United States Council of State and Territorial Epidemiologists defines a confirmed case of Legionnaires' disease as having both clinically compatible symptoms such as fever, myalgia, cough, and clinical or radiographic pneumonia and a confirmatory laboratory test such as either 1) isolation of *Legionella* from lower respiratory tract specimens, 2) positive urine antigen test specific for *L. pneumophila* serogroup 1, or 3) seroconversion for antibodies targeting *L. pneumophila* serogroup 1 showing at least 4-fold increase between acute and convalescent sera. Suspect cases are defined by clinically compatible symptoms combined with either seroconversion to other Legionella strains (i.e. a 4-fold rise in antibody titers to species or serogroups other than *L. pneumophila* serogroup 1), detection of *Legionella* antigens in respiratory secretions or tissue using techniques such as direct fluorescent antibody (DFA) staining, or detection of *Legionella* using a validated nucleic acid assay.

The incubation period for Legionnaires' disease is approximately two to 14 days (Cunha et al., 2016). Patients who develop Legionnaires' disease during hospitalization and who spent the entirety of an incubation period within the hospital are considered confirmed cases of hospital-acquired legionellosis. In these cases, public health measures to identify and remedy potential environmental sources are critical to stop further transmission to high numbers of susceptible individuals present in healthcare settings. The healthcare-associated case fatality rate is often higher than that for community-acquired disease. A study of 1,938 laboratory-confirmed cases occurring in Catalonia, Spain revealed that the case-fatality rate of hospital-acquired disease was nearly three-fold higher than that for sporadic disease (Dominguez et al., 2009).

A milder form of legionellosis, called Pontiac fever, is characterized by flu-like symptoms that resolve without specific treatment (Glick et al., 1978). Pontiac fever can be identified in individuals exposed to contaminated water sources either as clusters of only Pontiac fever cases or clusters containing both Legionnaires' disease and Pontiac fever cases. Some laboratory tests, such as the indirect fluorescent antibody (IFA) assay of patient serum and/or the urinary antigen test, may be useful in confirming or establishing a diagnosis of Pontiac fever (Burnsed et al., 2007).

Transmission and environmental sources

Although *Legionella* can be found in natural water sources, outbreaks of Legionnaires' disease typically occur when the organism amplifies and aerosolizes in complex man-made water systems such as hotels, hospitals, long-term care facilities, and cruise ships. Various conditions resulting in amplification of the organism in plumbing systems include water temperatures below 60°C, lack of adequate disinfectant, stagnation of water (especially in sections of pipe where water does not flow often termed "dead legs"), or association with biofilms and/or amoeba (Rogers et al., 1994; Wadowsky et al., 1985). Aerosol generating devices such as showerheads, hot tubs, decorative fountains, and cooling towers can also promote transmission of the organism to susceptible individuals.

Deficiencies in water system maintenance were evaluated among 23 Legionnaires' disease outbreak investigations conducted by the CDC between 2000 and 2014 (Garrison et al., 2016). Environmental source assessments revealed that 48% of the outbreaks were associated with two or more deficiencies, most often involving process failures and human errors resulting in permissive conditions for *Legionella* growth. Less commonly, equipment failures and unmanaged external changes (e.g. water main breaks) were factors associated with Legionnaires' disease outbreaks.

Incidence and trends in diagnostic testing

Recent active surveillance for legionellosis in ten states in the United States between 2011 and 2013 revealed an average incidence of 1.3 cases per 100,000 population (Dooling et al., 2015) which is similar to that reported in Europe for 2010 which was 1.2 cases per 100,000 inhabitants (Beauté et al, 2013). In the United States, legionellosis cases are reported to CDC through the National Notifiable Diseases Surveillance System (NNDSS). In 2012, a total of 3,688 cases of legionellosis occurring in the US were reported to NNDSS. However, it is believed that legionellosis is significantly underreported since an etiologic diagnosis may not always be sought for pneumonia patients. Legionellosis incidence is not uniformly distributed among the United States, as revealed by recent active surveillance from 2011 to 2013 which found the highest frequency in New York State (4.0 cases per 100,000 population) (Dooling et al., 2015). Regionally, the Northeastern United States has the highest rates of disease followed by the East North Central states.

During the past decade, legionellosis incidence has increased both in the United States and Europe by more than two-fold (Hicks et al., 2011; ECDC, 2016). The reasons for this increase are not fully understood. However, the explanation seems to be multi-factorial including increases in susceptible populations, improved diagnostic testing, and potentially new niches for *Legionella* amplification in building water systems. Cases tend to increase during summer months likely reflecting suitable conditions for *Legionella* growth and dispersal such as increased premise plumbing water temperatures and cooling tower usage) (Hicks et al., 2011).

Despite extensive research on outbreaks of *Legionella,* roughly 96% of legionellosis cases in the United States are not associated with an outbreak or cluster (Hicks et al., 2011). Despite the overwhelming majority, most sporadic legionellosis cases are not fully investigated. However, they are likely to be associated with similar environmental sources as those identified during outbreaks. Clearly, there is a need for more research in this area. Legionellosis cases and clusters are sometimes identified in travelers who resided in resorts, hotels, or on cruise ships. Approximately one-quarter of reported legionellosis cases in the United States are travel-associated (Hicks et al., 2011; Smith et al., 2007). In Europe, approximately 14% of Legionnaires' disease cases reported in 2014 were associated with travel (ECDC, 2016).

Historical perspective on selected *L. pneumophila* reference strains

L. pneumophila strain Philadelphia was first sequenced in 2004, revealing many important metabolic pathways and potential virulence genes (Chien et al., 2004). Genome sequencing has become a major focus of modern microbiology, and over 500 isolates of *L. pneumophila* have been sequenced as of mid-2016. Only a small subset of these sequences have been completed (i.e. closed genomes with no gaps). Many of these completely sequenced strains, such as Philadelphia-1, Pontiac-1, and OLDA were selected due to their historical significance to the field of *Legionella* research. Strains such as Paris, Lens, and Lorraine were also associated with historically important outbreaks occurring in Europe. Finally, a few strains belonging to the less commonly known *L. pneumophila* subspecies *fraseri* and *pascullei* have recently been sequenced in their entirety (Kozak-Muiznieks et al., 2016; Raphael et al., 2017). Many of the strains described in the following paragraphs were recovered during extensive investigations of specific outbreaks, and the results demonstrate the incredible diversity of this species.

OLDA

OLDA was the first documented *L. pneumophila* isolate. It was obtained in March 1947 from a sick guinea pig inoculated with the blood of a patient with undiagnosed mild febrile illness (Jackson et al., 1952, Bozeman et al., 1968). The organism did not grow on any bacteriological media available at that time, and it was not virulent for any laboratory animal model tested except guinea pigs and embryonated eggs (Jackson et al., 1952). The bacterium was then presumed to be a pathogen of guinea pigs rather than humans, because the patient's sera did not react in the complement fixation assay with OLDA antigen (Jackson et al., 1952, McDade et al., 1979). Based on observations of the bacterium from yolk sac smears, the OLDA bacillus appeared both intra- and extracellularly and resembled rickettsiae in color and morphology despite being somewhat larger (Bozeman et al., 1968). The strain was initially called a "rickettsia-like organism", although it was not serologically related to rickettsiae (Jackson et al., 1952). The three other oldest *Legionella* strains Tatlock, Wiga, and Heba were isolated in 1944-1959 and were grouped along with the OLDA strain as "rickettsia-like agents" (Bozeman et al., 1968). However, these strains have since been classified as different *Legionella* species. The identification of *L. pneumophila* as the agent of the 1976 outbreak in Philadelphia and the development of *Legionella*-specific media together permitted the proper classification of OLDA among legionellae (McDade et al., 1979). Both a strain isolated from the Philadelphia outbreak (Philadelphia-1) and OLDA grew on Mueller-Hinton agar, F-G agar, and charcoal-yeast extract agar but not on any other bacteriological media. These two strains also reacted with convalescent sera from LD patients in indirect fluorescent antibody tests. Finally, DNA-DNA hybridization studies confirmed the genetic relatedness of OLDA and Philadelphia-1 strains (McDade et al., 1979). It is suggested that the original presumption of the guinea-pig origin of strain OLDA was due to the low sensitivity of the complement fixation test used for analysis of the original patient's convalescent sera (McDade et al., 1979).

Pontiac-1

In July and August 1968, at least 144 people working at or visiting the Oakland County Health Department building in Pontiac, Michigan developed a self-limited acute febrile illness without pneumonia termed Pontiac fever (Glick et al., 1978). The patients experienced a disease course with a one to two-day incubation period before the illness lasted two to five days consisting of chills, fever, headache, and myalgia. There were 95 cases among 100 employees and 49 cases among 170 visitors to the building (Glick el al., 1978). None of the cases were fatal and only one person was hospitalized. The implication of a disease disseminated at a health department building, with extraordinarily high attack rate, and possible airborne spread precipitated intensive and prolonged epidemiological, environmental, and laboratory investigation. The inquiry involved a multidisciplinary team of epidemiologists, engineers, industrial toxicologists, and bacteriologists. Several of the first outbreak investigators who entered the building before the investigation was focused on an airborne source fell ill of Pontiac fever themselves (Glick et al., 1978). Clinical specimens including throat swabs, stool, and sera were collected from 280 people including case patients, their families, and employees from another office building. Specimens were inoculated into several cell lines, embryonated eggs, and numerous bacteriologic and fungal media in an attempt to culture the infectious organism. Even though Herpes simplex virus and *Mycoplasma* were isolated from several throat swabs, there was no indication of a common nor a widespread pathogen. Serological tests performed on 16 pairs of acute and convalescent sera with 31 viral and bacterial antigens were also uninformative (Glick et al., 1978).

Samples of nearby air, soil, drinking water, and sewage were collected and tested for the presence of microorganisms. Yet, again, none of the isolated organisms were predominant nor widespread (Glick et al., 1978). When a defective air-conditioning system was identified as a possible source of spread of the infectious agent, multiple swabs of the system's surfaces as well as water at the base of evaporative condenser were taken for culture and isolation. *Corynebacterium*, *Bacillus*, *Pseudomona*s, and *Flavobacterium* species were isolated from the water, but these organisms were not serologically connected to the illness. Extracts from air-conditioning filters were obtained and underwent spectrophotometric and chromatographic analyses, which did not reveal any abnormalities (Glick et al., 1978).

Luckily, exposure of Hartely guinea pigs to the building and aerosols of evaporative condenser water eventually led to the isolation and characterization of the Pontiac Fever pathogen. At first, six guinea pigs, along with mice, rabbits, and rhesus monkeys, were placed in the health department building in mid-July of 1968 to test whether they were susceptible to the "mysterious" agent (Kaufmann et al., 1981). Only the guinea pigs developed fever and bronchopneumonia whereas none of the other animals exposed to the building fell ill. After the identifying the pathogen, controlled exposure experiments with guinea pigs were conducted in the health department building and later at the CDC laboratory with aerosolized evaporative condenser water collected in August 1968 (Kaufmann et al., 1981). These controlled guinea pig experiments found that animals given streptomycin and tetracycline were partially protected from the infection. Subsequent laboratory studies conducted at CDC between 1968 and 1970 showed that the animals did not develop bronchopneumonia when the water was filtered or autoclaved. These observations suggested that the infectious agent was most likely of bacterial origin, and small Gram-variable bacterial rods were observed in the nodules that developed in the lungs of ill animals (Kaufmann et al., 1981). However, all attempts to culture this microorganism either from the evaporative condenser water or from lungs of sick guinea pigs were unsuccessful, and the attempts were discontinued in 1971 (Kaufmann et al., 1981). Several factors contributed to the eventual isolation of the Pontiac fever agent in 1977 including the proper storage of key specimens (e.g. patient paired sera and guinea pig lung tissue) from the outbreak and the successful investigation of a Legionnaires' disease outbreak in 1976 that led to the development of culture media for *L. pneumophila*. One of the first clues suggesting a connection between Pontiac fever and Legionnaires' disease was the positive results of indirect fluorescent antibody tests where sera from 37

Pontiac fever patients reacted with the newly isolated *L. pneumophila* Philadelphia strain (Kaufmann et al., 1981).

A complex scheme involving inoculation of guinea pigs, eggs, and Mueller-Hinton agar supplemented with 1% hemoglobin and IsoVitaleX was developed to culture *Legionella* and used to isolate a *L. pneumophila* strain named "Pontiac" from the previously frozen lung tissue of guinea pigs exposed either to the health department building in 1968 or to the untreated evaporative condenser water during aerosolization experiments in 1969 (Kaufmann et al., 1981). Though both outbreaks were caused by the same *Legionella* species and serogroup, it remains unclear why the 1968 outbreak caused Pontiac fever in such a large number of individuals while the 1976 outbreak in Philadelphia caused Legionnaires' disease. The infectious agents in both outbreaks were *L. pneumophila* serogroup 1, which is the most common cause of legionellosis in nearly every country where it has been identified. Since the description of these first Pontiac fever and Legionnaires' disease outbreaks, mixed outbreaks with both forms of legionellosis have been described (Ambrose et al., 2014; Burnsed et al., 2007; Euser et al., 2010; Thomas et al., 1993). However, the exact pathogenic mechanism whereby *Legionella* causes Pontiac fever (e.g. inhalation of *Legionella* or exposure to its lipopolysaccharide) remains to be determined.

Philadelphia

The term 'Legionnaires' disease' was first coined to describe a large-scale, explosive outbreak of respiratory disease in the summer of 1976 in Philadelphia, Pennsylvania. Two hundred twenty-one individuals, mostly American Legionnaires, fell ill with a severe form of pneumonia shortly after attending their annual convention from July 21-24, 1976. Earlier that year a novel H1N1 influenza A virus caused a cluster of respiratory illness, including one fatality, at nearby Fort Dix. Thus, initial efforts and fears were focused on detection of a "swine flu" outbreak and put the nation on high alert for a possible repeat of the 1918 pandemic (Gaydos et al., 2006). Epidemiologists quickly ruled out influenza, but efforts to identify the true cause continued for more than six months as scientists' attempts to cultivate the pathogen in every known virologic host and bacteriologic media were unsuccessful. Initial laboratory reports noted bacteria in thin section microscopy of lung tissue from many patients. However, these were not considered to be the primary agent of disease. In mid-December 1976, researchers at the Rocky Mountain Laboratory in Montana failed to find any significant antibody titers in the blood of 12 affected Legionnaires against a panel of rickettsial species, the cause of deadly spotted fevers and typhus. Shortly thereafter, Joseph McDade and colleagues in the CDC Leprosy and Rickettsia Branch attempted to cultivate their "bacterial contaminants" as they would rickettsia. By early 1977, a previously unknown, acid fast, Gram negative bacillus, which reacted with Legionnaires' disease patient sera, was recovered from the frozen lung tissues of four Legionnaires' disease case patients and given the strain names Philadelphia-1, -2, -3 and -4 in the order they were isolated. Two of these case patients were Legionnaire delegates (isolates Philadelphia-2 and -4), one was the spouse of a Legionnaire (Philadelphia-3), and one was a non-Legionnaire case patient with "Broad Street pneumonia" (i.e. a person who never entered the convention hotel but did pass its entrance on Broad Street) (Philadelphia-1) (Mercante et al., 2016a).

CDC researchers also discovered that the Philadelphia epidemic of 1976 was not the first salvo of an emerging disease, but simply the latest. Similar "rickettsia-like" bacteria, now understood to be various *Legionella* species, caused outbreaks of Legionnaires' disease and Pontiac fever in the United States from 1957, and possibility prior (Glick et al., 1978; Kaufmann et al., 1981; Osterholm et al., 1983; Thacker et al., 1978). Early work employing the original Philadelphia strains formed the basis of microbiological and epidemiological studies of this new genus and included comparisons of antigenic cross-reactivity (Cherry et al., 1978;

McKinney et al., 1979), cellular fatty acid composition (Moss et al., 1977), enzyme electrophoretic mobility profiles (Selander et al., 1985), and genome similarity by the traditional DNA-DNA hybridization method (Brenner et al., 1978).

Isolate Philadelphia-1 became the type strain for *L. pneumophila* serogroup 1. Philadelphia-1 was also the first *Legionella* strain to be fully sequenced in 2004, which has contributed greatly to our genetic understanding of the organism. The remaining Philadelphia-2, -3, and -4 strains were recently sequenced and found to be genetically related, yet distinct from strain Philadelphia-1 (Mercante et al, 2016a). Surprisingly, Philadelphia strains -2 and -4 harbor a growth phase-dependent, self-replicating element that encodes a homologue of the *Agrobacterium lvh/lvr* virulence-associated type IVA secretion locus as well as a CRISPR system. The size, composition, and chromosomal integration site of this element is highly similar to the pP36 element found in *L. pneumophila* strain Paris (Doléans-Jordheim et al., 2006; Cazalet et al., 2004). Apart from this element, the Philadelphia-1 strain diverged from its sister strains at two large genomic loci that were acquired, ostensibly, through horizontal gene transfer. These two loci, variable genomic region (VGR)-1 and VGR-2, encode diverse gene products including virulence determinates such as components of the Dot/Icm Type IVB secretion apparatus. These "Philadelphia" VGR alleles were conserved in other *L. pneumophila* strains of the same sequence type (ST36) (Mercante et al., 2016a).

The number of polymorphisms present among the historical Philadelphia strains suggests they diverged from a common ancestor years before the 1976 outbreak and may have persisted at "hotel A" in downtown Philadelphia for a substantial period. Supporting this hypothesis, a smaller scale respiratory disease outbreak with clinical and epidemiological features compatible with Legionnaires' disease was documented at the same hotel two years prior to the 1976 Legionnaires' convention in 1974. Several members of a fraternal organization known as the Independent Order of Odd Fellows sickened during that 1974 outbreak demonstrated high convalescent *Legionella* antibody titers in 1977 (Terranova et al, 1977). Despite the extensive environmental investigation in 1976, the original source of *Legionella* dispersal was not uncovered. However, the hotel cooling system was considered the most likely culprit.

Interestingly, the Philadelphia sequence type (ST36), or strains matching five or more ST36 alleles, together represent ~42% of the outbreak-related STs in the US between 1982 and 2012 where CDC investigators matched clinical and environmental isolates (Kozak-Muiznieks et al., 2014). These STs form a "clonal complex" and, at least with ST36, appear to share a common genomic structure (Mercante et al, 2016a). Conversely, similar ST36-based complexes only represent ~12% of sporadic cases of legionellosis in the United States over the same time period, and they comprise a much smaller percentage of sampled environmental isolates with no disease association.

Paris

One-third of *L. pneumophila* serogroup 1 clinical isolates (N=75) from individuals with either community-acquired or hospital-associated Legionnaires' disease admitted to hospitals in Paris, France between 1986 and 1997 were indistinguishable by both arbitrarily primed PCR and PFGE methods (Lawrence et al., 1999). The same strain was identified among environmental samples collected from cooling towers and water systems at hospitals, as well as various buildings and swimming pools in locations throughout Paris. Aurell and colleagues (2003) showed that the Paris strain accounted for ~12% of clinical *Legionella* isolates in France examined between 1998 and 2002. Further studies demonstrated that the strain is also present in several other countries (Cazalet et al., 2008; Aurell et al., 2003; Harrison et al., 2006; Borchardt et al., 2008; Harrison et al.,

2009). Interestingly, the Paris strain has been associated with various monoclonal antibody types (Edelstein and Metlay, 2009) suggesting additional genetic variation among seemingly related isolates. In 2004, the genome of the Paris strain was sequenced by Cazalet et al. (2004) and shown to contain genes for a type V secretion system as well as an *lvh-lvr* type IV secretion apparatus which has also been recently described as part of the Philadelphia-2 and -4 genomes (Mercante et al., 2016a). As an ST1 representative, strain Paris is frequently used as reference in comparative genomic analysis studies involving this sequence type.

Lens

Between November 2003 and January 2004, a total of 86 laboratory confirmed cases of Legionnaires' disease were identified in the Pas-de-Calais District located in Northern France (Nguyen et al., 2006). The case fatality rate of this outbreak was 21%. Twenty-three *L. pneumophila* serogroup 1 isolates were obtained from patient sputa. PFGE analysis revealed that these isolates were indistinguishable from each other and distinct from previously defined PFGE patterns. In addition, the outbreak strain, termed "Lens" also detected at the suspected environmental source of the outbreak which was a cooling tower at an industrial plant producing petrochemical solvents. This strain was also identified in water collected during environmental sampling from two other cooling towers and a car wash station, none of which were linked to the outbreak (Nguyen et al., 2006). Comparative analysis of the complete genome sequence of strain Lens revealed a high level of unique gene content implying the significant genomic plasticity which is typical of this species. Also detected in the Lens genome was a ~60 kb plasmid encoding proteins associated with plasmid transfer functions (Cazalet et al., 2004).

Lorraine

A new endemic serogroup 1 strain (Lorraine) spread throughout France and has dramatically increased in prevalence since 2002 (Ginerva et al.,

2008). The Lorraine strain appears to have displaced another French strain, Paris. It was found that Lorraine strain accounted for 10.5% of all clinical isolates in 2005, while the prevalence of strain Paris decreased from 10% in the late 1990s to 6.5% between 2000 and 2006. Most of the isolates associated with the Lorraine strain belong to the Allentown monoclonal antibody (MAb) subgroup and are ST47. Ginerva and colleagues (2008) reported that the Lorraine strain was associated with two Legionnaires' disease outbreaks in France: one in Lyon in 2005 involving 34 cases and another in Paris in 2006 involving 12 cases. Soon after this first report, it became apparent that the Lorraine strain, or ST47 strains in general, were the most frequently isolated strains from clinical specimens in the Netherlands, Belgium and the UK (Euser et al., 2013; Vekens et al., 2012; Harrison et al., 2009).

Interestingly, the Lorraine strain is rarely recovered during routine environmental sampling. For example, in a study of the distribution of *L. pneumophila* in the UK (Harrision et al., 2009), ST47 accounted for 25.7% of clinical *L. pneumophila* isolates. However, only one ST47 isolate (0.4%) was identified among environmental isolates obtained during routine monitoring unrelated to Legionnaires' disease outbreak investigations. In the Netherlands, ST47 accounted for 41% of *L. pneumophila* serogroup 1 clinical isolates (Euser et al., 2013), but similar to the UK, ST47 was recovered only three times between 2002 and 2013 from environmental sources, all of which involved outdoor whirlpool spas (Schalk et al., 2014). One of these outdoor whirlpool spas was implicated in a combined Legionnaires' disease and Pontiac fever outbreak that affected four members of the same family and involved one fatal case of Legionnaires' disease (Euser et al., 2010). Even though no clinical *Legionella* isolates were recovered during the outbreak, an ST47 strain was recovered from an outdoor whirlpool spa used by three out of four cases as well as a garden shower and two garden hoses. In a follow up study, an ST47 strain was isolated from soil in that family's garden. It was then hypothesized that the strain was introduced into the whirlpool spa either by wind or by people entering the spa with soil on their feet (Schalk et al., 2014) and suggesting that soil could be an alternative source of this pathogenic strain.

In a comparative study of host factors associated with community-acquired Legionnaires' disease due to either the Paris or Lorraine strains, the Lorraine strain was linked to smoking and was less lethal (Ginevra et al., 2009). In contrast, infection with the Paris strain was associated with steroid therapy and a history of cancer, but it was negatively associated with smoking. The genetic features of the Paris and Lorraine strains were strikingly different in yet another study which examined SNPs in the genomes of temporary and geographically separated strains (Underwood et al., 2013). In this study, two ST47 isolates from the UK and France isolated in different years differed from each other by just four SNPs. Similarly, a recent study demonstrated that ST47 isolates selected from various regions in the United Kingdom differed by a maximum of one SNP (David et al., 2016). In contrast, two ST1 strains isolated two years apart differed by 280 SNPs (Underwood et al., 2013). These results indicate that the ST47 lineage is highly clonal and less diverse than the ST1 lineage. The authors speculated that the ST47 strain may occupy a niche with strong purifying selection, or that it is subject to different selective pressures when inside a human host. Perhaps the ST47 ancestor was not formerly efficient at causing human infection until the lineage acquired this ability more recently (Underwood et al., 2013).

An examination of the ST47 isolates deposited into the European Society of Clinical Microbiology and Infectious Diseases (ESCMID) Study Group for Legionella Infections (ESGLI) SBT database revealed several other interesting features of this enigmatic strain. Whereas there are multiple ST47 entries from Belgium, France, the Netherlands, and the UK, there are few submissions from other European countries. Only four originate from Germany, with the first entry in 2010, and one entry each from Austria, Greece, Italy, and Turkey. There are no ST47 records from Asian countries and only one deposit from North America, specifically from British Columbia, Canada. The uneven geographical distribution of the Lorraine strain, as deduced from SBT database, could be partially explained by differences among labs in either conducting SBT or submitting strain information to the database. Interestingly, the Lorraine strain has not been identified among serogroup 1 isolates submitted to or isolated by the CDC *Legionella* reference laboratory during routine SBT analysis of clinical isolates or during a retrospective analysis of serogroup 1 clinical strains received by the lab between 1982 and 2006.

Detroit-1

The Detroit-1 strain of *L. pneumophila* is a clinical isolate recovered in December 1977 from the lung biopsy material of a 32-year-old female who developed pneumonia after rejecting a kidney transplant (Neblett, et al., 1979; Saravolatz et al., 1979). This sporadic case displayed a very high convalescent serum antibody titer of 8;192, and the patient received antibiotic treatment one week prior to biopsy and culture isolation. Strain Detroit-1 is serogroup 1 and was the first of three initial "Detroit" isolates recovered from separate immunocompromised patients at the Henry Ford Hospital in Detroit, Michigan (Saravolatz et al., 1979). These cases occurred at a time when the scientific community was beginning to recognize both the trend of hospital-acquired Legionnaires' disease infections and the strong disease connection with underlying immunocompromise (Kirby et al., 1978; Beaty et al., 1978).

In early nutritional growth studies, the Detroit-1 strain was found to be fastidious on solid defined, minimal media compared to other *L. pneumophila* strains and serogroups (George et al., 1980). It also displayed higher comparable susceptibility to several antibiotics, including penicillin and rifampicin (Saravolatz et al., 1980). Detroit-1 does not appear to harbor extrachromosomal plasmids (Johnson et al., 1982). Various genetic relatedness studies employing Infrequent Restriction Site (IRS)-PCR, PFGE, and DNA-DNA hybridization suggested strain Detroit-1, as well as Detroit strains -2 and -3 in the hybridization studies, are distinct from serogroup 1 strains Philadelphia-1 and Pontiac (Brenner et al., 1988; Riffard et al., 1998). In fact, the Detroit-1 strain belongs to the *L. pneumophila* subspecies *fraseri* (Selander et al., 1985; Brenner et al., 1988). In particular, *L. pneumophila* strain Detroit-1 should not be confused with the *L. micdadei* strain "Detroit", which is also a human disease-associated isolate.

Dallas 1E

The Dallas 1E *L. pneumophila* strain was obtained from a cooling tower sampled during a Legionnaires' disease outbreak investigation in September 1978 in Dallas, Texas (England et al., 1980). The 79th Annual National Veterans of Foreign Wars (VFW) Convention took place in Dallas during August 1978 with 36,000 attendees including 19,000 VFW members (Berry et al., 1978). The first case was diagnosed in a 58-year-old New Jersey man who had onset of symptoms on August 21st after traveling to the VFW convention in Dallas and arriving there on August 12st, 1978. There were five confirmed Legionnaires' disease cases diagnosed by a four-fold or greater rise in antibody titer from paired sera with *L. pneumophila* serogroup 1 used as an antigen, and two presumptive cases were diagnosed by a single convalescent titer of ≥256 (Berry et al., 1978). None of the identified cases were fatal. As part of the outbreak investigation, water samples were collected from 13 cooling towers on September 20, 1978. In addition, two isolates, Dallas 1E and Dallas 2E, were obtained from locations two blocks apart which were not epidemiologically associated with the outbreak cases (England et al., 1980). Neither isolate stained with fluorescent antibodies when prepared against the four *L. pneumophila* serogroups known at that time. In fact, Dallas 1E and Dallas 2E were the first strains representing a fifth *L.*

pneumophila serogroup. The antibodies prepared against these new strains reacted with previously non-typable strains Dallas 3E, Burlington-1E and Cambridge-2, placing them as members of *L. pneumophila* serogroup 5. Dallas 3E was obtained from the same cooling tower as Dallas 2E, only three months later. Burlington-1E was an environmental isolate obtained at the site of a previous outbreak in Burlington, Vermont. The Cambridge-2 strain was recovered from the lung tissue of a patient from Cambridge, England. Despite their close proximity, neither Dallas 1E nor Dallas 2E strains were implicated in the VFW outbreak. Dallas 1E, however, is considered the type strain for *L. pneumophila* serogroup 5. Based on a comparison of electrophoretic motilities of 22 enzymes as well as DNA-DNA hybridization studies conducted in 1980s, both Dallas 1E and Dallas 2E were assigned to the *L. pneumophila* subspecies *fraseri* (Selander et al., 1985; Brenner et al., 1988).

MICU-B, D7119, F4185

L. pneumophila subspecies *pascullei* is represented by only a few strains (Brenner et al., 1988). Currently, all known representatives of this subspecies are connected to a health care facility (HCF-A) located in Pittsburgh, Pennsylvania. HCF-A was opened in 1954 and started to identify Legionnaires' disease cases in 1979 when *Legionella*-specific culture methods were introduced into the hospital's microbiology laboratory (Brown et al., 1980). Between February 1979 and March 1980, 25 confirmed and nine presumptive cases of Legionnaires' disease were identified, and the disease incidence rate remained as significant for the next several years (Brown et al., 1980, Stout et al., 1998). In 1981, the HCF-A began monthly sampling of the potable water system, which resulted in the recovery of *L. pneumophila* and *L. micdadei* isolates from multiple sites including faucets, showers, and hot water storage tanks (Best et al., 1983). The *L. pneumophila* isolates were 85% serogroup 1 and 15% of serogroup 5, and *L. micdadei* at the time was known as the

"Pittsburgh pneumonia agent". During subsequent sampling in 1982, three *L. pneumophila* serogroup 5 strains were recovered from the "shower heads, sink drains, and spigots" (Garrity et al., 1982). The strains were presumably named based on the location of the water sampling site as"U7W" was named from the 7th floor of the west wing, "U8W" was named from the 8th floor of the west wing, and "MICU-B" was named from the medical intensive care unit (Stout et al., 1982). Based on the results of DNA-DNA hybridization and electrophoretic mobility of 22 enzymes, these three strains substantially differed from the rest of *L. pneumophila* isolates used in the analyses. Due to the difference, they were grouped together as the *L. pneumophila* subspecies, *pascullei,* with U8W as the type strain (Garrity et al., 1982; Selander et al., 1985; Brenner et al., 1988). Interestingly, legionellosis cases identified at HCF-A between March 1979 and December 1982 were due only to *L. pneumophila* serogroup 1, *L. micdadei,* or coinfection of these agents, but not *L. pneumophila* serogroup 5 (Best et al., 1983).

Despite a well-established Legionnaires' disease prevention program and implementation of multiple *Legionella* control measures such as a copper-silver ionization system installed in 1994, a large outbreak took place at the HCF-A in 2012 (Demirjian et al., 2015). There were five confirmed and 17 probable Legionnaires' disease cases including six deaths. The majority of environmental isolates obtained from the HCF-A potable water supply samples and three clinical isolates associated with this outbreak were characterized as *L. pneumophila* serogroup 1, ST1395. Interestingly, the 2012 outbreak strain and the three environmental strains obtained from the HCF-A in 1982 shared alleles at five out of seven sequence type loci (Kozak-Muiznieks, unpublished data). To better understand the relationship of the 2012 isolates to those from 1982, complete genomes of MICU-B (1982 environmental isolate), D-7119 (2012 clinical isolate), and F-4185 (2012 environmental isolate) were sequenced (Kozak-Muiznieks et al., 2016). Surprisingly, these isolates which are associated with the

Table 1. Genomic characteristics of *L. pneumophila* strains described in this chapter. Reference describes the initial genome sequence report of the indicated strain. Pontiac-1 strain was isolated from tissues inoculated in 1968. Detroit-1 and Dallas-1E strains are classified as *L. pneumophila* subspecies *fraseri*. MICU-B, D7119 and F4185 strains are classified as *L. pneumophila* subspecies *pascullei*.

Strain	Serogroup	ST	Source	Origin	Year Isolated	Reference
OLDA	1	1	Guinea pig	Unknown	1947	Mercante et al., 2016b
Pontiac-1	1	62	Guinea pig	Pontiac, Michigan	1977	Mercante et al., 2016b
Philadelphia-1	1	36	Lung	Philadelphia, Pennsylvania	1977	Chien et al., 2004
Paris	1	1	Unknown	France	Unknown	Cazalet et al., 2004
Lens	1	15	Unknown	France	~2003	Cazalet et al., 2004
Lorraine	1	47	Clinical specimen	France	2004	Gomez-Valero et al., 2011
Detroit-1	1	2206	Lung	Detroit, Michigan	1977	Raphael et al., 2017
Dallas 1E (ATCC33216)	5	1300	Cooling tower	Dallas, Texas	1978	Raphael et al., 2017
MICU-B	5	1335	Water	Pittsburgh, Pennsylvania	1982	Kozak-Muiznieks et al., 2016
D7119	1	1395	Sputum	Pittsburgh, Pennsylvania	2012	Kozak-Muiznieks et al., 2016
F4185	1	1335	Water	Pittsburgh, Pennsylvania	2012	Kozak-Muiznieks et al., 2016

same facility and separated by nearly 30 years, are from different serogroups. However, they are more related to each other than to other serogroup 1 or 5 strains.

Summary

The genomic era has provided great insight into the level of diversity among *L. pneumophila* isolates. Strains selected for sequencing projects often also have historical significance. Some strains selected from large outbreaks have been subsequently shown to be abundant in the environment or among those frequently causing clinical disease in a particular region. In other cases, strains were selected due to association with severe or mild clinical courses. For instance, Pontiac-1, which is associated with the first Pontiac fever outbreak, actually shares a great deal of gene content with other *L. pneumophila* strains but was not associated with the pneumonic form of legionellosis. Clearly, much remains to be discovered about the differences in virulence among *L. pneumophila* strains and the susceptibility to legionellosis among humans. As more strains are selected for genome sequencing, maintaining detailed metadata will be crucial for understanding the unique history of each isolate and may help associate novel genomic features with specific characteristics of these isolates in the future such as pathogenicity profiles. Obtaining a greater number of *L. pneumophila* isolates for sequencing from clinical specimens and environmental sources, both associated and unassociated with disease, will undoubtedly contribute to a more complete understanding of the diversity, ecology, and pathogenic potential of this species.

Acknowledgements

We thank Laura Cooley and Cynthia Whitney for critical review of the manuscript and their expertise on the epidemiology of legionellosis. The statements and conclusions in this chapter are those of the authors and do not necessarily represent the official position of the Centers for Disease Control and Prevention.

References

Ambrose J, Hampton LM, Fleming-Dutra KE, Marten C, McClusky C, Perry C, Clemmons NA, McCormic Z, Peik S, Mancuso J, Brown E, Kozak N, Travis T, Lucas C, Fields B, Hicks L, Cersovsky SB. (2014). Large outbreak of Legionnaires' disease and Pontiac fever at a military base. Epidemiol Infect 142:2336-2346.

Aurell H, Etienne J, Forey F, Reyrolle M, Girardo P, Farge P, Decludt B, Campese C, Vandenesch F, Jarraud S. (2003). *Legionella pneumophila* serogroup 1 strain Paris: endemic distribution throughout France. J Clin Microbiol. 41:3320-2.

Beaty HN, Miller AA, Broome CV, Goings S, Phillips CA. (1978). Legionnaires' disease in Vermont, May to October 1977. JAMA 240:127-131.

Beauté J, Zucs P, de Jong B; European Legionnaires' Disease Surveillance Network. (2013). Legionnaires disease in Europe, 2009-2010. Euro Surveill. 18:20417.

Berry EL, Dewlett H, Munson H, Webb C, Altman R, Taylor JW, Lyman DO, Donnell Jr HD. (1978). Convention-associated Legionnaires' disease. MMWR 27:387-388

Best M, Yu VL, Stout J, Goetz A, Muder RR, Taylor F. (1983). *Legionella*ceae in the hospital water-supply. Epidemiological link with disease and evaluation of a method for control of nosocomial Legionnaires' disease and Pittsburgh pneumonia. Lancet 2:307-310.

Borchardt J, Helbig JH, Lück PC. (2008). Occurrence and distribution of sequence types among *Legionella pneumophila* strains isolated from patients in Germany: common features and differences to other regions of the world. Eur J Clin Microbiol Infect Dis. 27:29-36.

Bozeman FM, Humphries JW, Campbell JM. (1968). A new group of rickettsia-like agents recovered from guinea pigs. Acta Virol. 12:87-93.

Brenner DJ, Steigerwalt AG, Weaver RE, McDade JE, Feeley JC, Mandel M. (1978). Classification of the Legionnaires' disease bacterium: An interim report. Current Microbiology 1:71-75.

Brenner DJ, Steigerwalt AG, Epple P, Bibb WF, McKinney RM, Starnes RW, Colville JM, Selander RK, Edelstein PH, Moss CW. (1988). *Legionella pneumophila* serogroup Lansing 3 isolated from a patient with fatal pneumonia, and descriptions of *L. pneumophila* subsp. pneumophila subsp. nov., *L. pneumophila* subsp. fraseri subsp. nov., and *L. pneumophila* subsp. pascullei subsp. nov. J Clin Microbiol 26:1695-1703.

Brown A, Yu VL, Elder EM, Magnussen MH, Kroboth F. 1980. Nosocomial outbreak of Legionnaire's disease at the Pittsburgh Veterans Administration Medical Center. Trans Assoc Am Physicians 93:52-59.

Burnsed LJ, Hicks LA, Smithee LM, Fields BS, Bradley KK, Pascoe N, Richards SM, Mallonee S, Littrell L, Benson RF, Moore MR, Legionellosis Outbreak Investigation Team. (2007). A large, travel-associated outbreak of legionellosis among hotel guests: utility of the urine antigen assay in confirming Pontiac fever. Clin Infect Dis. 44:222-8.

Cazalet C, Rusniok C, Brüggemann H, Zidane N, Magnier A, Ma L, Tichit M, Jarraud S, Bouchier C, Vandenesch F, Kunst F, Etienne J, Glaser P, Buchrieser C. (2004). Evidence in the *Legionella pneumophila* genome for exploitation of host cell functions and high genome plasticity. Nat Genet. 36:1165-73.

Cazalet C, Jarraud S, Ghavi-Helm Y, Kunst F, Glaser P, Etienne J, Buchrieser C. (2008). Multigenome analysis identifies a worldwide distributed epidemic *Legionella pneumophila* clone that emerged within a highly diverse species. Genome Res. 18:431-41.

Cherry WB, Pittman B, Harris PP, Hebert GA, Thomason BM, Thacker L, Weaver RE. (1978). Detection of Legionnaires disease bacteria by direct immunofluorescent staining. Journal of Clinical Microbiology 8:329-338.

Chien M, Morozova I, Shi S, Sheng H, Chen J, Gomez SM, Asamani G, Hill K, Nuara J, Feder M, Rineer J, Greenberg JJ, Steshenko V, Park SH, Zhao B, Teplitskaya E, Edwards JR, Pampou S, Georghiou A, Chou IC, Iannuccilli W, Ulz ME, Kim DH, Geringer-Sameth A, Goldsberry C, Morozov P, Fischer SG, Segal G, Qu X, Rzhetsky A, Zhang P, Cayanis E, De Jong PJ, Ju J, Kalachikov S, Shuman HA, Russo JJ. (2004). The genomic sequence of the accidental pathogen *Legionella pneumophila*. Science. 305:1966-8.

Cunha BA, Burillo A, Bouza E. (2016). Legionnaires' disease. Lancet. 387:376-85.

David S, Rusniok C, Mentasti M, Gomez-Valero L, Harris SR, Lechat P, Lees J, Ginevra C, Glaser P, Ma L, Bouchier C, Underwood A, Jarraud S, Harrison TG, Parkhill J, Buchrieser C. (2016). Multiple major disease-associated clones of *Legionella pneumophila* have emerged recently and independently. Genome Res. 26:1555-1564.

Demirjian A, Lucas CE, Garrison LE, Kozak-Muiznieks NA, States S, Brown EW, Wortham JM, Beaudoin A, Casey ML, Marriott C, Ludwig AM, Sonel AF, Muder RR, Hicks LA. (2015). The importance of clinical surveillance in detecting legionnaires' disease outbreaks: a large outbreak in a hospital with a *Legionella* disinfection system-Pennsylvania, 2011-2012. Clin Infect Dis. 60:1596-602.

Doléans-Jordheim A, Akermi M, Ginevra C, Cazalet C, Kay E, Schneider D, Buchrieser C, Atlan D, Vandenesch F, Etienne J, Jarraud S. (2006). Growth-phase-dependent mobility of the lvh-encoding region in *Legionella pneumophila* strain Paris. Microbiology. 152:3561-8.

Dominguez A, Alvarez J, Sabria M, Carmona G, Torner N, Oviedo M, Cayla J, Minguell S, Barrabeig I, Sala M, Godoy P, Camps N. (2009). Factors influencing the case-fatality rate of Legionnaires' disease. Int J Tuberc Lung Dis. 13:407-12.

Dooling KL, Toews KA, Hicks LA, Garrison LE, Bachaus B, Zansky S, Carpenter LR, Schaffner B, Parker E, Petit S, Thomas A, Thomas S, Mansmann R, Morin C, White B, Langley GE. (2015). Active bacterial core surveillance for legionellosis - United States, 2011-2013. MMWR. 64:1190-3.

Edelstein PH, Metlay JP. *Legionella pneumophila* goes clonal--Paris and Lorraine strain-specific risk factors. Clin Infect Dis. (2009). Jul 15;49(2): 192-4.

England AC 3rd, McKinney RM, Skaliy P, Gorman GW. (1980). A fifth serogroup of *Legionella pneumophila*. Ann Intern Med. 93:58-9.

European Centre for Disease Prevention and Control (ECDC). (2016). Legionnaires' disease in Europe, 2014. Stockholm, ECDC.

Euser SM, Pelgrim M, den Boer JW. (2010). Legionnaires' disease and Pontiac fever after using a private outdoor whirlpool spa. Scand J Infect Dis. 42:910-6.

Euser SM, Bruin JP, Brandsema P, Reijnen L, Boers SA, Den Boer JW. (2013). *Legionella* prevention in the Netherlands: an evaluation using genotype distribution. Eur J Clin Microbiol Infect Dis. 32:1017-22.

Farnham A, Alleyne L, Cimini D, Balter S. (2014). Legionnaires' disease incidence and risk factors, New York, New York, USA, 2002-2011. Emerg Infect Dis. 20:1795-1802.

Garrison LE, Kunz JM, Cooley LA, Moore MR, Lucas C, Schrag S, Sarisky J, Whitney CG. (2016). Vital signs: deficiencies in environmental control identified in outbreaks of Legionnaires' disease - North America, 2000-2014. MMWR. 65:576-584.

Garrity GM, Elder EM, Davis B, Vickers RM, Brown A. (1982). Serological and genotypic diversity among serogroup 5-reacting environmental *Legionella* isolates. J Clin Microbiol.15:646-53.

Gaydos JC, Top FH, Hodder RA, Russell PK. (2016). Swine influenza A outbreak, Fort Dix, New Jersey, 1976. Emerg Infect Dis. 12: 23-28.

George JR, Pine L, Reeves MW, Harrell WK. (1980). Amino acid requirements of *Legionella pneumophila*. J Clin Microbiol 11:286-291.

Ginevra C, Forey F, Campèse C, Reyrolle M, Che D, Etienne J, Jarraud S. (2008). Lorraine strain of *Legionella pneumophila* serogroup 1, France. Emerg Infect Dis. 14:673-5.

Ginevra C, Duclos A, Vanhems P, Campèse C, Forey F, Lina G, Che D, Etienne J, Jarraud S. (2009). Host-related risk factors and clinical features of community-acquired Legionnaires disease due to the Paris and Lorraine endemic strains, 1998-2007, France. Clin Infect Dis. 49:184-91.

Glick TH, Gregg MB, Berman B, Mallison G, Rhodes WW, Jr., Kassanoff I. (1978). Pontiac fever. An epidemic of unknown etiology in a health department: I. Clinical and epidemiologic aspects. American Journal of Epidemiology 107:149-160.

Gomez-Valero L, Rusniok C, Buchrieser C. (2009). *Legionella pneumophila*: population genetics, phylogeny, and genomics. Infect Gen Evol 9:727-739.

Gomez-Valero L, Rusniok C, Jarraud S, Vacherie B, Rouy Z, Barbe V, Medigue C, Etienne J, Buchrieser C. (2011). Extensive recombination events and horizontal gene transfer shaped the *Legionella pneumophila* genomes. BMC Genomics. 12:536.

Harrison TG, Fry NK, Afshar B, Bellamy W, Doshi N, Underwood AP. (2006). Typing of *Legionella pneumophila* and its role in elucidating the epidemiology of Legionnaires' disease. In: Cianciotto NP, Abu Kwaik Y, Edelstein PH, Fields BS, eds. *Legionella* - state of the art 30 years after its recognition. Washington, DC: ASM Press, p 94-9.

Harrison TG, Afshar B, Doshi N, Fry NK, Lee JV. (2009). Distribution of *Legionella pneumophila* serogroups, monoclonal antibody subgroups and DNA sequence types in recent clinical and environmental isolates from England and Wales (2000-2008). Eur J Clin Microbiol Infect Dis. 28:781-91.

Hicks LA, Garrison LE, Nelson GE, Hampton LM. (2011). Legionellosis-United States, 2000-2009. MMWR 60:1083-1086.

Jackson EB, Crocker TT, Smadel JE. (1952). Studies on two rickettsia-like agents probably isolated from guinea pigs. Bacteriol Proc 52:119.

Johnson SR, Schalla WO. (1982). Plasmids of serogroup 1 strains of *Legionella pneumophila*. Curr Microbiol 7:143-146.

Joly JR, McKinney RM, Tobin JO, Bibb WF, Watkins ID, Ramsay D. (1986). Development of a standardized subgrouping scheme for *Legionella pneumophila* serogroup 1 using monoclonal antibodies. J Clin Microbiol 23:768-771.

Kaufmann AF, McDade JE, Patton CM, Bennett JV, Skaliy P, Feeley JC, Anderson DC, Potter ME, Newhouse VF, Gregg MB, Brachman PS. (1981). Pontiac fever: isolation of the etiologic agent (*Legionella* pneumophilia) and demonstration of its mode of transmission. American Journal of Epidemiology 114:337-347.

Kirby BD, Snyder KM, Meyer RD, Finegold SM. (1978). Legionnaires' disease: clinical features of 24 cases. Ann Intern Med 89:297-309.

Kozak-Muiznieks NA, Lucas CE, Brown E, Pondo T, Taylor TH, Jr., Frace M, Miskowski D, Winchell JM. (2014). Prevalence of sequence types among clinical and environmental isolates of *Legionella pneumophila* serogroup 1 in the United States from 1982 to 2012. Journal of Clinical Microbiology 52:201-211.

Kozak-Muiznieks NA, Morrison SS, Sammons S, Rowe LA, Sheth M, Frace M, Lucas CE, Loparev VN, Raphael BH, Winchell JM. (2016). Three genome sequences of *Legionella pneumophila* subsp. *pascullei* associated with colonization of a health care facility. Genome Announc. 4: e00335-16.

Lawrence C, Reyrolle M, Dubrou S, Forey F, Decludt B, Goulvestre C, Matsiota-Bernard P, Etienne J, Nauciel C. (1999). Single clonal origin of a high proportion of *Legionella pneumophila* serogroup 1 isolates from patients and the environment in the area of Paris, France, over a 10-year period. J Clin Microbiol. 37:2652-5.

Lück PC, Helbig JH, Ehret W, Marre R, Witzleb W. (1992). Subtyping of *Legionella pneumophila* serogroup 1 strains isolated in Germany using monoclonal antibodies. Zentralbl. Bakteriol. 277:179-187.

Lück C, Fry NK, Helbrig JH, Jarraud S, Harrison TG. (2013). Typing methods for *Legionella.* Methods Mol Biol 954:119-148.

Marston BJ, Plouffe JF, File TM Jr, Hackman BA, Salstrom SJ, Lipman HB, Kolczak MS, Breiman RF. (1997). Incidence of community-acquired pneumonia requiring hospitalization. Results of a population-based active surveillance study in Ohio. The community-based pneumonia incidence study group. Arch Intern Med. 157:1709-18.

McDade JE, Brenner DJ, Bozeman FM. (1979). Legionnaires' disease bacterium isolated in 1947. Ann. Internal Med. 90:659-661.

McKinney RM, Thacker L, Harris PP, Lewallen KR, Hebert GA, Edelstein PH, Thomason BM. (1979). Four serogroups of Legionnaires' disease bacteria defined by direct immunofluorescence. Annals of Internal Medicine 90:621-624.

Mercante JW, Morrision SS, Desai HP, Raphael BR, Winchell JW. (2016a). Genomic analysis reveals novel bacterial diversity among the 1976 Philadelphia Legionnaires' disease outbreak isolates. PLoS One. 11:e0164074.

Mercante JW, Morrison SS, Raphael BH, Winchell JM. (2016b). Complete genome sequences of the historical *Legionella pneumophila* strains OLDA and Pontiac. Genome Announc. 4: e00866-16

Mercante JW, Winchell JM. 2015. Current and emerging *Legionella* diagnostics for laboratory and outbreak investigations. Clin Microbiol Rev. 28:95-133.

Moss CW, Weaver RE, Dees SB, Cherry WB. (1977). Cellular fatty acid composition of isolates from Legionnaires disease. Journal of Clinical Microbiology 6:140-143.

Neblett TR, Riddle JM, Dumoff M. (1979). Surface topography and fine structure of the Legionnaires' disease bacterium. A study of six isolates from hospitalized patients. Ann Intern Med 90:648-651.

Nguyen TM, Ilef D, Jarraud S, Rouil L, Campese C, Che D, Haeghebaert S, Ganiayre F, Marcel F, Etienne J, Desenclos JC. (2006). A community-wide outbreak of legionnaires disease linked to industrial cooling towers--how far can contaminated aerosols spread? J Infect Dis. 193:102-11.

Osterholm MT, Chin TD, Osborne DO, Dull HB, Dean AG, Fraser DW, Hayes PS, Hall WN. (1983). A 1957 outbreak of Legionnaires' disease associated with a meat packing plant. American Journal of Epidemiology 117:60-67.

Raphael BH, Kozak-Muiznieks NA, Mercante JM, Winchell JM. (2017). Complete genome sequences of *Legionella pneumophila* subsp. *fraseri* strains Detroit-1 and Dallas 1E. Genome Announc (in press).

Riffard S, Lo Presti F, Vandenesch F, Forey F, Reyrolle M, Etienne J. (1998). Comparative analysis of infrequent-restriction-site PCR and pulsed-field gel electrophoresis for epidemiological typing of *Legionella pneumophila* serogroup 1 strains. J Clin Microbiol 36:161-167.

Rogers J, Dowsett AB, Dennis PJ, Lee JV, Keevil CW. (1994). Influence of temperature and plumbing material selection on biofilm formation and growth of *Legionella pneumophila* in a model potable water system containing complex microbial flora. Appl Environ Microbiol . 60:1585-1592.

Saravolatz LD, Burch KH, Fisher E, Madhavan TOM, Kiani D, Neblett T, Quinn EL. (1979). The compromised host and Legionnaires' disease. Ann Intern Med 90:533-537.

Saravolatz LD, Pohlod DJ, Quinn EL. (1980). Antimicrobial susceptibility of *Legionella pneumophila* Serogroups I--IV. Scand J Infect Dis 12:215-219.

Schalk JA, Euser SM, van Heijnsbergen E, Bruin JP, den Boer JW, de Roda Husman AM. (2014). Soil as a source of *Legionella pneumophila* sequence type 47. Int J Infect Dis. 27:18-9.

Selander RK, McKinney RM, Whittam TS, Bibb WF, Brenner DJ, Nolte FS, Pattison PE. (1985). Genetic structure of populations of *Legionella pneumophila*. J Bacteriol. 163:1021-37.

Smith P, Moore M, Alexander N, Hicks L, O'Loughlin R. (2007). Surveillance for travel-associated legionnaires disease--United States, 2005-2006. MMWR. 56:1261-3.

Stout J, Yu VL, Vickers RM, Zuravleff J, Best M, Brown A, Yee RB, Wadowsky R. (1982). Ubiquitousness of *Legionella pneumophila* in the water supply of a hospital with endemic Legionnaires' disease. N Engl J Med. 306:466-8.

Stout JE, Lin YS, Goetz AM, Muder RR. (1998). Controlling *Legionella* in hospital water systems: experience with the superheat-and-flush method and copper-silver ionization. Infect Control Hosp Epidemiol 19:911-914.

Terranova W, Cohen M, Fraser D. 1978. (1974). Outbreak of Legionnaires' disease diagnosed in 1977: clinical and epidemiological features. Lancet 312:122-124.

Thacker SB, Bennett JV, Tsai TF, Fraser DW, McDade JE, Shepard CC, Williams KH, Jr., Stuart WH, Dull HB, Eickhoff TC. (1978). An outbreak in 1965 of severe respiratory illness caused by the Legionnaires' disease bacterium. J Infect Dis 138:512-519.

Thomas DL, Mundy LM, Tucker PC. (1993). Hot tub legionellosis. Legionnaires' disease and Pontiac fever after a point-source exposure to *Legionella pneumophila*. Arch Intern Med 153:2597-2599.

Underwood AP, Jones G, Mentasti M, Fry NK, Harrison TG. (2013). Comparison of the *Legionella pneumophila* population structure as determined by sequence-based typing and whole genome sequencing. BMC Microbiol. 13:302.

Vekens E, Soetens O, De Mendonça R, Echahidi F, Roisin S, Deplano A, Eeckhout L, Achtergael W, Piérard D, Denis O, Wybo I. (2012). Sequence-based typing of *Legionella pneumophila* serogroup 1 clinical isolates from Belgium between 2000 and 2010. Euro Surveill. 17:20302.

Vergis EN, Akbas E, Yu VL. (2000). *Legionella* as a cause of severe pneumonia. Semin Respir Crit Care Med. 21:295-304.

von Baum H, Ewig S, Marre R, Suttorp N, Gonschior S, Welte T, Lück C. (2008). Competence network for community acquired pneumonia study group. Community-acquired *Legionella* pneumonia: new insights from the German competence network for community acquired pneumonia. Clin Infect Dis. 46:1356-64.

Wadowsky RM, Wolford R, McNamara AM, Yee RB. (1985). Effect of temperature, pH, and oxygen level on the multiplication of naturally occurring *Legionella pneumophila* in potable water. Appl Environ Microbiol 49:1197-1205.

Yu VL, Plouffe JF, Pastoris MC, Stout JE, Schousboe M, Widmer A, Summersgill J, File T, Heath CM, Paterson DL, Chereshsky A. (2002). Distribution of *Legionella* species and serogroups isolated by culture in patients with sporadic community-acquired legionellosis: an international collaborative survey. J Infect Dis 186:127-128.

Chapter 5

Clinical Symptoms and Treatment of Legionellosis

Giancarlo Ceccarelli[1]*, Mario Venditti[1], Maria Scaturro[2] and Maria Luisa Ricci[2]*

[1]Department of Public Health and Infectious Diseases. University of Rome "Sapienza", Azienda Policlinico Umberto I Rome, Italy
[2]Department of Infectious Diseases, National Reference Laboratory for *Legionella*. National Institute for Health, Rome, Italy

*giancarlo.ceccarelli@uniroma1.it and marialuisa.ricci@iss.it

DOI: https://doi.org/10.21775/9781913652531.05

Abstract

Legionellosis causes a unique clinical course which can pose challenges for prompt diagnosis and treatment. This chapter discusses the clinical symptoms of the two forms of legionellosis, Legionnaires' disease and Pontiac fever, along with the current antibiotic treatment regimens for Legionnaires' disease. *Legionella* pneumonia does not have clear distinguishing features apart from other cases of bacterial pneumonia. Clinical scores and predicting tools have been developed to predict the likelihood of *Legionella* infection and determine the severity of infection. Antibiotics effective against *Legionella* must be active and concentrated in the intracellular space as well as distributed to infected tissues. Currently, fluoroquinolones and macrolides are the first line treatment with preference for fluoroquinolones and particularly, levofloxacin. However, immunocompromised patients may have unique clinical features requiring a more intensive antibiotic regimen. Lastly, the nonspecific symptoms of legionellosis may lead to misdiagnosis and treatment with antibiotics such as, beta-lactams which may cause a brief clinical improvement followed by a severe deterioration. Due to this dangerous phenomenon and the rise of antibiotic resistance, diagnosis and treatment regimens must be carefully considered.

Clinical Symptoms

"Legionellosis" is the term given to all diseases caused by the Gram-negative bacteria belonging to the genus, *Legionella*. There are two distinct forms of legionellosis, Pontiac Fever and Legionnaires' Disease. Although other species of *Legionella* can cause legionellosis, the *L. pneumophila* species is the etiologic agent in the vast majority of cases. Despite this fact, the clinical history of legionellosis is often polymorphic and characterized by multiple pathophysiological patterns and varying degrees of impairment.

Pontiac Fever is the milder of the two forms of legionellosis and causes a self-limited, non-pneumonic infection. It appears after an incubation period of 24-48 hours as an acute flu-like syndrome without pulmonary involvement and resolves within two to five days. The prodromal symptoms include malaise, myalgias and headache which is quickly followed by fever and sometimes cough and sore throat. Accompanying symptoms of diarrhoea, nausea and mild neurological symptoms such as vertigo and photophobia may also be present. The first outbreak of Pontiac Fever was caused by *L. pneumophila* serogroup 1 while subsequent epidemics have also been attributed to *L. feeleii, L. anisa* and *L. micdadei* species.

In contrast, symptoms of Legionnaires' disease arises after an incubation period ranging from two to 14 days (Den Boer et al., 2015). The illness persists for an average of five to six days appearing primarily as a pneumonia with nonspecific clinical and radiologic findings and sometimes accompanied by extra-pulmonary features. Radiological findings are not pathognomonic, and no specific radiological findings are associated with an increased risk of mortality (Poirier et al., 2017). However, the most common radiologic variations include focal alveolar type thickenings, unilateral consolidations and disseminated infiltrations. Blood tests are nonspecific for Legionnaires' disease, and common laboratory findings are elevated transaminases, hypophosphatemia and hyponatraemia. Possible gastrointestinal are diarrhea, nausea, vomiting and abdominal pain which can occur along with neurological findings such as mental status changes and cardiac symptoms such as relative bradycardia. Though infrequent, pulmonary complications of Legionnaires' disease have been described including pleural effusion, pneumothorax, lung abscess, empyema and respiratory failure. Further severe and systemic complications are rare but encompass shock, disseminated intravascular coagulation, purpura thrombocytopenic and renal failure. The severity of Legionnaires' disease ranges from a mild cough and low fever to rapidly progressive pneumonia with extra-pulmonary disease, coma and death. Cases of Legionnaires' disease are classified according to the Pneumonia Severity Index (PSI) score (Fine et al., 1997).

In general, infection with *Legionella* manifests differently across individuals and populations. Severity ranges from a mild cough and low fever to rapidly progressive pneumonia or extra-pulmonary disease, coma, and death. Among the complications of Legionnaires' disease, lung abscess, empyema, respiratory failure, shock, disseminated intravascular coagulation, purpura thrombocytopenic and renal failure are well described.

In immunocompromised populations, such as those with malignancies, recent transplantation, human immunodeficiency virus and other primary or secondary immunodeficiencies, infection with *Legionella* can cause unusual clinical presentations. Prominent extrapulmonary disease may occur such as severe neurologic and gastrointestinal manifestations, and the mortality rate is higher for these patients despite early appropriate treatment. Atypical pulmonary presentation such as nodular pneumonia and complications such as cavitation and empyema may occur (del Castillo et al., 2016; Lanternier et al., 2017).

Legionella pneumonia has no distinguishing clinical features apart from other atypical or bacterial forms of community acquired pneumonia. However, repeated efforts have been made to both identify predictors and create tools to increase the probability of a correct and timely diagnosis of *Legionella* infection. Table 1 shows the principal tools evaluated to improve legionellosis diagnosis.

Despite the availability of these tools, the reduced sensitivity of these tests and the severity of legionellosis require *L. pneumophila* to always be suspected in cases of community or nosocomial acquired lung infections. For this reason, the American Thoracic Society guidelines recommend antibiotics active towards *Legionella* both for community and nosocomial acquired pneumonia at least until laboratory investigations exclude Legionnaires' disease (Mandell et al., 2007; Kalil et al., 2016). Finally, from a prognostic point of view, hyponatraemia under 136 mEq/l, persistently increased serum PCT levels and APACHE II scores >15 upon admission are independent prognostic factors related to unfavorable clinical outcomes in legionellosis while WBC counts, CRP levels and CURB-65 scores are unable to effectively predict prognosis in legionellosis (de Jager et al., 2013).

Table 1. Clinical scores and tools proposed to predict *L. pneumophila*

Tool	Setting	Reliable scoring system	Notes
Community-Based Pneumonia Incidence Study (CBPIS) Score (Fernández-Sabé et al., 2003)	Identifies differences in presentation between *L. pneumophila* and bacterial pneumococcal pneumonia	No	Male sex, previous receipt of beta-lactam therapy and temperature 39°C positively associated with *L. pneumophila*. Purulent sputum, pleuritic chest pain and previous upper respiratory tract infection negatively associated with *L. pneumophila*.
"Fiumefreddo" score (Fiumefreddo et al., 2009)	Compares 82 patients with *Legionella* CAP with 368 patients with non-*Legionella* CAP	Yes (needs large clinical studies for validation)	Six parameters including high body temperature, absence of sputum production, low serum sodium concentrations and elevated levels of lactate dehydrogenase, C-reactive protein (CRP) and low platelet counts all identified as independent predictors of *Legionella* CAP.
Winthrop-University Hospital weighted point score system (modified) (Cunha, 2008)	Weighted clinical syndromic approach to predict the probability of Legionnaires' disease in patients with CAP	Yes (limited sensitivity and specificity for predicting *L. pneumophila* (Miyashita et al., 2017)	Based on a point score system using the following clinical and laboratory findings; fever, headache, confusion or lethargy, ear pain, non-exudative pharyngitis, hoarseness, purulent sputum, hemoptysis, chest pain, loose stools or watery diarrhea, abdominal pain, renal failure, shock or hypotension, splenomegaly, lack of response to beta-lactam antibiotics, microscopic hematuria, lymphopenia, acute hypoxemia, hyponatremia, hypophosphatemia, and elevated SGOT/SGPT, total bilirubin, lactate dehydrogenase (LDH), creatinine phosphokinase (CPK), CRP, cold agglutinins and ferritin.
Score by Saraya et al (Saraya et al., 2018)	Differentiating *L. pneumophila* pneumonia from *S. pneumoniae* pneumonia	Yes	Found four markers (relative bradycardia, LDH levels ≥292 IU/L, CRP levels ≥21 mg/dL, and sodium levels ≤137 meq/L) may be useful for differentiating pneumonia from *Legionella*-group from *Streptococcus*-group.

Antibiotic treatment

Bacteria of the *Legionella* genus are intracellular microorganisms. Therefore, antimicrobial agents effective for the treatment of Legionnaires' disease must be able to concentrate and remain active in intracellular spaces (Horwitz, 1983). Furthermore, these drugs must distribute and persist with adequate concentration properly in tissues infected with *Legionella*. Pontiac Fever has a benign clinical course in the absence of specific antibiotic treatment. However, other *Legionella* infections require specific treatment to reduce the probability of fatal outcomes.

The antibiotics meeting the requirements for Legionnaires' disease treatment are fluoroquinolones and macrolides in addition to tetracyclines which have lower performance. In contrast, beta lactams, carbapenems, aminoglycosides and chloramphenicol are all useless for the treatment of Legionnaires' disease, as they do not reach intracellular concentrations capable of carrying out an antibacterial effect (Edelstein and Cianciotto, 2005). Based on a number of *in vitro* studies measuring the anti-*Legionella,* particularly anti-*L. pneumophila* serogroup 1, activity in pulmonary alveolar macrophages of guinea pigs and human monocytes and other cellular lines fluoroquinolones, especially levofloxacin, were superior to macrolides. Among the macrolides, azithromycin appeared superior to clarithromycin and both were superior to erythromycin (Edelstein and Cianciotto, 2005; Pedro-Botet and Yu, 2006).

Currently the only clinical data available in literature is from observational studies, as there is a lack of prospective randomized comparative studies evaluating the different therapeutic approaches. Overall, the data obtained from the three larger studies available showed that patients treated with fluoroquinolones had a) more rapid defervescence with an average of 66 hours compared to 97 hours with macrolides, b) lower incidence of complications such as lung cavitation, pleural empyema and septic shock, c) less need for respiratory support with mechanical ventilation with 8.4% versus 18.5% with macrolides, d) lower mortality with 2.1% against 4.5% with macrolides, e) shorter duration of hospital stays with an average of 6.6 days compared to nine days with macrolides and f) lower incidence of side effects with 12.5% versus 23.4% with macrolides (Garrido et al.,

2005; Mykietiuk et al., 2005; Sabrià et al., 2005). The results were further confirmed by a study published in 2017 conducted on 176 cases of *L. pneumophila* between 2006 and 2011 in Japan which reported that among the oral antibiotics, the efficacy rates were 100% each for ciprofloxacin, moxifloxacin and garenoxacin, 91.7% for levofloxacin and only 81.8% for clarithromycin (Miyashita et al., 2017).

An important limitation in these studies, however, is usage of a standard macrolide. The standard fluoroquinolone used in the studies was levofloxacin. However, regarding macrolides, both erythromycin and clarithromycin were used in most cases. Interestingly, both erythromycin and clarithromycin were found to be less effective than azithromycin against intracellular *Legionella*, and azithromycin is the only macrolide that has shown the same anti-*Legionella in vitro* efficacy as fluoroquinolones (Pedro-Botet and Yu, 2006). Further supporting the superiority of fluoroquinolones is their application to the broader antimicrobial spectrum including against organisms which may co-infect immunocompromised patients vulnerable to legionellosis including penicillin-macrolide-resistant *Streptococcus*, methicillin-sensitive *Staphylococcus aureus*, *Pseudomonas aeruginosa* and Enterobacteriaceae (Edelstein and Cianciotto, 2005). A large study with six clinical trials conducted for the approval of the levofloxacin for *Legionella* CAP recorded high efficacy and negligible mortality associated with the drug (Yu et al., 2004).

A retrospective analysis focusing on azithromycin for the treatment of *L. pneumophila* reported that this macrolide specifically has clinical efficacy similar to fluoroquinolones not only in mild but also in severe settings (Nagel et al., 2014). This was confirmed by a study published in 2017 on a cohort of 446 patients with *L. pneumophila* treated with levofloxacin or azithromycin and found no significant difference in time to defervescence, time to achieve clinical stability, length of intravenous therapy, length of hospital stay and mortality rate. However, the same study reported that subjects treated with clarithromycin showed longer intravenous antibiotic treatment and longer hospital stay when compared with those treated with a fluoroquinolone (Garcia-Vidal et al., 2017).

Based on the current evidence, the primary regimens for treating *L. pneumophila* are either levofloxacin 750 mg IV/po daily, moxifloxacin 400 mg IV/po daily or azithromycin 1000 mg IV day one than 500 mg IV/ po daily. The initial intravenous azithromycin therapy recommended in hospitalized patients could be switched to oral once the patient has stabilized. Alternative choices are oral use of doxycycline 100 mg po twice daily or azithromycin 500 mg po day one then 250 mg once daily for four days in the case of immunocompetent, ambulatory patients with CAP (Gilbert et al., 2018). Despite insufficient evidence supporting the use of fluoroquinolone combination therapy along with azithromycin or rifampin, these therapeutic choices are sometimes recommended in immunocompromised patients with severe disease.

Italian guidelines for the prevention and treatment of legionellosis published in 2015 proposed a useful algorithm in which various therapeutic options are indicated as a first, second or third choice for the treatment of Legionnaires' disease according to severity and immunocompetence (Table 2) (Cagarelli et al., 2015).

The suggested duration of antibiotic therapy for legionellosis is seven to ten days with longer periods of 14-21 days in immunocompromised patients or for those with severe disease (Sharma et al., 2017). Infections with complications such as lung abscesses, pleural empyema, endocarditis or other extrapulmonary involvements require prolonged treatment. Related pulmonary radiologic findings can regress slowly and positive antigenuria can persist for months after initiation of treatment. Therefore, these tests should not be considered while determining the duration of antibiotic therapy (Edelstein and Cianciotto, 2005).

Table 2. Italian guidelines for legionellosis treatment in 2015. The immunocompetent and immunocompromised are further each divded into mild and severe cases. Mild cases are defined by PSI class I-111 and CURB-65 class I, and severe cases are defined by PSI class IV-V and CURB-65 class II-III. The regimen is distinct with treatments per oral (PO) only for immunocompetent patients with mild disease, and the regimens are the same with IV and PO treatments for immunocompetent patients with severe disease and all immunocompromised patients (Cagarelli et al., 2015).

Immuno-competent	Mild		Severe	
	First line	Levofloxacin 500 mg PO q24h	First line	Levofloxacin 500-750 mg IV q24h
		Moxifloxacin 400 mg PO q24h		Azithromycin 500 mg IV q24h
		Ciprofloxacin 500 mg PO q24h	Second line	Ciprofloxacin 400 mg PO q8h
		Azithromycin 500 mg PO q24h		Moxifloxacin 400 IV po q24h
		Clarithromycin 500 mg PO q12h	Third line	Erythromycin 750-1000 IV q6h and after 500 mg IV q6h Rifampicin 300-600 mg IV/PO q12h
	Second line	Eryhromycin500 mg PO q24h		
		Doxycycline 200 mg (1 dose) then 100 mg PO q12h		
Immuno-compromised	**Mild**		**Severe**	
	First line	Levofloxacin 500-750 mg IV q24h	First line	Levofloxacin 500-750 mg IV q24h
		Azithromycin 500 mg IV q24h		Azithromycin 500 mg IV q24h
	Second line	Ciprofloxacin 400 mg PO q8h	Second line	Ciprofloxacin 400 mg PO q8h
		Moxifloxacin 400 IV po q24h		Moxifloxacin 400 IV po q24h
	Third line	Erythromycin 750-1000 IV q6h and after 500 mg IV q6h Rifampicin 300-600 mg IV/PO q12h	Third line	Erythromycin 750-1000 IV q6h and after 500 mg IV q6h Rifampicin 300-600 mg IV/PO q12h

New treatment strategies with a variety of antibiotics are currently under evaluation. For example, evidence on the use of tigecycline as a second-line agent in monotherapy or in association with azithromycin or fluoroquinolones suggests a compelling option for severely ill patients with Legionnaires' disease refractory to conventional first-line agents (Slawek et al., 2017). These results, however, are burdened by limitations due to the small number of cases treated and necessitates larger clinical trials to properly assess the effectiveness.

The nonspecific nature of legionellosis symptoms may lead clinicians to misdiagnosis and prescription of antibiotics non-reactive to *Legionella* such as beta-lactams, cephalosporins or carbapenems. In these cases, an unexpected and initial clinical improvement can occasionally occur despite that the therapy does not effectively reach the causal organisms. However, improvement is temporary in the absence of prompt introduction macrolide or fluoroquinolone therapy. Furthermore, transient improvement can be followed by a severe clinical picture characterized by a sepsis and multi-organ failure. Tan et al. described a series of six patients with community-acquired pneumonia caused by common organisms such as *Streptococcus pneumoniae*, *Streptococcus pyogenes* and *Enterobacter spp,* and all patients were coinfected with *Legionella*. Legionellosis was only diagnosed and treated in two of the six patients. Those patients treated only with beta-lactam molecules showed a clinical course characterized by an initial improvement followed by a severe resurgence of the disease (Tan et al., 2002).

Despite clinical evidence against the use of beta-lactams, several *in vitro* studies reported that cephalosporins and carbapenems could be active against *L. pneumophila* and *Legionella spp* in assays using culture media supporting the growth of the *Legionella* species (Lewis et al., 1978; Fu and Neu, 1979; Edelstein and Meyer, 1980; Ruckdeschel et al., 1984). In addition, *in vivo* models of *Legionella* infection carried out on embryonated eggs demonstrated that cephalothin, a 1st generation cephalosporin, could delay the death of the embryos by 48 hours at doses of 2.5mg/embryo and prevent 100% of embryo deaths at maximum doses of 20 mg/embryo (Lewis et al., 1978). This unexpected result may be due to the fact that beta-lactams are partially active on *Legionella* in the initial stages of infection in extracellular space before penetration of the alveolar macrophages and pneumocytes. The subsequent severe relapse likely results from progression of the infectious process driven by the intracellular amount of *Legionella* which are unresponsive to beta-lactam antibiotics. Therefore, clinicians should always consider legionellosis with caution in cases with declining respiratory function supported by a clinical and radiological picture attributable to infectious pneumonia.

References

Cagarelli, R., Caraglia, A., La Mura, S., Giammarco, M., and Ottaviani, M. (2015). Linee guida per la prevenzione ed il controllo della legionellosi.

del Castillo, M., Lucca, A., Plodkowski, A., Huang, Y.-T., Kaplan, J., Gilhuley, K., Babady, N.E., Seo, S.K., and Kamboj, M. (2016). Atypical presentation of Legionella pneumonia among patients with underlying cancer: A fifteen-year review. J. Infect. *72*, 45-51.

Cunha, B.A. (2008). Severe *Legionella pneumonia*: rapid presumptive clinical diagnosis with Winthrop-University Hospital's weighted point score system (modified). Heart Lung J. Crit. Care *37*, 311-320.

Den Boer, J.W., Euser, S.M., Brandsema, P., Reijnen, L., and Bruin, J.P. (2015). Results from the National *Legionella* Outbreak Detection Program, the Netherlands, 2002-2012. Emerg. Infect. Dis. *21*, 1167-1173.

Edelstein, P.H., and Cianciotto, N.P. (2005). *Legionella*. In Principles and Practice Of Infectious Diseases, (Philadelphia: Churchill Livingstone Elsevier), pp. 2239-2251.

Edelstein, P.H., and Meyer, R.D. (1980). Susceptibility of *Legionella pneumophila* to twenty antimicrobial agents. Antimicrob. Agents Chemother. *18*, 403-408.

Fernández-Sabé, N., Rosón, B., Carratalà, J., Dorca, J., Manresa, F., and Gudiol, F. (2003). Clinical diagnosis of *Legionella pneumonia* Revisited: Evaluation of the Community-Based Pneumonia Incidence Study Group Scoring System. Clin. Infect. Dis. *37*, 483-489.

Fine, M.J., Auble, T.E., Yealy, D.M., Hanusa, B.H., Weissfeld, L.A., Singer, D.E., Coley, C.M., Marrie, T.J., and Kapoor, W.N. (1997). A prediction rule to identify low-risk patients with community-acquired pneumonia. N. Engl. J. Med. *336*, 243-250.

Fiumefreddo, R., Zaborsky, R., Haeuptle, J., Christ-Crain, M., Trampuz, A., Steffen, I., Frei, R., Müller, B., and Schuetz, P. (2009). Clinical predictors for Legionella in patients presenting with community-acquired pneumonia to the emergency department. BMC Pulm. Med. *9*, 4.

Fu, K.P., and Neu, H.C. (1979). Inactivation of beta-lactam antibiotics by *Legionella pneumophila*. Antimicrob. Agents Chemother. *16*, 561-564.

Garcia-Vidal, C., Sanchez-Rodriguez, I., Simonetti, A.F., Burgos, J., Viasus, D., Martin, M.T., Falco, V., and Carratalà, J. (2017). Levofloxacin versus azithromycin for treating L*egionella* pneumonia: a propensity score analysis. Clin. Microbiol. Infect. *23*, 653-658.

Garrido, R.M.B., Parra, F.J.E., Frances, L.A., Guevara, R.M.R., Sanchez-Nieto, J.M., Hernandez, M.S., Martinez, J.A.S., and Huerta, F.H. (2005). Antimicrobial Chemotherapy for Legionnaires' disease: Levofloxacin versus Macrolides. Clin. Infect. Dis. *40*, 800-806.

Gilbert, D.N., Chambers, H.F., Eliopoulos, G.M., Saag, M.S., and Pavia, A. (2018). Sanford Guide to Antimicrobial Therapy 2018 (Sperryville, VA: Antimicrobial Therapy, Inc).

Horwitz, M.A. (1983). Formation of a novel phagosome by the Legionnaires' disease bacterium (*Legionella pneumophila*) in human monocytes. J. Exp. Med. *158*, 1319-1331.

de Jager, C.P.C., Gemen, E.F.A., Leuvenink, J., Hilbink, M., Laheij, R.J.F., van der Poll, T., and Wever, P.C. (2013). Dynamics of Peripheral Blood Lymphocyte Subpopulations in the Acute and Subacute Phase of Legionnaires' Disease. PLoS ONE *8*, e62265.

Kalil, A.C., Metersky, M.L., Klompas, M., Muscedere, J., Sweeney, D.A., Palmer, L.B., Napolitano, L.M., O'Grady, N.P., Bartlett, J.G., Carratalà, J., et al. (2016). Management of Adults With Hospital-acquired and Ventilator-associated Pneumonia: 2016 Clinical Practice Guidelines by the Infectious Diseases Society of America and the American Thoracic Society. Clin. Infect. Dis. *63*, e61-e111.

Lanternier, F., Ader, F., Pilmis, B., Catherinot, E., Jarraud, S., and Lortholary, O. (2017). Legionnaire's Disease in Compromised Hosts. Infect. Dis. Clin. North Am. *31*, 123-135.

Lewis, V.J., Thacker, W.L., Shepard, C.C., and McDade, J.E. (1978). In vivo susceptibility of the Legionnaires' disease bacterium to ten antimicrobial agents. Antimicrob. Agents Chemother. *13*, 419-422.

Mandell, L.A., Wunderink, R.G., Anzueto, A., Bartlett, J.G., Campbell, G.D., Dean, N.C., Dowell, S.F., File, T.M., Musher, D.M., Niederman, M.S., et al. (2007). Infectious Diseases Society of America/American Thoracic Society Consensus Guidelines on the Management of Community-Acquired Pneumonia in Adults. Clin. Infect. Dis. *44*, S27-S72.

Miyashita, N., Higa, F., Aoki, Y., Kikuchi, T., Seki, M., Tateda, K., Maki, N., Uchino, K., Ogasawara, K., Kiyota, H., et al. (2017). Clinical presentation of Legionella pneumonia: Evaluation of clinical scoring systems and therapeutic efficacy. J. Infect. Chemother. *23*, 727-732.

Mykietiuk, A., Carratala, J., Fernandez-Sabe, N., Dorca, J., Verdaguer, R., Manresa, F., and Gudiol, F. (2005). Clinical Outcomes for Hospitalized Patients with *Legionella* Pneumonia in the Antigenuria Era: The Influence of Levofloxacin Therapy. Clin. Infect. Dis. *40*, 794-799.

Nagel, J.L., Rarus, R.E., Crowley, A.W., and Alaniz, C. (2014). Retrospective analysis of azithromycin versus fluoroquinolones for the treatment of *Legionella* pneumonia. P T Peer-Rev. J. Formul. Manag. *39*, 203-205.

Pedro-Botet, L., and Yu, V.L. (2006). Legionella: macrolides or quinolones? Clin. Microbiol. Infect. *12*, 25-30.

Poirier, R., Rodrigue, J., Villeneuve, J., and Lacasse, Y. (2017). Early Radiographic and Tomographic Manifestations of Legionnaires' Disease. Can. Assoc. Radiol. J. *68*, 328-333.

Ruckdeschel, G., Ehret, W., and Ahl, A. (1984). Susceptibility of Legionella spp. to quinolone derivatives and related organic acids. Eur. J. Clin. Microbiol. *3*, 373.

Sabrià, M., Pedro-Botet, M.L., Gómez, J., Roig, J., Vilaseca, B., Sopena, N., and Baños, V. (2005). Fluoroquinolones vs Macrolides in the Treatment of Legionnaires' disease. Chest *128*, 1401-1405.

Saraya, T., Nunokawa, H., Ohkuma, K., Watanabe, T., Sada, M., Inoue, M., Honda, K., Oda, M., Ogawa, Y., Tamura, M., et al. (2018). A Novel Diagnostic Scoring System to Differentiate between *Legionella pneumophila* Pneumonia and *Streptococcus pneumoniae* Pneumonia. Intern. Med. *57*, 2479-2487.

Sharma, L., Losier, A., Tolbert, T., Dela Cruz, C.S., and Marion, C.R. (2017). Atypical Pneumonia. Clin. Chest Med. *38*, 45-58.

Slawek, D., Altshuler, D., Dubrovskaya, Y., and Louie, E. (2017). Tigecycline as a Second-Line Agent for Legionnaires' Disease in Severely Ill Patients. Open Forum Infect. Dis. *4*.

Tan, M.J., Tan, J.S., and File, Jr., T.M. (2002). Legionnaires' disease with Bacteremic Coinfection. Clin. Infect. Dis. *35*, 533-539.

Yu, V.L., Greenberg, R.N., Zadeikis, N., Stout, J.E., Khashab, M.M., Olson, W.H., and Tennenberg, A.M. (2004). Levofloxacin Efficacy in the Treatment of Community-Acquired Legionellosis. Chest *125*, 2135-2139.

(2005). Guidelines for the Management of Adults with Hospital-acquired, Ventilator-associated, and Healthcare-associated Pneumonia. Am. J. Respir. Crit. Care Med. *171*, 388-416.

Chapter 6

Laboratory Diagnosis of Legionellosis

Giancarlo Ceccarelli[1]*, Mario Venditti[1], Maria Scaturro[2] and Maria Luisa Ricci[2]*

[1]Department of Public Health and Infectious Diseases
University of Rome "Sapienza", Azienda Policlinico Umberto I Rome, Italy
[2]Department of Infectious Diseases, National Reference Laboratory for *Legionella*. National Institute for Health, Rome, Italy

*giancarlo.ceccarelli@uniroma1.it and marialuisa.ricci@iss.it

DOI: https://doi.org/10.21775/9781913652531.06

Abstract

Despite efforts to find improved clinical and laboratory predictors for diagnosing cases of legionellosis, the clinical picture of Legionnaires disease, in particular, remains nonspecific. Due to the severity of *Legionella* pneumonia and the distinct antibiotic treatment regimen, prompt and accurate diagnosis is important to improve patient prognosis. For this reason, laboratory diagnostic methods are most essential tools for diagnosing legionellosis. This chapter discusses the diagnostic criteria along with an explanation of each laboratory method for the diagnosis of legionellosis including culture, urinary antigen, antibody detection and direct immunofluorescence. As there are benefits and disadvantages to each technique, this chapter describes how certain methods can be used in conjunction with one another in order to gain the most sensitive and specific diagnostic strategy.

Introduction

Although several attempts have been made to find specific clinical predictors identifying Legionnaires disease, the severe pneumonic disease has symptoms often indistinguishable from pneumonia caused by other microorganisms. Therefore, laboratory diagnosis of Legionnaires' disease must be considered an indispensable component for clinical diagnostic procedures (Boer and Yzerman, 2004; ; WHO, 2007; Fiumefreddo et al., 2009; Miyashita et al., 2017). In fact, according to the European Union Legionnaires' disease case definition (European Commission, 2018, https://eur-lex.europa.eu/legal-content/EN/TXT/PDF/?uri=CELEX:32018D0945) a confirmed case of Legionnaires' disease must have at least one of the following three in addition to clinical criteria of pneumonia:

1. Isolation of *Legionella* spp. from respiratory secretions or any normally sterile site.
2. Detection of *Legionella pneumophila* antigen in urine.
3. Significant rise in specific antibody level to *L. pneumophila* serogroup 1 in paired serum samples.

While probable cases of Legionnaires' disease must have at least one of the following four:

1. Detection of *L. pneumophila antigen* in respiratory secretions or lung tissue e.g. by DFA staining using monoclonal-antibody derived reagents.
2. Detection of *Legionella* spp. nucleic acid in respiratory secretions, lung tissue or any normally sterile site.
3. Significant rise in specific antibody level to *L. pneumophila* other than serogroup 1 or other.
   Legionella spp. in paired serum samples.
4. Single high level of specific antibody to *L. pneumophila* serogroup 1 in serum.

A scoring system predicting positive or negative *Legionella* urinary antigen in community acquired pneumonia has been proposed in order to direct clinicians suspicion of a Legionnaires' disease diagnosis. However, currently laboratory tests should ideally be performed in all of the following cases of pneumonia (WHO, 2007; Roed et al., 2015):

1. Patients with severe disease requiring hospitalization in an intensive care unit.
2. Patients who report risk factors for acquiring Legionnaires' disease.
3. Patients who have been exposed to *Legionella* during an outbreak.
4. Patients in whom no other aetiology is probable.

As mentioned before, diagnosis of confirmed and probable case of *Legionella* infection currently use the following methods:

1. Isolation of the bacterium by culture.
2. Detection of antibodies on sera in acute and convalescent phases of disease.
3. Detection of the urinary antigen.
4. Detection of the bacterium in tissues or in the body fluids by immunofluorescence.
5. Detection of bacterial DNA by PCR (not yet validated and discussed in further chapters).

The diagnostic methods for *L. pneumophila* serogroup 1 are relatively sensitive and specific compared to those of other serogroups of *L. pneumophila* and other *Legionella* species. Therefore, with the exception of culture, positive diagnostic tests for non-*L. pneumophila* serogroup 1 infections are only deemed as probable cases of Legionnaires' disease. In addition, since no diagnostic is 100% sensitive and specific, specialists internationally share the opinion that using a greater number of diagnostic methods leads to a greater likelihood of a correct diagnosis, especially in cases of nosocomial Legionnaires' disease (Luck et al., 2002; Helbig et al., 2003; Boer and Yzerman, 2004; Svarrer et al., 2012; Rota et al. 2014). Therefore, a negative result in one or all of the diagnostic tests used and validated cannot completely exclude the possibility of Legionnaires' disease.

Culture

Isolation by culture is currently the method of choice for *Legionella* detection and Legionnaires' disease diagnosis (Fields et al., 2002; Mercante and Winchell, 2015). Culture must be carried out before antibiotic therapy in order to prevent altering the results. Although, rare cases *Legionella* have been isolated from respiratory tract secretions and lung tissue a few days after initiation of antibiotic therapy (Luck et al., 2002; Stout et al., 2003; Scaturro et al. 2011). The analysis of clinical specimens by culture is extremely important, because it is the most specific diagnostic method. Culture allows for the isolation of all *Legionella* species and serogroups providing useful data for comparative studies with *Legionella* strains isolated from the environment and presumably associated with the infection. However, it has been demonstrated that sequence-based typing may also be performed without *Legionella* isolation directly on clinical samples (Luck et al., 2007; Scaturro et al., 2011). Culture is particularly important for diagnosis in some particular cases such as patients with severe pneumonia and respiratory failure, immunocompromised patients, nosocomial infections and cases in which the cause is a non-*L. pneumophila* serogroup 1 microorganism.

Handling and processing of *Legionella* specimens requires unique safety precautions including a class II biological safety cabinet. Additionally, specimens from the respiratory tract such as bronchoalveolar lavage, trans-tracheal aspirate, pleural fluid and sputum should be taken before antibiotic treatment. These kinds of samples do not need holding media such as buffer solution or saline which contrarily can inhibit *Legionella* growth (Fallon, 1990). A pre-treatment with a fluidifying solutions based on dithiothreitol is often required for respiratory samples to facilitate the inoculum on appropriate agar media. It is also important a heat treatment of a portion of the sample at 50°C for 30 minutes or use acid treatment (diluting 1:10 the sample in KCl-HCl, pH 2.2 and incubated at room temperature for four minutes) solution to remove interfering commensal bacteria that can cover or inhibit *Legionella* growth (Edelstein, 1985). In particular situations when the contamination of the specimen is very high and DFA examination and/or PCR demonstrates the presence of *Legionella,* both pre-treatments could be necessary for *Legionella* to be recovered (personal observation). The specimens should be stored avoiding desiccation and refrigerated at 4.0°C if not processed in a timely manner. After adequate homogenising treatment, specimens such as lung parenchyma are valuable for isolation of *Legionella* and detection by direct immune-fluorescence analysis (see below). In some cases, *Legionella* can be found in samples from extra-pulmonary sites, especially in autopsy specimens from the liver, spleen, pericardial fluid, kidney and skin abscess. All of these samples should be promptly cultivated to increase the possibility of *Legionella* isolation or frozen at -80 °C if delays are expected. Blood cultures are not usually considered as a sample for Legionnaires' disease diagnosis. However, in the absence of identifying any other respiratory pathogen and when cultivated on appropriate *Legionella* media, *Legionella* can be isolated even after a few days of antibiotic treatment (Edelstein et al. 1979; Durand, 2015).

Legionella does not grow on standard laboratory media, as it requires a medium containing yeast extract, L-cysteine, ferric pyrophosphate, alfa-ketoglutarate, charcoal, glycine, named buffered charcoal yeast extract (BCYE). L-Cystine and iron are essential for growth, charcoal prevents the formation of superoxide and peroxide during autoclaving and toxic compound present in agar (Hoffman et al., 1983; Rogers et al., 1993), glycine acts against other Gram-negative organisms and alfa-ketoglutarate is a growth stimulant (Wadowsky et al., 1981; Pine et al., 1986). These components present in media can be used with or without the addition of antibacterial and anti-fungal supplements. A few species such as *L. oackdrigensis* and *L. spiritensis* and *L. nagasakiensis* do not require cystine after primary isolation. On primary isolation, *Legionella* colonies may appear after two to three days in aerobic conditions at 36±2°C creating humid atmosphere to prevent desiccation. Growth is generally faster for agar-adapted strains. However, depending on the species and bacterial stress conditions, longer growth times of four to ten days could be necessary (Fallon, 1990). The morphology of *Legionella* colonies is circular and convex with a shining, smooth surface and typical frosted glass appearance when observed with a stereomicroscope under lateral white light (Figure 1). Under UV light (365 nm) some *Legionella* species have a blue-white brilliant aspect such as *L. bozemanii* (Figure 2), *L. cherri, L. dumofii, L.gormanii, L. parisiensis, L. steigerwaltii* and *L. tucsonensis.* Some species have a yellow-green aspect such as *L. pneumophila, L. longbeachae, L. birgminghamensis* and *L. jordanis*, and others have a red aspect such as *L. rubrilucens* (Edelstein and Cianciotto, 2006). On BCYE agar plates *Legionella* colonies are 1-2 mm in diameter, have grey-white, green or pink colour and grow larger and more opaque with more days of incubation (Fallon, 1990). Presumptive *Legionella* colonies can be confirmed by sub-cultivation on BCYE agar with and without L-cysteine, considering the *Legionella* colonies able to grow on BCYE agar only with L-cystine present. A number of commercial tests and methods such as agglutination, immunofluorescence, Maldi-Tof, genome sequencing of the *mip* gene and immune-chromatography allow for identification of *Legionella* at the species level and *L. pneumophila* serogroups.

Figure 1. Colony of *Legionella pneumophila* serogroup 1 with typical "frosted glass" margin viewed under a stereomicroscope with lateral light.

While the specificity is 100%, the sensitivity is extremely variable ranging from 5% to 99% (Edelstein, 1987; Edelstein et al., 1993; Lindsay et al., 2015). This wide range of values could be ascribed to the numerous disadvantages of this method. Despite culture being the preferred method, it is not time-efficient requiring at best four days to give a result. Clinical specimens are also difficult to obtain, as the type of respiratory secretion could affect *Legionella* isolation. Legionnaires' disease patients have a non-productive cough, and sputum, when available, is less suitable for *Legionella* recovery due to the possibility that the sample does not contain bacteria from low respiratory tract. In addition, the inhibitory growth effect of oral flora can impact *Legionella* isolation.

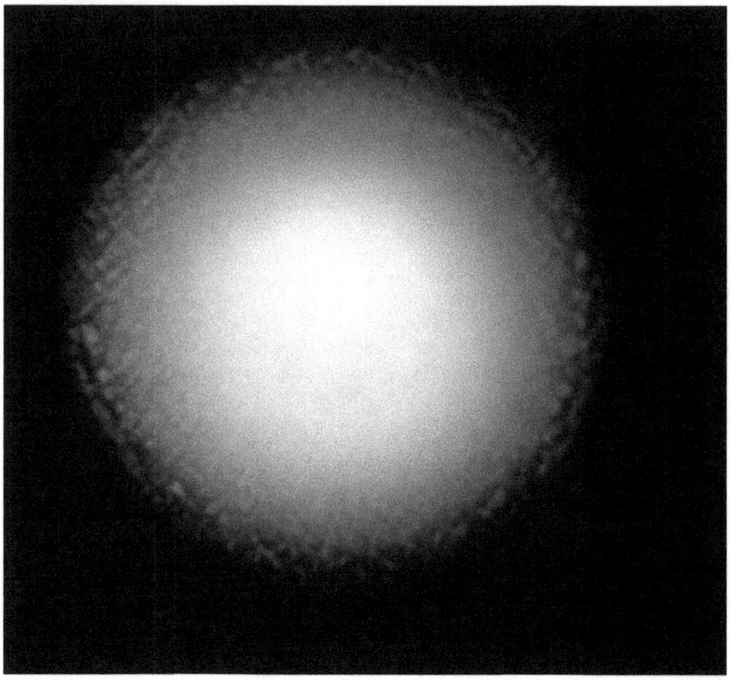

Figure 2. Culture of *Legionella bozemanii* (autofluorescent) in Buffered Charcoal Yeast Extract (BCYE) medium.

In order to improve isolation, a recent study using an amoebic co-culture procedure, the amoeba plate test, has shown both an increase of recovery and reduction of the time for identification of *Legionella* (Descours et al., 2018). Other respiratory secretions such as bronchoalveolar lavage (BAL) or transtracheal-aspirate are more appropriate for *Legionella* isolation but are more difficult to obtain due to the invasive procedures required. For this reason BAL and transtracheal aspirate are not carried out when the diagnosis has been already made with another method, for example by urinary antigen test. Other drawbacks of culture affecting viability of the organism are vulnerability to time lapses between collection and culture (Fields et al., 2002), the unique and possibly more difficult to acquire type of culture media required (Descours et al., 2014), the specific sample treatments necessary soon or after few days of incubation, the presence of commensal interfering microorganisms (Edelstein, 1985) and finally the general requirement of notable laboratory expertise.

Urinary antigen

In 1979 an ELISA technique was created to identify the *Legionella* antigen in urine for the first time (Tilton, 1979). In 1995 Pouffle et al. proposed to include the presence of urinary antigen (UA) in the case definition of Legionnaires' disease. Later, UA detection became the most utilized method for diagnosis of Legionnaires' disease in both Europe with 88.9% of cases using UA in 2015 and in the USA with 97.0% of cases from 2005-2009 (CDC, 2001-2009; J Beauté et al., 2017;). The lipopolysaccharide is the *Legionella pneumophila* molecular component of the UA present on the bacterium surface; it is highly diverse within this specie, and it is responsible for the subdivision of serogroups and subgroups (Luck and Helbig, 2013). Testing of UA presumably contributed to a significant increase of Legionnaires' disease diagnoses in Europe between 2011-2015 (Beaute et al., 2017).

The presence of soluble *Legionella pneumophila* antigen in urine can be intermittent, but it is generally detectable in infected patients one to three days after the onset of symptoms with a peak at five to ten days. In legionellosis, UA may persist for weeks to months, especially in immunocompromised patients where it may even persist up to one year (Kohler et al., 1984). UA has also been found during antibiotic therapy, and UA can be detected in Pontiac Fever (Burnsed et al., 2007). In suspected cases of legionellosis with severe clinical signs of pneumonia, UA should be performed along with one or more additional diagnostic tests such as culture, serology and PCR. Though the sensitivity of the test is often positively associated with the severity of the disease, the same should be done for mild and moderate cases of pneumonia in order to reduce false negatives (Yzerman et al., 2002; Blazquez et al., 2005).

Attempts have been made to develop UA test based on a 19-kDa peptidoglycan-associated lipoprotein molecules able to detect all *Legionella* species (Kim et al., 2003). A urinary antigen test detecting both *Legionella pneumophila* and *Legionella longbeachae* infections has recently been developed (Badoux et al., 2020). Generally, commercially available tests have been developed mainly to detect *L. pneumophila* serogroup 1. However, it was found that in some cases that UA tests were positive as result of infections caused by other *Legionella* serogroups (Benson et al., 2000; Olsen et al., 2009). Therefore, a positive UA result does not necessarily imply that the etiologic agent is *L. pneumophila* serogroup 1 despite this being the most frequent pathogen. Over recent years many UA kits have become commercially available and some claiming the ability to detect other serogroups of *L. pneumophila*. However, their sensitivity was not always confirmed (Svarrer et al., 2012).

The UA detection is carried out by two methods, immune-enzymatic method (EIA) and immune-chromatographic method (ICT). The EIA is the method of choice for the diagnosis of *L. pneumophila* serogroup 1 infections, and it has a specificity of 80-85%, (Hackman et al., 1996; Kazandjian et al., 1997; Helbig et al 2003; Svarrer et al., 2012) and a sensitivity of 79% (Svarrer et al., 2012). The ICT method is a very rapid test taking 15 minutes to one hour, depending on the brand, and does not require special laboratory equipment. The interpretation of the results is based on the presence or absence of two different coloured bands, one representing the sample and the other the control. Any visible line gives a positive result. However, samples with low antigen concentration could give a weak sample line which can be considered positive with confidence if the intensity increases after 45 minutes from the first observation (Helbig et al., 2001). If the weak band does not increase in intensity, the report must be formulated as doubt and must be confirmed by other tests. This is especially likely in cases where the urine is collected at the onset of symptoms such as urinary infections and proteinuria, etc. (Helbig et al., 2001). In some cases, concentrating urine samples increases the antigen level available for testing. However, this should always be accompanied by boiling due to the possibility of proteins present which could lead to false positives (Kohler et al.,1981; Rota et al., 2014).

In order to increase sensitivity and overcome the subjective nature of the band interpretation, an automatic reader of ICT antigen test to detect *L. pneumophila* serogroup 1 has recently been developed based on combining lateral flow chromatography with fluorescent antibody detection technology. The reader automatically scans the codified patient card, interpreting the test in few minutes and storing data. For concentrated and non-concentrated urine samples before and after boiling, the sensitivity of this method is higher compared to the that of other non-automatic ICT tests only when using non-concentrated urine samples (Beraud et al., 2015).

Compared to other diagnostic methods, the UA test has obvious advantages including the easy accessibility of samples, detection at early stages of disease, speed and specificity. A systematic study using a EIA and ICT commercial kits tested numerous *Legionella* strains and a few non-*Legionella* species demonstrated a high sensitivity for *L. pneumophila* and complete non-reactivity of antigens of non-*L. pneumophila* species (Okada et al., 2002).

Despite the advantages of the UA test, it is not without disadvantages. An example of a disadvantage is the persistence of the UA test which causes difficulty in distinguishing between acute infection, convalescence or previous infection. Another limitation of the test is that it mainly detects the antigens of *L. pneumophila* serogroup 1 despite that culture-confirmed cases for other serogroups can also give positive results (Helbig et al., 2001; Svarrer et al., 2012).

As demonstrated by recent studies, the utility of the UA diagnostic method should be continually scrutinized. The ICT test only has 32-47% sensitivity compared to EIA test which is 79% sensitive (Helbig et al., 2001; Svarrer et al., 2012). Although the overall sensitivity of UA test found in a recent meta-analysis study was equal to 74-99% (Shimada et al., 2014), it should be noted that sensitivity might vary in particular in different subpopulations, i.e. patients with Legionnaires' disease associated with travel vs. hospital vs. community acquired infection. In fact, in these three categories the sensitivity is respectively equal to 94%, 76-87% and 44-46%. Explaining these differences is the detection of some *L. pneumophila* strains which are MAb 3/1 positive and have a virulence associated epitope recognised by the MAb3/1 antibody of the Dresden panel. This phenomenon is predominant in cases of Legionnaires' disease associated with travel compared to other *L. pneumophila* MAb 3/1 negative strains more frequently found in community-acquired and nosocomial Legionnaires' disease (Helbig et al., 2003). Authors using *L. pneumophila* purified LPS demonstrated the utility of three commercial test to detect all of the subgroups of *L. pneumophila* serogroup 1 (Mab 3/1 positive and Mab 3/1negative) and all of the other serogroups of *L. pneumophila*, with the exception of serogroup 15. It was also shown that a higher limit of detection for Mab 3/1 negative strains compared to hypothesizing that LPS structure of these strains affects its release from bacteria and/or excretion from the kidney leading to a lower concentration in urine (Ranc et al., 2017).

False positives are rare and mainly reported in patients with serum sickness (Deforges et al., 1999), infections ascribable to *Nocardia asteroides* (Bailleul et al., 2004) and in a pseudo-epidemic episodes related to batches of a fallacious immune-chromatographic test (Rota et al., 2014). It has been suggested that by boiling samples the thermo-stable *Legionella* antigen is maintained and avoids false positives plausibly due to interfering protein content (Kholer et al., 1981). Urine concentration improves the sensitivity of the test, although it could interfere with the specificity (Svarrer et al., 2012). Particular consideration should be given when defining a false positive UA test when other pathogens (i.e. *Pseudomonas* or *Klebsiella*) are identified in culture for pneumonia patients. This is due to the ability of these pathogens to inhibit or overpower *Legionella* growth. In these cases, before culture, as already mentioned, double treatment of the sample is recommended first with heat and then with acid solution to allow the growth of *Legionella*. Finally, the unique use of UA test to confirm a Legionnaires' disease case remains quite restrictive and microbiologists should be encouraged to always use additional tests to confirm a Legionnaires' disease case.

Antibody detection

The introduction of the UA test by the end of 1990's gradually replaced other Legionnaires' disease diagnostic methods particularly, serology testing. Antibody detection is a method almost abandoned representing 0.1% of Legionnaires' disease cases diagnosed between 2011−2015 in Europe (Beute, 2017). Also, a similar trend has occurred in USA and Canada (Segal and Shuman, 1999; Benin et al., 2002). The serological methods currently used to detect *Legionella* antibodies in patient serum are indirect immunofluorescence (IFA) and enzyme-linked immunosorbent assay (ELISA). Various in-house and commercial kits exist both for IFA and ELISA which may use entire inactivated *Legionella* bacterium or isolated antigen (Elverdal et al., 2015).

Both IFA and ELISA are useful for retrospective epidemiological investigations, but they are less valid for clinical diagnosis given the late appearance of specific antibodies and the need to control a further serum sample during convalescence. A significant increase in antibody titre occurs ten days to nine weeks after disease onset in about 75% of patients with positive culture for *L. pneumophila* serogroup 1 (Edelstein and Cianciotto, 2005). Generally, patients develop antibodies within two weeks. However, over 25% of seroconversions are not detected due to incorrect collection of the sera in the early and convalescent phases of the disease (Edelstein and Cianciotto 2006).

Determining the antibody class is not helpful in differentiating between an ongoing and previous infection. In some studies IgM is detected early, and other studies have shown the presence of both IgM and IgG in the early phase (Elverdal et al. 2013). In some patients, only IgG or only IgM were found. Additionally, for those with IgM present, the positive IgM reading may persist for months after the initial test (Krech and Krech, 1983). IgA may be present in recent infections but promptly undergoes degradation. For this reason, it is appropriate to use a test that detects all antibody classes.

According to the EU case definition, only a four-fold increase of antibody titre between two sera taken in both the acute and convalescent phase of disease has diagnostic value. A positive result (≥ 256) on a single serum has a presumptive diagnostic value. These stringent criteria help avoid false positives due to cross-reactions with other pathogens. False negatives may occur due to poor antibody responses of immunocompromised patients, late seroconversion, or simply advanced age which causes a natural decline of the immune response. Another cause of false negatives is when an antigen non-homologous (i.e belonging to a different subgroup of *L. pneumophila*) to the one responsible for the infection is used for testing.

Finally, it should be noted that the specificity and sensitivity of indirect IFA was evaluated only for *L. pneumophila* serogroup 1 while sensitivity and specificity for other serogroups or species are unknown (Luck et al., 2002;

Muder, 2000). Due to the formation of cross-reactive antibodies, about 50% of patients infected with *L. pneumophila* non-serogroup 1 show seroconversion with specific antigens of *L. pneumophila* serogroup 1 (Edelstein, 2002). A negative result, however, does not rule out the diagnosis of Legionnaires' disease. In addition, laboratories using in-house or commercial test kits have different antigenic preparations resulting in varying detection of antibody levels. Therefore, for some antigenic preparations the specificity could presumably be relatively high for certain samples and low others (Rose et al., 2002).

Additional factors complicating the diagnostic confirmation are the cross-reactivity between *Legionella* and other microorganisms such as *Campylobacter* and *Pseudomonas* species (Boswell, 1996; Marshall et al., 1994) and the difficulty of distinguishing between current or previous infection in the case of a single serum sample or a constant antibody titer. Instances of cross reaction using IFA have been reported especially in patients with tuberculosis, pneumococcal pneumonia, *Pseudomonas* pneumonia, tularemia, exacerbation of cystic fibrosis, *Bacteroides fragilis* bacteraemia and leptospirosis (Edelstein et al., 1980; Bornstein et al., 1987).

The ELISA is a more specific serological tests for *L. pneumophila* serogroup 1 than IFA (Edelstein, 2002). ELISA is more frequently used in diagnostic laboratories due to the diffusion of numerous commercial kits, despite that the concordance between the ELISA test and IFA is about 91% (Edelstein, 2002). A paper by Elverdal et al. in 2013, evaluated an in-house test ELISA using both IgG and IgM antibodies for Legionnaires' disease diagnosis and found a sensitivity of 61% for IgG, 36% sensitivity for IgM and 66% sensitivity when both immune-globulins were combined. The authors also demonstrated that the combination of in-house ELISA and in-house PCR or culture gave a sensitivity of 93% while a sensitivity 94% was reached when in-house PCR and culture were both used. This study highlights, again, that sensitivity of Legionnaires' disease diagnosis can be improved only when more than one test is performed.

Direct immunofluorescence (DFA)

DFA is an effective method with sputum, endotracheal and trans-tracheal aspirates and on lung biopsies (Stout et al., 2003). Despite the fast turn-around of two to three hours, DFA is not commonly used the clinical setting due to low sensitivity of 25-75% (WHO, 2007). The technique also requires specimen preparation and personnel experienced in microscope observation. DFA is both influenced by the specificity of the particular antisera used and by the size of the preparation which is smaller compared to that of culture.

DFA performed on sputum can give positive results up to two to four days after the start of antibiotic therapy and often for longer periods of time in cases of cavitary pneumonia (Luck et al., 2002). Generally, 25% to 70% of patients diagnosed with cultured-positive Legionnaires' disease also have a positive DFA result. However, the specificity of the test is greater than 99.9% when monoclonal antibodies are used. Therefore, a negative result does not rule out the diagnosis of Legionnaires' disease, but a positive result almost always has diagnostic value if the slide reading has been performed correctly (Edelstein and Cianciotto, 2006).

DFA requires a significant amount of caution to prevent false positives, especially when the biopsy samples have accidentally been in contact with tap water which can occur mainly during autopsy. DFA is a cheap and easy test for screening respiratory or biopsy samples. It is particularly useful to select positive samples that should be cultivated in a greater extent on *Legionella*-specific media in order to successfully isolate bacterium important in a cluster or an outbreak situation for molecular epidemiology studies.

References

Badoux, P., Kracht-Kosten, L., Herpers, B., Euser, S. (2020). Method Comparison of the ImmuView *L. pneumophila* and *L. longbeachae* Urinary Antigen Test with the BinaxNOW *Legionella* Urinary Antigen Card for Detection of *Legionella pneumophila* Serogroup 1 Antigen in Urine. J Clin Microbiol. 58(3).

Bailleul, E. (2004). False-positive result with BinaxNOW *Legionella* Antigen immunochromatographic (ICT) assay: response to Helbig et al. (2001). J Med Microbiol. *53*, 173-173.

Beauté, J., and on behalf of the European Legionnaires' Disease Surveillance Network (2017). Legionnaires' disease in Europe, 2011 to 2015. Euro Surveill *22*.

Benson, R.F., Tang, P.W., and Fields, B.S. (2000). Evaluation of the Binax and Biotest urinary antigen kits for detection of Legionnaires' disease due to multiple serogroups and species of Legionella. J Clin Microbiol. *38*, 2763-2765.

Beraud, L., Gervasoni, K., Freydiere, A.M., Descours, G., Ranc, A.G., Vandenesch, F., Lina, G., Gaia, V., and Jarraud, S. (2015). Comparison of Sofia Legionella FIA and BinaxNOW® *Legionella* urinary antigen card in two national reference centers. Eur J Clin Microbiol Infect Dis *34*, 1803-1807.

Blázquez, R.M., Espinosa, F.J., Martínez-Toldos, C.M., Alemany, L., García-Orenes, M.C., and Segovia, M. (2005). Sensitivity of urinary antigen test in relation to clinical severity in a large outbreak of *Legionella pneumonia* in Spain. Eur J Clin Microbiol Infect Dis *24*, 488-491.

Boer, J.W. and Yzerman, E.P.F. (2004). Diagnosis of *Legionella* infection in Legionnaires disease. Eur J Clin Microbiol Infect Dis.

Burnsed, L.J., Hicks, L.A., Smithee, L.M.K., Fields, B.S., Bradley, K.K., Pascoe, N., Richards, S.M., Mallonee, S., Littrell, L., Benson, R.F., et al. (2007). A Large, Travel-Associated Outbreak of Legionellosis among

Hotel Guests: Utility of the Urine Antigen Assay in Confirming Pontiac Fever. Clinical Infectious Diseases *44*, 222-228.

Cosentini, R., Tarsia, P., Blasi, F., Roma, E., and Allegra, L. (2001). Community-acquired pneumonia: role of atypical organisms. Monaldi Arch Chest Dis *56*, 527-534.

Cristovam, E., Almeida, D., Caldeira, D., Ferreira, J.J., and Marques, T. (2017). Accuracy of diagnostic tests for Legionnaires' disease: a systematic review. Journal of Medical Microbiology *66*, 485-489.

Deforges, L., Legrand, P., Tankovic, J., Brun–Buisson, C., Lang, P., and Soussy, C. (1999). Case of False–Positive Results of the Urinary Antigen Test for *Legionella pneumophila*. CLIN INFECT DIS *29*, 953-954.

Descours, G., Hannetel, H., Reynaud, J.V., Ranc, A.G., Beraud, L., Kolenda, C., Campese, C., Lina, G., Ginevra, C., and Jarraud, S. (2018). Adaptation of Amoeba Plate Test To Recover *Legionella* Strains from Clinical Samples. J Clin Microbiol *56*, e01361-17, /jcm/56/5/e01361-17.atom.

Durando, P., Orsi, A., Alicino, C., Tinteri, C., Di Bella, A., Parodi, M.C., Marchese, A., Gritti, P.R., Fontana, S., Rota, M.C., et al. (2015). A Fatal Case of Nosocomial Legionnaires' Disease: Implications From an Extensive Environmental Investigation and Isolation of the Bacterium From Blood Culture. Infect. Control Hosp. Epidemiol. *36*, 1483-1485.

Edelstein, P.H. (1987). The laboratory diagnosis of Legionnaires' disease. Semin Respir Infect *2*, 235-241.

Edelstein, P.H. (1993). Legionnaires' Disease. Clinical Infectious Diseases *16*, 741-749.

Edelstein, P.H. (2002). Detection of antibodies to Legionella spp. In Manual of Clinical Laboratory Immunology, (Washington DC: ASM Press), pp. 486-476.

Edelstein, P.H., and Cianciotto, N.P. (2005). Legionella. In Principles and Practice of Infectious Disease, (Churchill Livingstone, Philadelphia: Elsevier), pp. 2711-12730.

Edelstein, P.H., and Cianciotto, N.P. (2006). Legionella Species and Legionnaires' Disease. In The Prokaryotes, M. Dworkin, S. Falkow, E. Rosenberg, K.-H. Schleifer, and E. Stackebrandt, eds. (Springer New York), pp. 988-1033.

Edelstein, P., Finegold, S., and Meyer, R. (1979). Isolation of Legionella pneumophila from blood. The Lancet *313*, 750-751.

Elverdal, P.L., Jørgensen, C.S., Krogfelt, K.A., and Uldum, S.A. (2013). Two years' performance of an in-house ELISA for diagnosis of Legionnaires' disease: Detection of specific IgM and IgG antibodies against Legionella pneumophila serogroup 1, 3 and 6 in human serum. Journal of Microbiological Methods *94*, 94-97.

European Commission (2008). Commission Decision of 28 April 2008 amending Decision 2002/253/EC laying down case definitions for reporting communicable diseases to the Community network under Decision No 2119/98/EC of the European Parliament and of the Council. Brussels.

Fallon, R.J. (1990). *Legionella*. In Pinciples of Bacteriology, Virology and Immunity, pp. 275-287.

Fields, B.S., Benson, R.F., and Besser, R.E. (2002). Legionella and Legionnaires' Disease: 25 Years of Investigation. Clinical Microbiology Reviews *15*, 506-526.

Fiumefreddo, R., Zaborsky, R., Haeuptle, J., Christ-Crain, M., Trampuz, A., Steffen, I., Frei, R., Müller, B., and Schuetz, P. (2009). Clinical predictors for Legionella in patients presenting with community-acquired pneumonia to the emergency department. BMC Pulm. Med. *9*, 4.

Formica, N., Yates, M., Beers, M., Carnie, J., Hogg, G., Ryan, N., and Tallis, G. (2001). The impact of diagnosis by *Legionella* urinary antigen test on the epidemiology and outcomes of Legionnaires' disease. Epidemiol. Infect. *127*, 275-280.

Hackman, B.A., Plouffe, J.F., Benson, R.F., Fields, B.S., and Breiman, R.F. (1996). Comparison of Binax Legionella Urinary Antigen EIA kit with

Binax RIA Urinary Antigen kit for detection of Legionella pneumophila serogroup 1 antigen. J. Clin. Microbiol. *34*, 1579-1580.

Harrison, T.G., Uldum, S.A., Lück, P.C., and Helbig, J.H. (2001). Detection of Legionella pneumophila antigen in urine samples by the BinaxNOW immunochromatographic assay and comparison with both Binax Legionella Urinary Enzyme Immunoassay (EIA) and Biotest Legionella Urin Antigen EIA. Journal of Medical Microbiology *50*, 509-516.

Helbig, J.H., Uldum, S.A., Bernander, S., Lück, P.C., Wewalka, G., Abraham, B., Gaia, V., and Harrison, T.G. (2003). Clinical utility of urinary antigen detection for diagnosis of community-acquired, travel-associated, and nosocomial legionnaires' disease. J. Clin. Microbiol. *41*, 838-840.

Hoffman, P.S., Pine, L., and Bell, S. (1983). Production of superoxide and hydrogen peroxide in medium used to culture *Legionella pneumophila*: catalytic decomposition by charcoal. Appl. Environ. Microbiol. *45*, 784-791.

Kazandjian, D., Chiew, R., and Gilbert, G.L. (1997). Rapid diagnosis of *Legionella pneumophila* serogroup 1 infection with the Binax enzyme immunoassay urinary antigen test. J. Clin. Microbiol. *35*, 954-956.

Kim, M.J., Sohn, J.W., Park, D.W., Park, S.C., and Chun, B.C. (2003). Characterization of a Lipoprotein Common to Legionella Species as a Urinary Broad-Spectrum Antigen for Diagnosis of Legionnaires' Disease. Journal of Clinical Microbiology *41*, 2974-2979.

Kohler, R.B. (1981). Rapid Radioimmunoassay Diagnosis of Legionnaires' Disease: Detection and Partial Characterization of Urinary Antigen. Ann Intern Med *94*, 601.

Kohler, R.B., Winn, W.C., and Wheat, L.J. (1984). Onset and duration of urinary antigen excretion in Legionnaires disease. J. Clin. Microbiol. *20*, 605-607.

Krech, T.U., and Krech, U.H. (1983). Differentiation of recent and past infections with *Legionella pneumophila* by determination of specific IgM. Zentralbl Bakteriol Mikrobiol Hyg A *255*, 44-47.

Lindsay, D.S.J. (2004). Laboratory diagnosis of legionnaires' disease due to *Legionella pneumophila* serogroup 1: comparison of phenotypic and genotypic methods. Journal of Medical Microbiology *53*, 183-187.

Lück, C., and Helbig, J.H. (2013). Characterization of *Legionella* Lipopolysaccharide. In *Legionella*, C. Buchrieser, and H. Hilbi, eds. (Totowa, NJ: Humana Press), pp. 381-390.

Luck, P.C., Helbig, J.H., and Schuppler, M. (2002). Epidemiology and Laboratory Diagnosis of *Legionella* Infections. Laboratoriums Mediz *26*, 174-182.

Luck, P.C., Ecker, C., Reischl, U., Linde, H.-J., and Stempka, R. (2007). Culture-Independent Identification of the Source of an Infection by Direct Amplification and Sequencing of *Legionella pneumophila* DNA from a Clinical Specimen. Journal of Clinical Microbiology *45*, 3143-3144.

Mercante, J.W., and Winchell, J.M. (2015). Current and Emerging *Legionella* Diagnostics for Laboratory and Outbreak Investigations. Clin. Microbiol. Rev. *28*, 95-133.

Mykietiuk, A., Carratala, J., Fernandez-Sabe, N., Dorca, J., Verdaguer, R., Manresa, F., and Gudiol, F. (2005). Clinical Outcomes for Hospitalized Patients with *Legionella* Pneumonia in the Antigenuria Era: The Influence of Levofloxacin Therapy. Clinical Infectious Diseases *40*, 794-799.

Okada, C., Kura, F., Wada, A., Inagawa, H., Lee, G., and Matsushita, H. (2002). Cross-Reactivity and Sensitivity of Two *Legionella* Urinary Antigen Kits, Biotest EIA and Binax NOW, to Extracted Antigens from Various Serogroups of *L. pneumophila* and Other *Legionella* Species. Microbiology and Immunology *46*, 51-54.

Olsen, C.W., Elverdal, P., Jørgensen, C.S., and Uldum, S.A. (2009). Comparison of the sensitivity of the *Legionella* urinary antigen EIA kits from Binax and Biotest with urine from patients with infections caused by less common serogroups and subgroups of Legionella. Eur J Clin Microbiol Infect Dis *28*, 817-820.

Pine, L., Hoffman, P.S., Malcolm, G.B., Benson, R.F., and Franzus, M.J. (1986). Role of keto acids and reduced-oxygen-scavenging enzymes in the growth of *Legionella* species. J. Clin. Microbiol. *23*, 33-42.

Plouffe, J.F., File, T.M., Breiman, R.F., Hackman, B.A., Salstrom, S.J., Marston, B.J., Fields, B.S., and Community Based Pneumonia Incidence Study Group (1995). Reevaluation of the Definition of Legionnaires' Disease: Use of the Urinary Antigen Assay. Clinical Infectious Diseases *20*, 1286-1291.

Ranc, A.G., Carpentier, M., Beraud, L., Descours, G., Ginevra, C., Maisonneuve, E., Verdon, J., Berjeaud, J.M., Lina, G., Jarraud, S. (2017). Legionella pneumophila LPS to evaluate urinary antigen tests. Diagn Microbiol Infect Dis. 89(2):89-91.

Roed, T., Schønheyder, H.C., and Nielsen, H. (2015). Predictors of positive or negative *Legionella* urinary antigen test in community-acquired pneumonia. Infectious Diseases *47*, 484-490.

Rota, M.C., Fontana, S., Montano-Remacha, C., Scaturro, M., Caporali, M.G., Vullo, V., Scorzolini, L., Ercole, A., and Ricci, M.L. (2014). Legionnaires' Disease Pseudoepidemic Due to Falsely Positive Urine Antigen Test Results. Journal of Clinical Microbiology *52*, 2279-2280.

Scaturro, M., Fontana, S., and Ricci, M.L. (2011). Use of Nested Polymerase Chain Reaction Based on Sequence-Based Typing of Clinical Samples to Determine the Source of Infection for Hospital-Acquired Legionnaires' Disease. Infect. Control Hosp. Epidemiol. *32*, 510-512.

Segal, G., and Shuman, H.A. (1999). Legionella pneumophila utilizes the same genes to multiply within Acanthamoeba castellanii and human macrophages. Infect. Immun. *67*, 2117-2124.

Shimada, T., Noguchi, Y., Jackson, J.L., Miyashita, J., Hayashino, Y., Kamiya, T., Yamazaki, S., Matsumura, T., and Fukuhara, S. (2009). Systematic Review and Metaanalysis: urinary antigen tests for Legionellosis. Chest *136*, 1576-1585.

Stout, J.E., Rihs, J.D., and Yu, V.L. (2003). Legionella. In Manual of Clinical Microbiology, (Washington DC: ASM Press), pp. 809-823.

Svarrer, C.W., Luck, C., Elverdal, P.L., and Uldum, S.A. (2012). Immunochromatic kits Xpect Legionella and BinaxNOW Legionella for detection of *Legionella pneumophila* urinary antigen have low sensitivities for the diagnosis of Legionnaires' disease. Journal of Medical Microbiology *61*, 213-217.

Tilton, R.C. (1979). Legionnaires' Disease Antigen Detected by Enzyme-Linked Immunosorbent Assay. Ann Intern Med *90*, 697.

United States Center for Disease Control (CDC) (2011). Legionellosis --- United States, 2000--2009.

Wadowsky, R.M., and Yee, R.B. (1981). Glycine-containing selective medium for isolation of Legionellaceae from environmental specimens. Appl. Environ. Microbiol. *42*, 768-772.

Yzerman, E.P.F., Boer, J.W., Lettinga, K.D., Schellekens, J., Dankert, J., and Peeters, M. (2002). Sensitivity of Three Urinary Antigen Tests Associated with Clinical Severity in a Large Outbreak of Legionnaires' Disease in The Netherlands. Journal of Clinical Microbiology *40*, 3232-3236.

(2007). *Legionella* and the prevention of legionellosis (Geneva: World Health Organization).

Chapter 7

Clinical Significance of (non-*Legionella pneumophila*) *Legionella* Species

Diane S.J. Lindsay*

Scottish Microbiology Reference Laboratory, Glasgow, UK

*Diane.Lindsay@ggc.scot.nhs.uk

DOI: https://doi.org/10.21775/9781913652531.07

Abstract
Despite most human cases of clinical legionellosis being caused by *Legionella pneumophila,* there are many non-*L. pneumophila* species of the family, Legionellaceae, which have clinical, environmental and public health relevance. This chapter explores the history, notable features and geographical distribution of the most common non-*L. pneumophila* species of the *Legionella* genus with emphasis on *L. longbeachae, L. bozemanae, L. micdadei, L. dumoffii* and *L. anisa*. The *Legionella* species and *L. pneumophila* can be indistinguishable in terms of the clinical features of the disease that they cause and the respective treatments they require. However, infections caused by *Legionella* species predominate in the immunosuppressed and these infections can feature distinct clinical and radiologic findings such as an association with extra pulmonary sites of infection. Current diagnostic tools such as the urinary antigen test are biased towards detection of *L. pneumophila* serogroup 1 and increases the likelihood of cases caused by *Legionella* species being under or mis-

diagnosed. Therefore, the true incidence of legionellosis cases caused by non-*L. pneumophila* species is probably under reported.

History

The first *Legionella* species was identified in 1943 (McDade *et al.*, 1977). It was a rickettsial-like organism that only grew in guinea pig blood and embryonated hens' eggs and was initially classified as part of the *Tatlockia* genus (Hebert *et al.*, 1980a). After the first reported outbreak of Legionnaires' disease in 1976 and the subsequent formulation of selective agar, this organism was cultured and confirmed as a *Legionella* species and re-named *Legionella micdadei* in honour of Joseph McDade (Garrity *et al.*, 1980).

Non-*L. pneumophila Legionella* species have been implicated in sporadic cases of Legionnaires' disease and outbreaks of Pontiac Fever (Fields *et al.*, 2002, Khodr *et al.*, 2007, Hamilton *et al.*, 2018). The majority of Legionnaires' disease cases associated with large outbreaks are attributed to *L. pneumophila* serogroup (Sg) 1. In recent years, *L. pneumophila* Sg 1 has become the main focus of diagnostic testing. However, there are now over 50 species in the *Legionella* genus that have been identified from both clinical and environmental sources with approximately half of the species capable of causing human disease (Khodr *et al.,* 2016). A current list of non-*L. pneumophila* strains is detailed in Table 1. This is not exhaustive as new strains are identified regularly and clinical strains are also found in the environment. Legionnaires' disease cases caused by *Legionella* species, although less common than *L. pneumophila* Sg 1, have been reported in susceptible individuals particularly the immunosuppressed (Mercant and Winchell, 2015, Miyashita *et al.*, 2020).

Table 1. Examples of non-*L. pneumophila* species of the family Legionellaceae	
Source	Non-*L. pneumophila* strains
Clinical	*L. anisa L. birminghamensis L. bozemanae L. cardiaca L cincinnatiensis L.clemsonenesis L. donaldsonii L. dumoffii L. erythra L. feeleii L. gormanii L. hackeliae L. jamestownensis L. jordanis L. lansingensis L. londiniensis L. longbeachae L. lytica L.maceachernii L. micdadei L. nagasakiensis L. oakridgensis L. parisiensis L. rubrilucens L. sainthelensi L. steelei L. tucsonensis L. wadsworthii L. waltersii L. worsleiensis*
Environmental	*L. adelaidensis L.beliardensis L. brunensis L.busanensis L. cherrii L. drancourtii L. fairfieldensis L fallonii L. geestiae L. gratiana L. gresilensis L. israelensis L. impletisoli L. massiliensis L. moravica L. nautarum L. norrlandica L. quarteirensis L. quinlivanii L rowbothamii L. santicrucis L. saoudiensis L. shakespearei L spiritensis L. steigerwaltii L. tauriensis L.thermalis L.tunisiensis L. yabucchiae*

Disease

Legionellosis accounts for about 2-9% of community acquired pneumonia (CAP) worldwide. There are currently two distinct disease categories, Legionnaires' disease and Pontiac fever (Fields *et al.*, 2002, Murdoch, 2003, Diederen, 2007, Hamilton *et al.*, 2018). Legionnaires' disease can be a potentially fatal, atypical pneumonic illness that often requires hospitalisation. Non-specific laboratory findings include hyponatraemia, hypophosphatemia, elevated creatine kinase, myoglobinuria, leucocytosis with relative lymphopenia, high erythrocyte sedimentation rate and C-reactive protein levels, elevated serum ferritin levels and microscopic haematuria. Pulmonary infiltration is usually present in chest radiographs, and the most common pattern is a patchy, uni-lobar infiltrate progressing to consolidation of the lung tissue. Pleural effusion occurs in 15–50% of patients at hospital admission. In immunosuppressed patients, especially those on glucocorticoids, round nodular opacities can appear expanding and cavitating in about 10% of cases (Cunha and Cunha, 2016).

The majority of this chapter will concentrate on Legionnaires' disease, but Pontiac fever is often underdiagnosed and presents as a mild flu-like illness with a high attack rate. *L. pneumophila* (Mangione *et al.*, 1985), *L. micdadei* (Fallon and Rowbotham, 1990, Goldberg *et al.*, 1989), *L. anisa* (Fields *et al.*, 1990), *L. longbeachae* (Cramp *et al.*, 2010) and *L. feeleii* (Brenner *et al.*, 1984, Herwaldt *et al.*, 1984) have all been implicated in Pontiac fever outbreaks. The diagnosis is usually made on the basis of epidemiological, clinical, microbiology and environmental laboratory findings that include a common source with a short incubation and resulting in a non-fatal, non-pneumonic illness characterized by malaise, myalgia and fever (Hamilton *et al.*, 2018). A causative bacterium is rarely isolated from Pontiac fever patients suggesting that it is a sensitization of the respiratory mucosa or a toxin-related outcome of exposure instead of a systemic disease. However, in the Oklahoma City Pontiac fever outbreak in 2004 there was a high frequency of urine antigen positivity (36%) to *L. pneumophila* Sg 1. This provided the first tangible evidence that this outbreak was due to inhalation of live or dead *L. pneumophila* Sg 1 and not just from inhalation of *Legionella* bacterial endotoxin (Burnsed *et al.*, 2007).

A recent study compared the characteristics of *L. feeleii* strains that were isolated from a freshwater sample implicated in an outbreak of Pontiac fever (LfPF) and from a patient with *L. feeleii* Legionnaires' disease (LfLD). Growth of LfPF and LfLD after *in vitro* infection of rodents, human macrophage cell lines and human lung epithelial cell lines showed that LfLD had a higher cell infection rate, stronger internalization by host cells and greater cytotoxicity than that of LfPF strain. Large amounts of IL-6 and IL-8 were secreted by human host cells after infection with LfLD but not when infected with LfPF. Based on the above results, this is one of the first studies to show distinct pathogenic differences in a *Legionella* species that could ultimately determine disease outcome (Wang *et al.*, 2015).

Cases of asymptomatic carriage have also been described in two cases of *L. steelei* that occurred in the US and Australia (Edelstein *et al.*, 2012). The organism was isolated from respiratory secretions, but neither patient was thought clinically to have legionellosis. *In vitro*, the organisms were perfectly adapted to live in *Acanthamoeba castellani* (ATCC 30234) hosts. However, they were unable to establish a productive infection in human respiratory alveolar epithelial cell line A549 and it may have been isolated as a transient colonizer or perhaps a contaminant. If the European Centre for Disease Prevention and Control (ECDC) guidelines on *Legionella* infections were adhered to, both cases would be definitive of Legionnaires' disease due to the presence of respiratory symptoms and culture positive results. In 2011, an environmental isolate of *L. steelei* was isolated from a hospital water supply in Scotland and was deposited in the National Collection of Type Cultures (NCTC) as the first confirmed environmental isolate of *L. steelei* (personal communication, D. Lindsay).

Legionellaceae is mainly transmitted from the environment to humans by inhalation or aspiration of contaminated water and can result in lung infiltration and pneumonic illness. Cases of Legionnaires' disease may be sporadic or occur as part of an outbreak. Sporadic cases are reported throughout the year, but most cases occur in the summer and autumn. This trend presumably occurs because the warmer weather encourages proliferation of *Legionella* in water. The disease tends to occur more in middle-aged and elderly people as well as those with impaired respiratory and cardiac function, heavy smoking history or immunocompromised states (Fields *et al.*, 2002, Murdoch, 2003, Diederen, 2007). The majority of Legionnaires' disease involving non-*pneumophila Legionella* species occurs in highly immunosuppressed individuals with underlying comorbidities leading to a normal and commensal water micro-organism becoming an opportunistic pathogen (Muder and Yu, 2002).

Microbiology

The main diagnostic tests for identification of Legionnaires' disease are urinary antigen tests (UAT), serology, real time (RT) PCR and culture (Lindsay *et al.*, 2004, Gadsby *et al.*, 2015, Pierre *et al.*, 2017). However, to identify non-*L. pneumophila* Legionnaires' disease cases only serology, RT-PCR and culture are available as UAT is primarily specific to *L. pneumophila* Sg 1. Urinary antigen is detected in the acute phase of disease by either enzyme linked immuno-sorbent assays (ELISA) or immunochromatographic tests and is the most popular diagnostic test for the detection of Legionnaires' disease (see Chapter 6).

While the *Legionella* UAT is a valuable tool, reliance on UAT can result in undetected cases of Legionnaires' disease (Benin *et al.*, 2002, Fields *et al.*, 2002, St Martin *et al.*, 2013). In Europe, nearly 80% of Legionnaires' disease cases in 2014 were identified by UAT compared with only 11% identified by culture (ECDC, 2016). The majority of UAT are most sensitive for the detection of the *L. pneumophila* Sg 1 "Pontiac" or Dresden monoclonal antibody 2 positive strains (up to 90%), and less sensitive for other monoclonal antibody types of *L. pneumophila* Sg 1 (60%), and poorly sensitive (5%) for other *L. pneumophila* serogroups and other *Legionella* species (Harrison *et al.*, 1998, Dominguez *et al.*, 2001). All current UATs are reported to be specific for *L. pneumophila* Sg 1, and this may be a contributing factor in the late diagnosis of infections caused by non-*L. pneumophila* Sg1 and other species of *Legionella*.

Studies in the United States pre- and post-UAT popularity suggested that the proportion of Legionnaires' disease associated with *L. pneumophila* Sg 1 is variable from year to year and ranges from 50-91% in culture confirmed cases (Marston *et al.*, 1994, Benin *et al.*, 2002). International surveillance suggests that there has not been a decrease in the non-*L. pneumophila* Sg 1 disease as seen in countries like Denmark that utilise culture and *Legionella* species-specific PCR at levels above the rest of the European Union and US (Joseph and Ricketts, 2010, ECDC, 2013). In the period from 1996 to 2006, over a third of culture confirmed Legionnaires' disease cases in Denmark were non-*L. pneumophila* Sg 1 which is similar to the rates in the US before the rise of urinary antigen tests and the decline of culture techniques (Jespersen *et al.*, 2009, Svarrer *et al.*, 2011, St Martin *et al.*, 2013). The Danish research revealed that mortality rates in non-L. *pneumophila* Sg1 Legionnaires' disease patients were significantly higher than those for the *L. pneumophila* Sg 1 infected populations (St Martin *et al.*, 2013). The study demonstrated survival rates as low as is 73% in culture positive and urinary antigen negative cases, which is a pattern suggesting non-*L. pneumophila* Sg 1 infection. There are three possible explanations for the higher mortality rate in these patients including the presence of pre-existing co morbidity, misdiagnosis resulting in treatment delay and/or an increase in non-*L. pneumophila* Sg1 pathogenicity (Jespersen *et al.*, 2009). Therefore, it is reasonable to assume that significant under diagnosis of non-*L. pneumophila* Sg 1 Legionnaires' disease has occurred due to an over reliance on current generation UAT.

Serology can be very useful if acute and convalescent sera are available. However, serological assays tend to be performed in specialist laboratories using in house methods as current manufactured assays lack sensitivity and specificity. The reported sensitivities of serological assays vary substantially from 41% to 94% (den Boer *et al.*, 2004). A retrospective serological survey of patients with community-acquired pneumonia (CAP) of unknown aetiology found that 8% of undiagnosed infections were due to *L. bozemanae*. In the same multicenter German Competence Network for CAP study, 10 % of all cases were due to non-*L. pneumophila Legionella* species (von Baum *et al.*, 2008). Serology suffers from known cross-reactions with *E. coli*, *Campylobacter* and *Coxiella* antibodies and historically only dominated in the years after the original Philadelphia outbreak (Wilkinson *et al.*, 1979, 1981, Fallon and Abraham, 1992, Finidori *et al.*, 1992). However, serology can be invaluable if appropriately controlled and can detect titers to all clinically relevant *Legionella* species that could be beneficial in the diagnosis of non-*L. pneumophila* Legionnaires' disease cases. Currently, a four-fold rise in titer or single high titer to *Legionella* species is classified as probable of case of Legionnaires' disease (ECDC, 2013).

Culture is still the gold standard in diagnosis of *Legionella*. However, culture is rarely performed due to a lack of availability of lower respiratory tract samples, the fastidious nature of the organism and a general lack of experience in *Legionella* identification. When viewed under a plate microscope Legionellae are identified morphologically by a typical cut-glass appearance. Some *Legionella* species fluoresce blue white (e.g. *L. bozemanae*) or red (e.g. *L. erythra*). Identification can be confirmed by MALDI-TOF in a routine laboratory but only if platforms have access to a *Legionella* species database. If an organism is cultured, it can be identified by monoclonal sub-typing, sequence-based typing for *L. pneumophila* serogroups and macrophage infectivity potentiator (*mip*) speciation for *Legionella* species or whole genome sequencing (WGS) for all strains (Ratcliff *et al.*, 1998, McAdam *et al.*, 2014, Bacigalupe *et al.*, 2017). There is a *mip* speciation database (http://www.hpa-bioinformatics.org.uk/cgi-bin/legionella/mip/mip_id.cgi) currently hosted by Public Health England and initially created at IMVS in Australia (Ratcliff *et al.*, 1998). FASTA files of the macrophage infectivity potentiator (*mip*) gene of unknown organisms can be compared against all *Legionella* species and some as yet un-named new species for identification purposes.

WGS has many applications in the context of *Legionella*. The data generated by WGS can be used to identify species, subtypes and clades.

For outbreak investigations, WGS allows for definitive matching of patient and environmental isolates as well as depicting the diversity of isolates cultured from a point source outbreak (McAdam *et al*., 2014). In the future, WGS could replace *mip* speciation and sequence based typing (SBT) although most of the current work in the area relates to *L. pneumophila* Sg 1 (McAdam *et al*., 2014, David *et al*., 2016) although there has been extensive work done on the *L. longbeachae* genome (Cazalet *et al*., 2010, Bacigalupe *et al*., 2017, Haviernik *et al*., 2020).

qPCR on respiratory samples has been utilised as a screening tool for large numbers of respiratory pathogens including *Legionella*. Unfortunately, the targets in these assays for clinical samples tend to be specific to either *L. pneumophila* or *L pneumophila* Sg 1. Therefore, with qPCR a similar bias to that of UAT exists towards identifying only *L. pneumophila* and excluding *Legionella* species (Mentasti *et al*., 2012). However, the most recent publications on qPCR are reported to be specific for more *Legionella* species especially the clinically relevant ones such as *L. longbeachae, L. micdadei* and *L. bozemanae*. This expansion should increase case ascertainment if utilised correctly in severe CAP cases (Benitez and Winchell, 2016, Cross *et al*., 2016).

Immunosuppression

There are many reports of *Legionella* species infections in immunosuppressed patients. In some studies, between 50–70% of immunocompromised transplant recipients with post-operative CAP had non-*L. pneumophila Legionella* infections (Sivagnanamus *et al*., 2016). A review of 33 legionellosis cases in cancer patients identified 18 *L. pneumophila* cases, four strains of *L. donaldsonii*, three strains of *L. micdadei*, one each of *L. bozemanae, L. feeleii, L. gormanii, L. longbeachae, L. maceachernii, L. parisiensis*, *L. sainthelensi* and a new species *Legionella* sp. strain D5382. The five cases of pneumonia due to *L. donaldsonii* and *Legionella* sp. D5382 were the first reports of human infection with these organisms. All patients demonstrated the clinical picture of pneumonia except one asymptomatic carrier. The mortality rate in this group was very high at 27%, and 27 of the 33 patients had underlying hematologic malignancies. There were 23 patients with leukopenia. Six patients were recipients of allogeneic hematopoietic stem cell transplants and their respiratory infections were caused by five *Legionella* species (Han *et al.,* 2015). Often *Legionella* is the last organism to be tested in a case of atypical pneumonia when all other avenues have proved negative. *Legionella* can often co-infect with *Streptococcus pneumoniae,* and thus it can be missed in initial diagnosis as *S. pneumoniae* is treated with macrolides that inadvertently also cover *Legionella* species (Tan *et al.*, 2002, Jorgensen *et al.*, 2015).

An international collaborative study compared the most common causative agents in cases of sporadic community acquired legionellosis in the United States (72.2% of cases reviewed), Italy (12.6%), Switzerland (6.1%), New Zealand (4.3%) and Australia (4.7%) and cited *L. pneumophila* as responsible for nearly 92% of cases and *L. longbeachae* present in 4% of cases (Yu *et al.*, 2002).

In the USA, the aetiology of legionellosis cases was audited over a 15-year study period between 1999 and 2013 (del Castillo *et al.*, 2015). Forty cases of *Legionella* infection were identified, and nine were due to non-*pneumophila* species. The majority (89%) of patients with non-*pneumophila* infection had underlying hematologic malignancy compared to (58%) with *L. pneumophila* infections. The causative organisms of these infections included *L. micdadei*, *L. jordanis*, *L. maceachernii* and *L. bozemanae*. Radiographic findings showed varied nodular infiltrates mimicking an invasive fungal infection only seen among patients with hematologic malignancy and hematopoietic stem cell transplant recipients. The radiographic findings were also statistically significantly more associated with non-*L. pneumophila* infections (Kaul and Riddell, 2009). In a more recent Japanese study although most sporadic cases were *L. pneumophila*, *L. bozemanae*, *L. dumofii L. micdadei* and *L. longbeachae* were also isolated (Miyashita *et al.*, 2020).

Other species of the *Legionella* genus isolated from patients in Europe in 2014 were *L. longbeachae*, *L. micdadei*, *L. bozemanae*, *L. maceachernii*, *L. sainthelensi*, other *Legionella* species and *L. species* not identified (ECDC, 2016). A study of *L. maceachernii* infection identified six patients all with underlying disorders such as HIV, multiple myeloma, pulmonary fibrosis, systemic lupus erythematosus and autoimmune haemolytic anaemia or had undergone liver transplantation, and all but one patient died (Yu *et al.*, 2009). *L. sainthelensi* has previously caused an outbreak in two nursing homes that occurred in 29 elderly patients all who had received pureed food. The association of the illness with dietary characteristics indicative of swallowing disorders suggests that aspiration was the most likely mode of infection. Therefore, diagnosis of legionellosis should always be considered during outbreaks of respiratory infection in nursing homes (Loeb *et al.*, 1999). The other prominent *Legionella* species will be covered separately in this chapter.

Legionella longbeachae

L. longbeachae infection is a problem in Australia and New Zealand where it causes the majority of Legionnaires' disease cases (Steele *et al.*, 1990, Li *et al.*, 2002, Whiley and Bentham, 2011). The method of infection is also unusual as *L. longbeachae* is mostly associated with the manipulation and handling of soils and potting compost rather than acquired by inhalation or aspiration of contaminated water droplets (Chapter 2). A review of legionellosis in Southern Australia from 1996 to 2000 reported that 42% of cases were attributable to *L. longbeachae*, compared with 51% due to *L. pneumophila* (Li *et al.*, 2002). Between 1999 and 2010 in Western Australia, 87% of diagnosed cases of Legionnaires' disease were caused by *L. longbeachae*. Meanwhile, only 9% of cases were caused by *L.*

pneumophila (Commission for Occupational Safety and Health, 2010). Similarly, in New Zealand in 2011 the Ministry of Health found that *L. longbeachae* was responsible for more cases of legionellosis than *L. pneumophila*, with 42% and 30% of laboratory-reported cases of infection, respectively (IESRNZ, 2011). *L. longbeachae* infection has also been noted in the United States (Duchin *et al.*, 2000), Japan (Okazaki *et al.*, 1998) and in Thailand (Phares *et al.*, 2007) where they found that *L. longbeachae* was responsible for 5% of clinically defined cases of pneumonia in a rural district. This pattern differs dramatically to both Europe and the US where the incidence of infection with *L. longbeachae* has been historically low. However, the number of cases of infection appears to be increasing (Whiley and Bentham, 2011). In 2012, it was reported that *L. longbeachae* had been the causative agent in only eleven cases of infection in the United Kingdom since 1984, seven of which occurred in Scotland (Lindsay *et al.*, 2012). Further work revealed that from 2008 to date, 26 cases of *L. longbeachae* infection have been reported in Scotland. In most of these cases the patients had been in direct contact with commercially available growing media before the onset of symptoms (Cameron *et al.*, 2016). Most of the Scottish cases demonstrated severe community acquired pneumonia requiring intensive care treatment. However, the clinical characteristics of *L. longbeachae* Legionnaires' disease are broadly similar to those reported for all Legionnaires' disease cases. Additionally, the temporal distribution of infection is indistinguishable with more cases in the spring/summer season which is also the optimal time for compost use (Isenman *et al.*, 2016). Interestingly, the potting compost material used in Scotland comes from the same manufacturing bases as those for the rest of the UK and some of Europe. Therefore, the increase in Scotland cases would be expected to be seen in other parts of the UK and Europe although there was a recently reported case in Switzerland (Degranges *et al.* 2020) and New Zealand continues to see *L. longbeachae* as the predominant cause of Legionnaires' disease (Priest *et al.*, 2019)

A possible explanation for the increase in Scotland was the introduction of a *Legionella* species specific PCR into routine community acquired pneumonia screening (Lindsay *et al.*, 2012, Gadsby *et al.*, 2016). Similarly, a RT-PCR assay for detection of all *Legionella* spp. and simultaneous specific identification of four clinically relevant *Legionella* species, *L. anisa*, *L. bozemanae*, *L. longbeachae*, and *L. micdadei*, using 5'-hydrolysis probe RT-PCR has been reported that could increase overall detection of *Legionella* species (Cross *et al.*, 2016).

Legionella bozemanae

L. bozemanae infections in immunocompromised patients can produce cavitary lung lesions (Harris and Albrecht, 1998, Widmer *et al.*, 2007). However, patients with cystic lesions in the upper lobes frequently also suffer from tuberculosis. In addition, other diagnoses should be considered such as invasive Aspergillosis or an autoimmune disease (Taylor and Albrecht, 1995). Historically, there are reports of patients with AIDS succumbing to *L. bozemanae* infections after undergoing transplantation (Harris *et al.*, 1998, Muder *et al.*, 2000). Diagnosis of *L. bozemanae* can be difficult as the commonly used urinary antigen tests for *Legionella* and serological kits consistently fail to detect *L. bozemanae* (Muder & Yu, 2002). However, it should be included in the differential diagnosis of cystic lung lesions in the lower lobes when more frequently observed diagnoses such as tuberculosis and non-infectious diseases have been ruled out (Yu *et al.*, 2002).

Legionella micdadei

L. micdadei was identified in immunocompromised patients in Pittsburgh and later called Pittsburgh Pneumonia Agent (Myerowitz *et al.*, 1979, Hebert *et al.*, 1980b). It is the second most common *Legionella* species found in the US, after *L. pneumophila*, comprising about 9% of cases (Medarov *et al.*, 2004). Radiographic appearance in immunocompromised patients often show nodular infiltrates with a tendency to cavitate and enlarge (Lachant and Prasad, 2015). Pulmonary nodules are common in immunocompromised patients and account for significant morbidity and mortality (Jain *et al.*, 2004). The nodule aetiology varies by degree of immunosuppression, prophylaxis, exposure and medical history making them difficult to diagnose with an infectious differential including *Aspergillus*, *Cryptococcus*, *Zygomycetes*, *Nocardia*, *Legionella*, mycobacterial infection, CMV, adenovirus, septic emboli or endocarditis (Rogers *et al.*, 1979, Waldron *et al.*, 2015). If infiltration is identified, bronchial lavage should be undertaken to improve isolation and identification of the causative organism/s (Jain *et al.*, 2004). In a 2015 case report, *L. micdadei* was mis-diagnosed as an acid-fast bacillus (AFB) of *Mycobacterium* in an immunocompromised host after a stem cell transplantation who had the co-morbidities of familial Kaposi's sarcoma and pre-cursor T-cell acute lymphoblastic leukaemia (Waldron *et al.*, 2015) and continues to be reported (Huang *et al.*, 2019, Foissac *et al.*, 2019). As *Legionella* requires selective media to grow, and the commercially available UAT only detect *L. pneumophila* Sg 1, it is important to consider this organism in the differential diagnosis for AFB in the immuno-compromised host.

Legionella dumoffii

This organism was first identified in a fatal Legionnaires' disease case isolated from post-mortem lung (Lewallan *et al.*, 1979). It has continued to cause sporadic cases (Harris *et al.*, 1981) and was implicated in a nosocomial outbreak in Canada in the 1980's linked to contaminated distilled water that was being used in respiratory therapy equipment (Joly *et al.*, 1986). It has also been implicated in prosthetic valve endocarditis (Tompkins *et al.*, 1988). A case of community-acquired pneumonia caused by *L. dumoffii* in a patient with hairy cell leukemia has also been reported (Fang *et al.*, 1990). *Legionella* have been shown to multiply in monocytes and cell-mediated immunity appears to be the primary mechanism of the host defence in man. Hairy cell leukemia is characterized by monocyte dysfunction and such patients have a predilection for infection by microbes that are controlled by cell-mediated defences. PCR has been utilised to detect *Legionella spp* DNA in peripheral leukocyte, serum and urine in an Legionnaires' disease case, and the subsequent sequencing of PCR amplicon identified the causative agent as *L. dumoffii* (Murdoch and Chambers, 2000). *Legionella* pneumonia typically presents as a multi-lobar pneumonia, but *L. dumoffii* has been implicated along with *L. pneumophila*, *L. micdadei*, *L. bozemanae* and *L. maceachernii* in lung abscesses. Clinical factors that cause impaired cellular immunity and subsequent neutrophil accumulation like corticosteroid treatment may be a predisposing factor for the development of these abscesses (Yu *et al.*, 2009). *L. dumoffii* has also been shown to possess Type 2 and 4 secretory systems which may infer greater pathogenicity as these are more often associated with *L. pneumophila* Sg 1 (Qin et al., 2017)

Legionella anisa

This organism is a common contaminant of potable and industrial water systems (Gorman *et al.*, 1985, van der Mee-Marquet *et al.*, 2006). However, the first reported clinical case was isolated from pleural fluid of a patient suffering from nosocomial pleurisy, and this confirmed the potential pathogenicity of *Legionella anisa* (Bornstein *et al.*, 1989). The organism has continued to cause sporadic cases and outbreaks of Pontiac fever (Fenstersheib *et al.*, 1990 Fields *et al.*, 1990, Tanabe *et al.*, 2009, Vaccaro *et al.*, 2016). Similar to other *Legionella* species, *L. anisa* has been reported in unusual and extrapulmonary clinical presentations such as chronic endocarditis and osteomyelitis (Compain *et al.*, 2015, Sanchez *et al.*, 2013) and continues to cause infection in HIV associated pneumonia (Head *et al.*, 2019)

Extrapulmonary legionellosis

In general, reports of extrapulmonary *Legionella* spp. infections are scarce (Chapter 5). In a rare case, *L. micdadei* was isolated from a mass on the left side of the neck of a healthy 9-year-old girl with low grade fever (Qin *et al.*, 2002). *L. micdadei* has also been cultivated from a cerebral abscess in an immunocompromised patient with Waldenström macroglobulinemia (Charles *et al.*, 2013) and from a prosthetic heart valve following endocarditis (Fukuta *et al.*, 2012). A new species, *Legionella cardiaca* has been isolated from a native valve endocarditis (Pearce *et al.*, 2012). Similarly, *L. anisa* has been associated with an aortic aneurysm (Tanabe *et al.*, 2009). There are reports of cellulitis caused by *L. feeleii* (Loridant *et al.*, 2011) and *L. maceachernii* (Chee and Baddour, 2007) as well as a rare case of cutaneous *L. longbeachae* infection in a patient receiving long-term corticosteroids for immune thrombocytopenia (Grimstead *et al.*, 2015). *Legionella* species have also been implicated in arthritic conditions usually involving *L. bozemanae* in immunocompromised patients with dermatomyositis and anti-synthetase syndrome (Just *et al.*, 2012, Ibranosyan *et al.*, 2019). Without the use of *Legionella* species specific PCR screening or culture, the organisms in all of the cases above would not have been identified.

Pontiac fever

Pontiac fever was initially attributed to aerosol exposure to an unknown organism in 1968 (Glick *et al.*, 1978) and subsequently to *L. pneumophila* (Kaufmann *et al.*, 1981, Girod *et al.*1982, Burnsed *et al.*, 2007, Hamilton *et al.*, 2018). The resulting non-pneumonic and flu-like illness is hypothesised to be the result of a sensitisation of the lung mucosa by live and/or dead bacteria and has a possible link to high endotoxin levels detected in water samples causing pneumonitis (Fields *et al.*, 2001) (Chapter 5). There are also Pontiac fever outbreaks linked to non-*pneumophila Legionella*. In the Lochgoilhead Pontiac fever outbreak in 1989, 170 individuals presented with headache, fatigue, arthralgia, myalgia, cough and breathlessness after visiting the same leisure and hotel complex (Goldberg *et al.*, 1989). *L. micdadei* was isolated from a whirlpool spa in the leisure complex following amoebal co-cultivation and a high percentage of cases (>50%) showed *L. micdadei* seroconversion (Fallon and Rowbotham, 1990) and it has been implicated in more than one whirlpool spa outbreak (Fields *et al.*, 2001). Both a specific antibody response to the causative organism and high endotoxin levels in the whirlpool spa were identified suggesting endotoxin mediated sensitisation. More recently, there were ten documented cases associated with a Pontiac fever outbreak in horticultural workers where *L. longbeachae* Sg 2 was isolated from clinical samples from two of the cases along with an indistinguishable strain found in the compost (Cramp *et al.*, 2010). A possible reason for the less pathogenic nature of Pontiac fever relates to the *Legionella* species identified in these outbreaks along with the age and health status of the individuals affected.

Environmental overview

In various studies of potable, river and cooling tower water environments the preponderance of *Legionella* species over *L. pneumophila* is intriguing. *L. anisa* is the most frequently observed non-*L. pneumophila* species in the environment, as it is commonly isolated from cooling towers, drinking water, wastewater treatment plants and in hospital water distribution systems (Gorman *et al.*, 1985, Doleans *et al.*, 2004, van der Mee Marquet *et al.*, 2006 Huang *et al.*, 2009, Paranjape *et al.*, 2020). A meta-analysis of 28 studies comparing culture to qPCR from environmental sampling showed that 72% of environmental samples were positive for *Legionella* species by qPCR but only 34% were positive by culture (Whiley and Taylor, 2014). The study did not identify the percentage split of results into *L. pneumophila* and *Legionella* species, but other studies have shown a preponderance of *Legionella* species other than *L. pneumophila* in water samples (Doelans *et al.*, 2004, van der Mee-Marquet *et al.*, 2006, Lau *et al.*, 2013). A study of drinking water storage tank sediments in the USA detected high levels of *Mycobacterium* but also *L. pneumophila* and many *Legionella* species by qPCR. In addition, there was a significant correlation between *Legionella* and *Acanthamoeba* in these water systems (Lu *et al.*, 2015).

Molecular methods have also revealed the presence of pathogenic non-*L. pneumophila* species in environmental samples, which represents a risk factor for *Legionella* infections (Doleans *et al.*, 2004, Craun *et al.*, 2010, Huang *et al.*, 2011). In Spain, *L. feeleii, L. anisa, L. donaldsonii, L. bozemanae, L. dumoffi* and *L jordanis* have been isolated from drinking water treatment plants, water distribution networks and cooling towers (Rivera *et al.*, 2007, Magnet *et al.*, 2015). These non-*L. pneumophila* species have been previously described as etiological agents of respiratory tract infections. Although *Legionella* species are ubiquitous in the aquatic environments, they are rarely reported to cause Legionnaires' disease outbreaks (Muder and Yu, 2002, Siegel *et al.*, 2010, Potts *et al.*, 2013).

Some *Legionella* species have a less pathogenic nature and are better adapted for survival in water biofilms and amoeba than in humans. This probably relates to the Lag operon, T4SS secretory pathways present in *L. pneumophila* Sg 1 that allow the legionella to evade phagocystosis by macrophages in the lungs and cause systemic disease. This may also be another reason why large Pontiac fever outbreaks tend to be associated with non-Pontiac strains of *L. pneumophila* Sg 1 or *Legionella* species. Differences in pathogenicity of different *Legionella* species was noted by Gomez-Valero *et al.* in 2014 who found that genes specific to *L. pneumophila* and *L. longbeachae* increased the successful infection of mammalian cells when compared to *L. micdadei*, *L. hackeliae* and *L. fallonii*.

When the genomes of *L. pneumophila* and *L. longbeachae* are compared there are a number of similarities which maybe the reason why they are the most successful pathogens in the *Legionella* genus. However, unlike *L. pneumophila*, *L. longbeachae* lacks flagella and produces a capsule that, along with a chemotaxis system and sequences for cellulolytic enzymes, likely improves survival in potting soil and evasion of host defences. In some of the Scottish cases, *L. longbeachae* was isolated from composted material that patients had been working with prior to becoming ill. Whole genome sequencing was performed on all patient and environmental strains and not once was the same strain isolated from the patient found in the suspect compost (Bacigalupe *et al.*, 2017). This is probably a result of the difficulty in isolating *Legionella* from the microbial rich potting composts and the diversity of *Legionella* species found in composted material (Whiley and Bentham, 2011, Currie *et al.*, 2015).

Similar to the picture seen in environmental waters, the range and diversity of *Legionella* present in potting compost is large and the pasteurisation process PAS-100 treatment may kill sessile free-living *Legionella* but not those found inside amoebal cysts (BSI PAS 100: 2018). Therefore, purely testing of composted material post PAS-100 may be insufficient to ensure eradication of these micro-organisms, as they can survive within an encysted amoebal host and re-infect the substrate once conditions improve for the amoeba and *Legionella.*

Treatment

Infection with *Legionella* species are treated in the same way as *L. pneumophila* with macrolides and quinolones (Chapter 5). There is very little information on anti-microbial resistance for all *Legionella* due to the difficulties faced in cultivating this fastidious organism. Growth of *Legionella* within pulmonary macrophages is a prerequisite for disease, as the organism does not grow extracellularly in the human body due to high sodium concentrations in extracellular compartments that have been shown to inhibit bacterial replication. Specifically, antimicrobials that show high intracellular penetration, such as quinolones and later-generation macrolides, are effective treatments (Chiaraviglio and Kirby, 2015).

Among the macrolides, azithromycin, roxithromycin, josamycin and clarithromycin have shown similar potent effects on intracellular growth. In contrast, erythromycin showed relatively poor intracellular activity. These results are consistent with earlier observations of poor clinical outcomes associated with erythromycin, a macrolide no longer recommended for *Legionella* therapy (Miller, 1981).

In intracellular assays of drug susceptibility, quinolones were significantly more active against *L. micdadei* and *L. bozemanae* than against *L. pneumophila* (Stout *et al.*, 1998). In a later study, the rank order of intracellular activity against *L. pneumophila* Sg 1 was quinolones > ketolides > macrolides (Stout *et al.*, 2005). Therefore, due to the potent intracellular activity of quinolones and later-generation macrolides, these are now recommended as first-line therapies for *Legionella*. Among the quinolone class, the intracellular dose-response curve for levofloxacin was compared to the curve for ciprofloxacin. It was found that levofloxacin was the most efficient antibiotic for intracellular bacterial killing (Mills *et al.*, 2005, Pedro-Botet and Yu, 2006, Chiaraviglio and Kirby, 2015). Other studies have documented clinical failure with ciprofloxacin, suggesting the potential clinical superiority of levofloxacin that is consistent with other in vitro findings (Kurz *et al.*, 1988, Unertl *et al.*, 1989). In a study of intensive care cases of non-pneumococcal CAP, dual therapy and early administration of antibiotics was associated with increased survival. In the case of severe Legionnaires' disease, a third-generation cephalosporin such as cefotaxime or ceftriaxone and a later generation fluorquinolone like levofloxacin were the preferred treatment options (Gattarello *et al.*, 2015). However, in patients with multiple antibiotic allergies, the third generation glycylcycline, tigecycline has been used successfully. (Arget *et al.*, 2019)

Conclusion

The true clinical significance of non-*L. pneumophila* infections will only become evident when urinary antigen and PCR tests that detect all *Legionella* species become both widely available and utilised correctly. At the present time, infection with non-*L. pneumophila* species seems to cause only sporadic cases of Legionnaires' disease, and it is often only detected in severely ill patients with co-morbidities. However, any *Legionella* species can be potentially fatal. Sporadic hospital, community and travel associated cases can occur when this normally aquatic micro-organism comes into contact within a susceptible or immunosuppressed host. Depending on the literature, the aetiology of CAP is never confirmed in 40-60% of cases (Yu *et al.*, 2002, den Boer *et al.*, 2004). Therefore, pneumonia caused by *Legionella* species other than *L. pneumophila* may be missed because of the diagnostic bias towards *L. pneumophila* Sg 1 inherent in current testing methods.

References

Arget, M., Kosar, J., Suen, B and Peermohamed, S. (2019). Successful Treatment of Legionnaires' Disease with Tigecycline in an Immunocompromised Man with a Legion of Antibiotic Allergies. *Cureus.* 11(4):e4577. Published 2019 Apr 30. doi:10.7759/cureus.4577

Bacigalupe, R., Lindsay, D., Edwards, G and Fitzgerald, J. R. (2017). Population Genomics of *Legionella longbeachae* and Hidden Complexities of Infection Source Attribution. Emerg Infect Dis. 23, 750-757.

Benin, A.L., Benson, R. F and Besser, R. E. (2002). Trends in Legionnaires' disease, 1980-1998: declining mortality and new patterns of diagnosis. Clin Infect Dis 35,1039–1046. http://dx.doi.org/10.1086/342903.

Benitez, A. J and Winchell, J. M. (2016). Rapid detection and typing of pathogenic non-pneumophila *Legionella* spp. isolates using a multiplex real-time PCR assay. Diagn Microbiol Infect Dis. 84, 298-303.

Bornstein, N., Mercatello, A., Marmet, D., Surgot, M., Deveaux, Y and Fleurette J. (1989). Pleural infection caused by *Legionella anisa*. J Clin Microbiol. 27, 2100-2101.

Brenner, D. J., Steigerwalt, A. G. and Broome, C. V. (1984). A new *Legionella* species, *Legionella feeleii* species nova, causes Pontiac fever in an automobile plant, *Ann. Intern. Med. 100*, 333-338. BSI PAS 100:2018.

Burnsed, L. J., Hicks, L. A., Smithee, L. M., Fields, B. S., Bradley, K. K., Pascoe, N., *et al.* (2007). A large travel-associated outbreak of legionellosis among hotel guests: utility of the urine antigen assay in confirming Pontiac fever. Clin Infect Dis. 44, 222-228.

Cameron, R. L., Pollock, K. G. J., Lindsay, D. S. J and Anderson, E. (2016). Comparison of *Legionella longbeachae* and *Legionella pneumophila* cases in Scotland; implications for diagnosis, treatment and public health response. J Med Micro. 65, 142-6

Cazalet, C., Gomez-Valero, L., Rusniok, C., Lomma, M., Dervins-Ravault, D., Newton, H. J. et al. (2010). Analysis of the *Legionella longbeachae* genome and transcriptome uncovers unique strategies to cause Legionnaires' disease. PLoS Genetics, 6.

Charles, M., Johnson, E., Macyk-Davey, A., Henry, M., Nilsson, J. E., Miedzinski, L and Zahariadis, G. (2013). *Legionella micdadei* brain abscess. J Clin Microbiol. 51, 701-704.

Chee, C. E and Baddour, L. M. (2007). *Legionella maceachernii* soft tissue infection. Am J Med Sci. 334, 410-413.

Chiaraviglio, L and Kirby, J. E. (2015). High-throughput intracellular antimicrobial susceptibility testing of *Legionella pneumophila*. Antimicrob Agents Chemother 59: 7517–7529.

Commission for Occupational Safety and Health. (2010). Code of Practice: Prevention and Control of Legionnaires' Disease. Report for

Government of Western Australia, Department of Commerce, Department of Mines and Petroleum. Available at: https://www.commerce.wa.gov.au/sites/default/files/atoms/files/legionnaires-code.pdf

Compain, F., Bruneval, P., Jarraud, S., Perrot, S., Aubert, S., Napoly, V., Ramahefasolo, A., Mainardi, J. L and Podglajen I. (2015). Chronic endocarditis due to *Legionella anisa*: a first case difficult to diagnose. New Microbes New Infect. 8, 113-115.

Cramp, G. J., Harte, D., Douglas, N.M., Graham, F., Schousboe, M. and Sykes, K. (2010). An outbreak of Pontiac fever due to *Legionella longbeachae* serogroup 2 found in potting mix in a horticultural nursery in New Zealand, Epidemiol. Infect. 138, 15-20.

Craun, G. F., Brunkard, J. M., Yoder, J. S., Roberts, V. A., Carpenter, J., Wade, T. *et al.* (2010). Causes of outbreaks associated with drinking water in the United States from 1971 to 2006. Clin Microbiol Rev. 23, 507-512.

Cross, K. E., Mercante, J. W., Benitez, A. J., Brown, E. W., Diaz, M. H and Winchell, J. M. (2016). Simultaneous detection of *Legionella* species and *L. anisa*, *L. bozemanii*, *L. longbeachae* and *L. micdadei* using conserved primers and multiple probes in a multiplex real-time PCR assay. Diagn Microbiol Infect Dis. 85, 295-301.

Cunha, C. B and Cunha, B. A. (2016). Legionnaire's Disease since Philadelphia: lessons learned and continued progress. Infect Dis Clin 16, 30087-93.

Currie, S and Beattie, T. K., (2015). Compost and *L. longbeachae*: an emerging infection. Perspectives in Public Health. 135, 309-315.

David, S., Mentasti, M., Tewolde, R., Aslett, M., Harris, S. R., Afshar, B., Underwood, A., Fry, N. K., Parkhill, J and Harrison, T. G. (2016). Evaluation of an optimal epidemiological typing scheme for *Legionella pneumophila* with whole genome sequence data using validation guidelines. J Clin Microbiol. 54, 2135-48.

Desgranges F., Coste, A. T., Wernly, D., Dufour, J., Opota, O and Meylan S. (2020) Immunosuppressed gardener pricked by roses grows *Legionella longbeachae*. Lancet. 2020 Feb 22;395(10224):604. doi: 10.1016/S0140-6736(20)30112-4.

del Castillo, M., Lucca, A., Plodkowski, A., Huang, Y-T., Kaplan, J., Gilhuley, K., Babady, N. E., Seo, S. K. and Kamboj, M. (2015). Atypical presentation of *Legionella* pneumonia among patients with underlying cancer: A fifteen-year review. J Inf. 72, 45-51.

den Boer JW, Yzerman EP. (2004). Diagnosis of *Legionella* infection in Legionnaires' disease. Eur J Clin Micro Infect Dis. 23, 871-878.

Doleans, A., Aurell, H., Reyrolle, M., Lina, G., Freney, J., Vandenesch, F. *et al.* (2004). Clinical and environmental distributions of *Legionella* strains in France are different. J Clin Microbiol. 42, 458–60.

Dominguez, J., Gali, N., Blanco, S., Pedroso, P., Prat, C., Matas, L., *et al.* (2001). Assessment of a new test to detect *Legionella* urinary antigen for the diagnosis of Legionnaires' disease. Diagn Microbiol Infect Dis. 41,199-203.

Diederen, B. M. W. (2007). *Legionella* spp. and Legionnaires' disease J Infect. doi:10.1016/ j.jinf.2007.09.010.

Duchin, J. S., Koehler, J., Kobayashi, J. M., Rakita, R. M., Olson, K., Hampson, N. B. *et al.* (2000). Legionnaires' diseases associated with potting soil: California, Oregon, and Washington, May–June 2000. MMWR Morbidity and Mortality Weekly Report; 49, 777–780.

Edelstein, P. H., Edelstein, M. A., Shephard, L. J., Ward, K. W and Ratcliff, R. M. (2012). *Legionella steelei* sp. nov., isolated from human respiratory specimens in California, USA and South Australia. Int J Syst Evol Microbiol. 62, 1766-1771.

European Centre for Disease Prevention and Control. (2013). Legionnaires disease in Europe. 2011. ECDC, Stockholm, Sweden. https://www.ecdc.europa.eu/en/publications-data/legionnaires-disease-europe-2011.

European Centre for Disease Prevention and Control. (2016). Legionnaires' Disease in Europe, 2016. https://www.ecdc.europa.eu/en/publications-data/legionnaires-disease-annual-epidemiological-report-2016.

Fang, G. D, Stout, J. E., Yu, V. L., Goetz, A., Rihs, J. D and Vickers, R. M. (1990). Community-acquired pneumonia caused by *Legionella dumoffii* in a patient with hairy cell leukemia. Infection. 18, 383-385.

Fallon, R. J and Rowbotham T. J. (1990). Microbiological investigations into an outbreak of Pontiac fever due to *L. micdadei* associated with use of a whirlpool. J. Clin Pathol. 43, 479-483

Fallon, R. J and Abraham, W. H. (1992). Crossreactions between *Legionella* and *Campylobacter* spp. Lancet 340,551-552.

Fenstersheib, M. D., Miller, M., Diggins, C., Liska, S., Detwiler, L., Werner, S. B., Lindquist, D., Thacker, W. L and Benson, R. F. (1990). Outbreak of Pontiac fever due to *Legionella anisa*. Lancet. 336, 35-37.

Fields, B. S., Barbaree, J. M., Sanden, G. N and Morrill, W. E. (1990). Virulence of a *Legionella anisa* strain associated with Pontiac fever: an evaluation using protozoan, cell culture, and guinea pig models. *Infect. Immun*. 58, 3139-3142.

Fields, B. S., Haupt, T., Davis, J. P., Arduino, M. J., Miller, P.H and Butler., J. C. (2001). Pontiac Fever Due to *Legionella micdadei* from a whirlpool spa: possible role of bacterial endotoxin. J Inf Dis. 184:1289-92

Fields, B. S., Benson, R. F and Besser, R. E. (2002). *Legionella* and Legionnaires' disease: 25 years of investigation, *Clin. Microbiol. Rev. 15*, 506-526.

Finidori, J. P., Raoult, D., Bornstein, N and Fleurette J. (1992). Study of cross-reaction between *Coxiella burnetii* and *Legionella pneumophila* using indirect immunofluorescence assay and immunoblotting. Acta Virol. 36, 459-465.

Foissac, M., Bergon, L., Vidal, J., Cauquil, P., Mainar, A and Mourguet, M. (2019). Pneumonia and pulmonary abscess due to *Legionella micdadei* in an immunocompromised patient. Germs. 9, 89-94.

Fukuta, Y., Yildiz-Aktas, I. Z., William Pasculle, A and Veldkamp, P. J. (2012). *Legionella micdadei* prosthetic valve endocarditis complicated by brain abscess: case report and review of the literature. Scand. J. Infect. Dis. 44,414–418.

Gattarello S., Lagunes L., Vidaur L., Solé-Violán J., Zaragoza R., Vallés J., Torres., A, Sierra R., Sebastian R, Rello J. (2015). Improvement of antibiotic therapy and ICU survival in severe non-pneumococcal community-acquired pneumonia: a matched case-control study. Crit Care. 19:335.

Gadsby, N. J., Helgason, K. O., Dickson, E. M., Mills, J. M., Lindsay, D. S. J., Edwards, G. F., Hanson, M. F and Templeton K. E. (2016). Molecular diagnosis of *Legionella* infections - clinical utility of front-line screening as part of a pneumonia diagnostic algorithm. J Inf. 72, 161-170.

Garrity, G.M., Brown, A and Vickers, R. M. (1980). *Tatlockia* and *Fluoribacter*: two new genera of organisms resembling *Legionella pneumophila*. Int. J. Syst. Bacteriol. 30, 609-614.

Girod, J. C., Reichman, R. C., Winn, W. C Jr., Klaucke, D. N., Vogt, R. L and Dolin R. (1982). Pneumonic and non-pneumonic forms of legionellosis. The result of a common-source exposure to *Legionella pneumophila*. Arch Intern Med. 142, 545-7.

Goldberg, D. J., Wrench, J. G., Collier, P.W., Emslie, J. A., Fallon, R. J., Forbes, G. I. *et al.* (1989). Lochgoilhead fever – outbreak of non-pneumonic legionellosis due to *Legionella micdadei*. The Lancet. 1, 316–8.

Gomez-Valero, L., Rusniok, C., Rolando, M., Neou, M., Dervins-Ravault, D., Demirtas, J., Rouy, Z., Moore, R. J., Chen, H., Petty, N. K., Jarraud, S., Etienne, J., Steinert, M. and Heuner, K., Gribaldo, S., Médigue, C., Glöckner, G., Hartland, E. L. and Buchrieser, C. (2014). Comparative analyses of *Legionella* species identifies genetic features of strains causing Legionnaires' disease. Genome Biol. 15, 505-526.

Gorman, G. W., Feeley, J. C., Steigerwalt, A., Edelstein, P. H., Moss, C. W and Brenner, D. J. (1985). *Legionella anisa*: a new species of *Legionella* isolated from potable waters and a cooling tower. Appl Environ Microbiol. 49, 305–309.

Glick, T. H., Gregg, M. B., Berman, B., Mallison, G., Rhodes, W. W and Kassanoff, I. (1978). Pontiac fever. An epidemic of unknown etiology in a health department: Clinical and epidemiologic aspects. Am J Epidemiol. 107, 149-160.

Grimstead, D., Tucker, D., Harris, K and Turner, D. (2015). Cutaneous *Legionella longbeachae* infection in immuno-suppressed woman, United Kingdom. Emerg Infect Dis. 21, 1426–1428.

Hamilton, K, A., Prussin, A. J., Ahmed, W and Haas, C. N (2018) Outbreaks of Legionnaires' Disease and Pontiac Fever 2006-2017. Curr Environ Health Rep. 5(2):263-271.

Han, X. Y., Ihegword, A., Evans, S. E., Zhang, J., Li, L., Hongjing, C., Tarrand, J. J. and El-Kweifi, O. (2015). Microbiological and clinical studies of legionellosis in 33 patients with cancer. J Clin Microbiol 53, 2180 –2187. doi:10.1128/JCM.00380-15.

Harris, P. P., Aufdemorte, T., Ewing, E. P., Johnson, J. E., Tio, F. O. (1981). Fluorescent-antibody detection of *Legionella dumoffii* in a fatal case of pneumonia. J Clin Microbiol. 13, 778-80.

Harris, A., Lally and Albrecht M. (1998). *Legionella bozemanae* pneumonia in three patients with AIDS. Clin Infect Dis. 27, 97–99

Harrison T, Uldum S, Alexiou-Daniel S, Bangsborg J, Bernander S, Drasbrevear V, et al. (1998). A multicenter evaluation of the Biotest Legionella urinary antigen EIA. Clin Microbiol Infect. 4,359-65.

Haviernik, J., Dawson, K., Anderson, T., Murdoch, D., Chambers, S., Biggs, P., Cree, S and Slow, S. (2020). Complete Genome Sequence of a Legionella longbeachae Serogroup 2 Isolate Derived from a Patient with Legionnaires' Disease. Microbiol Resour Announc. 9:e01563-19. https://doi.org/10.1128/MRA01563-19.

Head, B. M., Trajtman, A., Bernard, K., Burdz, T., Vélez, L., Herrera, M., Rueda, Z.V and Keynan Y. (2019) Legionella co-infection in HIV-associated pneumonia. Diagn Microbiol Infect. 95, 71-76.

Hebert, G. A., Moss, C. W., McDougal, L. K., Bozeman, F. M., McKinney, R. M. and Brenner, D. J. (1980a). The rickettsia-like organisms TATLOCK (1943) and HEBA (1959): bacteria phenotypically similar, but genetically distinct from *Legionella pneumophila* and the WIGA bacterium. Ann. Intern. Med. 92, 45-52.

Hebert, G. A., Thomason, B. M., Harris, P. P., Hicklin, M. D and McKinney, R. M. (1980b). Pittsburgh Pneumonia Agent: a bacterium phenotypically similar to *Legionella pneurnophila* and identical to the TAT- LOCK bacterium. Ann. Intern. Med. 92, 53-54.

Herwaldt, L. A., Gorman, G. W., McGrath, T., Toma, S., Brake, B., Hightower, A. W., Jones, J., Reingold, A. L., Boxer, P. A., Tang, P. W., Moss, C. W., Wikinson, H. (1984). A new *Legionella* species, *Legionella feeleii* species nova, causes Pontiac fever in an automobile plant. Ann Intern Med. 100, 333-338.

Huang, S. W., Hsu, B. M., Ma, P. H and Chien, K. T. (2009). *Legionella* prevalence in wastewater treatment plants of Taiwan. Water Sci Technol. 60, 1303–1310.

Huang, S. W., Hsu, B. M., Chen, N. H., Huang, C. C., Huang, K. H., Chen, J. S. *et al.* (2011). Isolation and identification of *Legionella* and their host amoebae from weak alkaline carbonate spring water using a culture method combined with PCR. Parasitol Res. 109, 1233–1241.

Huang, Y., Ma, Y., Miao, Q., Pan, J., Hu, B., Gong, Y and Lin, Y. (2019). Arthritis caused by *Legionella micdadei* and *Staphylococcus aureus*: metagenomic next-generation sequencing provides a rapid and accurate access to diagnosis and surveillance. Ann Transl Med. Oct;7(20):589. doi: 10.21037/atm.2019.09.81.

Institute of Environmental Science and Research Ltd (IESRNZ). (2011). Notifiable and other diseases in New Zealand: Annual Report 2011.online at: http://www.surv.esr.cri.nz/surveillance/annual_surveillance.php?we_objectID=2963

Ibranosyan, M., Beraud, L., Lemaire, H., Ranc, A. G., Ginevra, C., Jarraud, S and Descours, G. (2019). The clinical presentation of

Legionella arthritis reveals the mode of infection and the bacterial species: case report and literature review. BMC Infect Dis.19, 864-869.

Isenman, H. L., Chambers, S. T., Pithie, A. D., MacDonald, S. L., Hegarty, J. M., Fenwick, J. L., Maze, M. J., Metcalf, S. C and Murdoch, D. R. (2016). Legionnaires' disease caused by *Legionella longbeachae*: Clinical features and outcomes of 107 cases from an endemic area. Respirology. 21, 1292-1299.

Jain, P., Sandur, S., Meli., Y., Arroliga, A. C., Stoller, J. K and Mehta, A. C. (2004). Role of flexible bronchoscopy in immunocompromised patients with lung infiltrates. Chest. 125, 712–722.

Jespersen, S., Sogaard, O. S., Fine, M. J. and Ostergaard, L. (2009). The relationship between diagnostic tests and case characteristics in Legionnaires' disease. Scand J Infect Dis 41, 425– 432.

Jørgensen, C. S., Uldum, S. A., Sørensen, J. F., Skovsted, I. C., Otte, S and Elverdal, P. L. (2015). Evaluation of a new lateral flow test for detection of *Streptococcus pneumoniae* and *Legionella pneumophila* urinary antigen. J Microbiol Methods. 116, 33-36.

Joly, J. R., Déry, P., Gauvreau, L., Coté, L and Trépanier, C. (1986). Legionnaires' disease caused by *Legionella dumoffii* in distilled water. CMAJ. 135, 1274-1277.

Joseph, C. A. and Ricketts, K. D. (2010) European Working Group for Legionella Infections. Legionnaires' disease in Europe 2007-2008. Euro Surveill 15(8): pii_19493. http://www.eurosurveillance.org/ViewArticle.aspx?ArticleId_19493.

Just, S. A., Knudsen, J. B., Uldum, S. A and Holt, H.M. (2012). Detection of *Legionella bozemanae*, a new cause of septic arthritis, by PCR followed by specific culture. J Clin Microbiol. 50, 4180-4182.

Kaufmann, A. F, McDade, J. E., Patton, C. M., *et al.* (1981). Pontiac fever: isolation of the etiologic agent (*Legionella pneumophila*) and demonstration of its mode of transmission. Am J Epidemiol. 114, 337–47.

Kaul, D. R and Riddell, J. (2009). Approach to the immuno-compromised patient with pulmonary nodules. *Current Fungal Infection Reports.* 3, 45–54.

Khodr, A., Kay, E., Gomez-Valero, L., Ginevra, C., Doublet, P., Buchrieser, C and Jarraud, S. (2016). Molecular epidemiology, phylogeny and evolution of *Legionella* Infection, Genetics and Evolution 43, 108–122.

Kurz, R. W., Graninger, W., Egger, T. P., Pichler, H and Tragl, K. H. (1988). Failure of treatment of Legionella pneumonia with ciprofloxacin. J Antimicrob Chemother 22:389–391.

Lachant, D and Prasad, P. (2015). *Legionella micdadei*: a forgotten etiology of growing cavitary nodules: a case report and literature review. Case Rep Pulm. http://dx.doi.org/10.1155/2015/535012.

Lau, R., Maqsood, S., Harte, D., Caughley, B and Deacon, R. (2013). Prevalence of *Legionella* strains in cooling towers and legionellosis cases in New Zealand. J Environ Health. 75, 82–89.

Lewallen, K. R., McKinney, R. M., Brenner, D. J., Moss, C. W., Dail, D. H., Thomason, B. M and Bright, R.A. (1979). A newly identified bacterium phenotypi-cally resembling but genetically distinct from *Legionella pneumophila* an isolate in a case of pneumonia. Ann. Intern. Med. 91, 831-834.

Li, J. S., O'Brien, E. D and Guest, C. (2002). A review of national legionellosis surveillance in Australia, 1991 to 2000. Communicable Diseases Intelligence Quarterly Report 26, 461–8.

Lu, J., Struewing, I., Yelton, S. and Ashbolt, N. (2015). Molecular survey of occurrence and quantity of *Legionella* spp., *Mycobacterium* spp., *Pseudomonas aeruginosa* and amoeba hosts in municipal drinking water storage tank sediments. J Appl Microbiol. 119, 278-88.

Lindsay, D. S., Abraham, W. H., Findlay, W., Christie, P., Johnston, F and Edwards, G. F. (2004). Laboratory diagnosis of legionnaires' disease due to *Legionella pneumophila* serogroup 1: comparison of phenotypic and genotypic methods. J Med Microbiol.53,183–187.

Lindsay, D. S. J., Brown, A.W., Brown, D. J., Pravinkumar, S. J., Anderson, E and Edwards, G. F. S. (2012). *Legionella longbeachae* serogroup 1 infections linked to potting compost. J Med Micro. 61, 218–22.

Loeb, M., Simor, A. E., Mandell, L., Krueger, P., McArthur, M., James, M., Walter, S., Richardson, E., Lingley, M., Stout, J., Stronach, D and McGeer, A. (1999). Two nursing home outbreaks of respiratory infection with *Legionella sainthelensi*. J Am Geriatr Soc. 47, 547-552.

Loridant, S., Lagier, J. C and La Scola, B. (2011). Identification of *Legionella feeleii* cellulitis. Emerg Infect Dis. 17,145–6.

Magnet, A., Peralta, R. H., Gomes, T. S., Izquierdo, F., Fernandez-Vadillo, C., Galvan, A. L et al. (2015). Vectorial role of Acanthamoeba in *Legionella* propagation in water for human use. Sci Total Environ. 2015; 505:889-895.

Mangione E. J., Remis, R. S., Tait, T.A., Gee, H. B., Gorman, G. W.,Wentworth, G. G., Baron, P. A., Hightower, A. W., Barbaree, J. M and Broome, C. V. (1985). An outbreak of Pontiac fever related to whirlpool use, Michigan 1982. JAMA *253*, 535-539.

Marston, B. J., Lipman, H. B. and Breiman, R. F. (1994). Surveillance for Legionnaires' disease. Risk factors for morbidity and mortality. Arch Intern Med 154,2417–2422.

Mentasti, M., Fry, N. K., Afshar, B., Palepou-Foxley, C., Naik, F. C and Harrison, T. G. (2012). Application *of Legionella pneumophila*-specific quantitative real-time PCR combined with direct amplification and sequence-based typing in the diagnosis and epidemiological investigation of Legionnaires' disease. Eur J Clin Microbiol Infect Dis. 31, 2017–2028.

Mercante, J. W and Winchell, J. M. (2015). Current and emerging *Legionella* diagnostics for laboratory and outbreak investigations. *Clin Microbiol Rev* 28, 80-118. doi:10.1128/CMR.00029-14.

Medarov, B. I., Siddiqui, A. K., Mughal, T., Moshiyakhov, M and Rossoff, L. (2004). *Legionella micdadei* infection presenting as severe secretory diarrhoea and a solitary pulmonary mass. Clin Infect Dis. 38, 63–65.

McAdam, P. R, Vander Broek, C. W., Lindsay, D. S. J., Ward, M. J., Stevens, J. M., Hanson, M., Gillies, M., Watson, M., Edwards, G. F. and Fitzgerald, J. R. (2014). Gene flow in environmental *Legionella pneumophila* leads to genetic and pathogenic heterogeneity within a Legionnaires' disease outbreak. Genome biology15, 504-14.

McDade, J. E., Shepard, C. C., Fraser, D. W., Tsai, T. R., Redus, M. A and Dowdle W. R. (1977). Legionnaires' Disease– isolation of a bacterium and demonstration of its role in other respiratory disease. New Eng J Med. 297, 1197–203.

McKinney, R.M., Porschen, R.K., Edelstein, P. H., Bissett, M. L., Harris, P. P., Bondell, S. P. et al. (1981). *Legionella longbeachae* species nova, another etiologic agent of human pneumonia. Annals Int. Med. 94, 739–43.

Miller, A. C. (1981). Erythromycin in Legionnaires' disease: a re-appraisal. J Antimicrob Chemother 7, 217–222.

Mills, G. D., Oehley, M. R and Arrol, B. (2005). Effectiveness of beta lactam antibiotics compared with antibiotics active against atypical pathogens in non-severe community acquired pneumonia: meta-analysis. BMJ. 330, 456.

Miyashita ,N., Higa, F., Aoki,Y., Kikuchi, T., Seki, M., Tateda, K., Maki, N., Uchino, K., Ogasawara, K,, Kiyota, H and Watanabe, A. (2020) Distribution of Legionella species and serogroups in patients with culture-confirmed Legionella pneumonia. J Infect Chemother. https://doi.org/10.1016/j.jiac.2019.12.016.

Muder, R. R, Stout J. E and Yu, V. L. (2000). Nosocomial *Legionella micdadei* infection in transplant patients: fortune favors the prepared mind. Am J Med. 108, 346–348.

Muder, R. R and Yu, V. L. (2002). Infection due to *Legionella* species other than L. pneumophila. Clin Infect Dis. 35, 990–998.

Murdoch, D. R and Chambers, S. T. (2000). Detection of *Legionella* DNA in peripheral leukocytes, serum, and urine from a patient with pneumonia caused by Legionella dumoffii. Clin Infect Dis.

Murdoch, D. R. (2003). Diagnosis of *Legionella* infection. Clin Infect Dis. 36, 64-69.

Myerowitz, R. L., Pasculle, A. W., Dowling, J. N et al. (1979). Opportunistic lung infection due to 'Pittsburgh pneumonia agent. New England J Med. 301, 953–958.

Okazaki, M., Umeda, B., Koide, M and Saito, A. (1998). *Legionella longbeachae* pneumonia in a gardener. Kansenshogaku Zasshi. 72, 1076–9.

Paranjape, K., Bédard, É., Whyte, L. G., Ronholm, J., Prévost, M and Faucher, S. P. (2020). Presence of *Legionella* spp. in cooling towers: the role of microbial diversity, *Pseudomonas*, and continuous chlorine application. *Water Res.* 2020;169:115252. doi:10.1016/j.watres.2019.115252

Pearce, M. M., Theodoropoulos, N., Mandel, M. J., Brown, E., Reed, K. D and Cianciotto, N. P. (2012). *Legionella cardiaca* sp. nov., isolated from a case of native valve endocarditis in a human heart. Int J Syst Evol Microbiol. 62, 2946–2954.

Pedro-Botet, L and Yu, V. L. (2006). Legionella: macrolides or quinolones? Clin Microbiol Infect 12, 25–30.

Phares, C. R., Wangroongsarb, P., Chantra, S., Paveenkitiporn, W., Tondella, M. L, Benson RF *et al.*, (2007). Epidemiology of severe pneumonia caused by *Legionella longbeachae, Mycoplasma pneumoniae*, and *Chlamydia pneumoniae*: 1-year, population-based surveillance for severe pneumonia in Thailand. Clinical Infectious Diseases. 45, 147–155.

Pierre, D. M., Baron, J., Yu, V. L and Stout, J. E. (2017) Diagnostic testing for Legionnaires' disease. Ann Clin Microbiol Antimicrob. 16, 59-62.

Potts, A., Donaghy, M., Marley, M., Othieno, R., Stevenson, J., Hyland, J., Pollock, K. G., Lindsay, D., Edwards, G., Hanson, M. F and Helgason, K.

O. (2013). Cluster of Legionnaires' disease cases caused by *Legionella longbeachae* serogroup 1, Scotland, August to September 2013. Euro Surveill. 2013;18(50):pii=20656. Available at: http://www.eurosurveillance.org/ViewArticle.aspx?ArticleId=20656

Priest, P. C., Slow, S., Chambers, S. T., Cameron, C. M., Balm, M. N., Beale, M. W., Blackmore,T. K.,Burns, A. D., Drinković, D., Elvy, J. A., Everts, R. J., Hammer, D. A., Huggan, P. J., Mansell, C. J., Raeder, V. M., Roberts, S. A., Robinson, M. C., Sathyendran, V., Taylor, S. L., Thompson, A. W., Ussher J. E., van der Linden, A. J., Williams, M. J., Podmore, R. G., Anderson, T. P., Barratt, K., Mitchell, J. L., Harte, D. J., Hope, V. T and Murdoch, D. R. (2019). The burden of Legionnaires' disease in New Zealand (LegiNZ): a national surveillance study. Lancet Infect Dis.19,770-777

Qin, X., Abe, P. M., Weissman, S. J and Manning, S. C. (2002). Extrapulmonary *Legionella micdadei* infection in a previously healthy child. Pediatr Infect Dis J. 21,1174–1176.

Qin, T., Zhou, H., Ren, H and Liu W. Distribution of Secretion Systems in the Genus Legionella and Its Correlation with Pathogenicity. Front Microbiol. 2017 Mar 14;8:388. doi: 10.3389/fmicb.2017.00388.

Ratcliff, R. M., Lanser, J. A., Manning, P. A and Heuzenroeder, M. W. (1998). Sequence-based classification scheme for the genus *Legionella* targeting the *mip* gene. J Clin Microbiol. 36, 1560-1567.

Rogers, B., Donowitz, G., Walker, G., Harding, S and Sande, M. (1979). Opportunistic pneumonia: a clinical pathological study of five cases caused by an unidentified acid-fast bacterium. New England J Med. 301, 959–961

Rivera, J. M, Aguilar, L., Granizo, J. J., Vos-Arenilla, A., Giménez, M. J., Aguiar, J. M and Prieto J. (2007). Isolation of *Legionella* species/serogroups from water cooling systems compared with potable water systems in Spanish healthcare facilities. J Hosp Infect. 67, 360-6.

Sanchez, M. C., Sebti, R., Hassoun, P., Mannion, C., Goy, A. H., Feldman, T., Mato, A and Hong T. (2013). Osteomyelitis of the patella caused by *Legionella anisa*. J Clin Microbiol. 51,2791-3.

Siegel, M. O., Fedorko, D. P., Drake, S. K., Calhoun, L. B. and Holland, S. M. (2010). *Legionella feeleii* serotype 2 pneumonia in a man with chronic lymphocytic leukemia: a challenging diagnosis. J Clin Microbiol. 48, 2294–7.

Sivagnanamus, S and Pergam, S. A. (2016). Legionellosis in Transplantation. Curr Infect Dis Rep 18, 9. doi: 10.1007/s11908-016-0517-x.

Steele, T. W., Lanser, J. and Sangster, N. (1990). Isolation of *Legionella longbeachae* serogroup-1 from potting mixes. *Appl Environ Microbiol* 56, 49–53.

Stout JE, Arnold B, Yu VL. (1998). Comparative activity of ciprofloxacin, ofloxacin, levofloxacin, and erythromycin against *Legionella* species by broth microdilution and intracellular susceptibility testing in HL-60 cells. Diagn Microbiol Infect Dis. 30,37–43

Stout, J. E., Sens, K., Mietzner, S., Obman, A and Yu, V. L. (2005). Comparative activity of quinolones, macrolides and ketolides against *Legionella* species using in vitro broth dilution and intracellular susceptibility testing. Int J Antimicrob Agents. 25, 302-307.

St-Martin, G., Uldum, S and Molbak, K. (2013). Incidence and prognostic factors for Legionnaires' disease in Denmark 1993-2006. ISRN Epidemiol 2013:8. http://dx.doi.org/10.5402/2013/847283.

Svarrer, C. W and Uldum, S. A. (2011). The occurrence of *Legionella* species other than *Legionella pneumophila* in clinical and environmental samples in Denmark identified by *mip* gene sequencing and matrix-assisted laser desorption ionization time-of-flight mass spectrometry. Clin Micro Inf. 18, 1004-1009.

Tan, M. J., Tan, J. S., File, T. M. (2002). Legionnaires' disease with bacteremic coinfection. Clin Infect Dis. 35, 533-539.

Tanabe, M., Nakajima, H., Nakamura, A., Ito, T., Nakamura, M., Shimono, T. *et al.* (2009). Mycotic aortic aneurysm associated with *Legionella anisa*. J Clin Microbiol. 47, 2340–2343.

Taylor, T. H and Albrecht, M. A. (1995). *Legionella bozemanae* cavitary pneumonia poorly responsive to erythromycin: case report and review. Clin Infect Dis. 20, 329–334.

Tompkins, L. S., Roessler, B. J., Redd, S. C., Markowitz, L. E and Cohen, M. L. (1988). Legionella prosthetic-valve endocarditis. N Engl J Med. 318, 530-535.

Unertl, K. E., Lenhart, F. P., Forst, H., Vogler, G., Wilm, V., Ehret, W., Ruckdeschel, G. (1989). Ciprofloxacin in the treatment of legionellosis in critically ill patients including those cases unresponsive to erythromycin. Am J Med 87,128–131.

Vaccaro, L., Izquierdo, F., Magnet, A., Hurtado, C., Salinas, M. A., Gomes, T. S. *et al.* (2016). First Case of Legionnaire's Disease Caused by *Legionella anisa* in Spain and the Limitations on the Diagnosis of *Legionella* non-*pneumophila* infections. PLoS one 11, 0159726. doi: 10.1371/journal.pone.0159726.

van der Mee-Marquet, N., Domelier, A. S., Arnault, L., Bloc, D., Laudat, P., Hartemann, P., *et al.* (2006). *Legionella anisa*, a possible indicator of water contamination by *Legionella pneumophila*. J Clin Microbiol. 44, 56–59.

von Baum, H., Ewig, S., Marre, R., Suttorp, N., Gonschior, S., Welte, T and Lück, C. (2008). Communnity-acquired *Legionella* pneumonia: new insights from the German competence network for community acquired pneumonia. Competence Network for Community Acquired Pneumonia Study Group. Clin Micro Dis. 46, 1356-1364.

Waldron, P. R., Martin, B. A and Ho, D. Y. (2015). Mistaken identity: *Legionella micdadei* appearing as an acid fast bacillus on lung biopsy of a hematopoietic stem cell transplant patient. Transpl. Infect. Dis. 17, 89-93.

Wang, C., Mitsumasa, S., Tamami, T., Kazunobu, Amako and Shin-ichi, Y. (2015). Comparative analysis of virulence traits between a *Legionella feeleii* strain implicated in Pontiac fever and a strain that caused Legionnaires' disease. Microbial Pathogenesis. 89,79-86.

Whiley, H. and Bentham, R. (2011). *Legionella longbeachae* and Legionellosis. EID. 17, 579–83.

Whiley, H and Taylor, M. (2014). *Legionella* detection by culture and qPCR: comparing apples and oranges. Crit Rev Microbiol. doi: 10.3109/1040841X.2014.885930.

Widmer, A., Hohl, P., Dirnhofer, S., Bassetti, S., Marsch, S and Frei, R. 2007. *Legionella bozemanae*, an elusive agent of fatal cavitary pneumonia Infection. 35, 180–181. doi:10.1007/s15010-007-6251-4.

Wilkinson,H.W., Farshy, C. E., Fikes, B. J., Cruce, D. D and Yealy, L. P. (1979). Measure of immuno-globulin G, M and A specific titers against *Legionella pneumophila* and inhibition of titers against nonspecific gram negative bacterial antigens in the indirect immunofluorescence test for legionellosis. J. Clin. Microbiol. 10,685-689.

Wilkinson H.W., Cruce, D. D and Broome, C. V. (1981). Validation of *Legionella pneumophila* indirect immunofluorescence assay with epidemic sera. J Clin Microbiol. 13, 139-146.

Yu, V. L., Plouffe, J. F., Pastoris, M. C., Stout, J. E., Schousboe, M., Widmer, A *et al.*, (2002). Distribution of *Legionella* species and serogroups isolated by culture in patients with sporadic community-acquired legionellosis: an international collaborative survey. J Infect Dis. 186, 127–128.

Yu, H., Higa, F., Koide, D., Haranaga, S., Yara, S and Tateyama, M. (2009). Lung abscess caused by *Legionella* species: Implication of the immune status of hosts. Intern Med. 48, 1997-2002.

Chapter 8

Regulatory and Risk Management Strategies for Control of *Legionella*

***Susanne* Surman-Lee,[1]* and James T. Walker[2]**

[1]Leegionella Ltd., Rockford, Ringwood BH24 3NA, UK
[2]Walker on Water, UK

*susannelee@leegionella.co.uk

DOI: https://doi.org/10.21775/9781913652531.08

Abstract
Despite four decades passing since the discovery of *Legionella*, Legionnaires' disease is the deadliest waterborne disease in the United States, and according to the WHO, *Legionella* still has the highest burden of all water borne pathogens. This is despite years of improved understanding of the microorganism, engineering practices and control measures. A number of large outbreaks in The Netherlands, UK, and USA resulted in changes in legislation to enable the publication of regulatory documents for the control of Legionella in water systems. Countries where regulatory documents have been published often lack resources to monitor the implementation across wide ranges of water systems that could pose health risks. The WHO advised that Water Safety Groups (WSGs) be initiated, and this has been adopted into guidance in the UK for large healthcare establishments. The advantage of this approach is that it enables teams with the required expertise to undertake or engage

with other professionals to carry out risk assessments to identify the hazards to vulnerable patients.

Introduction

The worldwide incidence of Legionnaires' disease is rising (Parr et al., 2015; European Centre for Disease Prevention and Contro (ECDC), 2017b; Amemura-Maekawa et al., 2018; ECDC, 2019). Within Europe, Legionnaires' disease is notifiable in all 31 European Union/European Economic Area (EU/EEA) countries where roughly 70% of all reported cases are community acquired (ECDC, 2019). Due to the increase in reported cases over recent years the World Health Organisation (WHO) has identified *Legionella* as causing "a significant health burden in the EU and the highest health burden of all waterborne pathogens" (WHO, 2017) and is, therefore, advocating that monitoring of *Legionella* be included in the revision of the EU Drinking Water Directive at the point of use (POU) in priority buildings such as public buildings and healthcare premises. There may be many reasons for the increasing numbers of cases including improved surveillance, ascertainment, reporting and diagnostic tools, but the impact of shifting weather patterns as a result of climate change cannot be ruled out (Walker, 2018; De Giglio et al., 2019).

From a regulatory perspective the determination of drinking water safety is universally based on the absence of faecal indicators *Escherichia coli*, coliforms, enterococci and where applicable, *Clostridium perfringens* (Official Journal of the European Communities, 1998). Whilst faecal contamination within buildings can occur due to cross contamination between drinking water and waste water systems or as a result of inadvertent cross connections, backflow or damage during maintenance or alterations etc. is usually a transient problem and appropriate remedial work and disinfection will effectively reduce the risk. Faecal indicators, however, are not useful indicators for the presence of waterborne opportunistic pathogens (OPs) such as *Legionella*, *Pseudomonas aeruginosa*, non-tuberculous *Mycobacteria* (NTMs) and other emerging waterborne pathogens associated with causing illness in susceptible persons as a result of deteriorating water quality within building water systems (Wang et al., 2017).

Where conditions exist within water systems and associated equipment encourage colonisation and growth of OPs, safe management of the system becomes a continuing challenge for the rest of the life of the system. It is not necessary for the whole system to be colonised to pose a significant risk. Experience has shown that just small deadlegs and blind ends can harbour millions of colony forming units (cfu) of legionellae leading to intermittent detection when routine monitoring. The potential for biofilm sloughing off and colonising other areas of the system increases the possibility of causing cases. While *Pseudomonas aeruginosa* and other similar OPs however, are more likely to be found close to outlets (Flemming and Bendinger, 2014) and particularly where outlets with inserts are used which increases the surface area available for microbial colonisation (Walker and Moore, 2015). *Legionella,* however, can be found both at the outlet associated with biofilms and also growing systemically throughout the system (van der Lugt et al., 2017). There is evidence that outbreaks of Legionnaires' disease associated with building water distribution can result in outbreaks which may last for years (Haley et al., 1979; Kool, 1998; Lepine, 1998; Colville, 1993).

Is regulation the answer?

Historically notorious outbreaks of Legionnaires' disease been followed by the introduction of measures to prevent such outbreaks into the regulatory framework. For example, the first UK guidelines produced by the Health and Safety Executive (Guidance Note EH48 (HSE)) were introduced in 1987 after the Stafford Hospital Outbreak in 1985 (O'Mahony et al., 1990). In the Netherlands, changes in legislation where introduced after an outbreak caused by a spa pool on display which resulted in 23 deaths

(Sonder et al., 2008). A recent review of regulations worldwide carried out by Van Kenhove et al., (2019) identified that whilst targets varied within legislation for both monitoring parameters and safe levels of *Legionella* there was general agreement on the main principles to control *Legionella* growth such as avoidance of temperatures which encourage *Legionella* growth and avoiding stagnation.

However, for such regulations to work there must be capacity for enforcement which requires considerable resources not only in terms of manpower but also training, sampling, laboratory services and judicial processes. At a time when financial resources are scarce the cost of operating and managing water systems and water using equipment such as evaporative cooling systems, industrial systems, spa pools etc., together with the resources needed for inspection and enforcement can all too easily drop down the list of priorities.

Evidence for whether compliance with regulation a positively impacts water systems is difficult to determine. Whilst both governments and regulators need to ensure that regulation achieves the policy objectives, rarely do they complete the full process including developing policy and developing regulations to enforce the policy, followed by an evaluation stage to determine the effectiveness of the regulation on public health. In the UK it is the courts who have the final say in the interpretation of the law. However, the regulators provide information on what is needed to comply with the law in the forms of Approved Codes of Practice (Health Safety and Executive (HSE), 2013) and associated guidance (Health and Safety Executive (HSE), 2013a, b, c) in order for all stakeholders including, those designing and building water distribution systems and associated systems and equipment such as public health engineers and building architects, those who commission, manage and operate systems, owners and managers, those outsourcing services and appointing contractors all understand the regulatory expectations. In many cases, supplementary guidance with additional information on the practical aspects of compliance is given. This may come from government bodies for example the Department of Health for healthcare buildings or various professional organisations. Targeted inspection programmes can assess compliance where further intervention is required to protect public health but also to raise awareness as demonstrated by an inspection programme carried out by the UK regulatory body, the Health and Safety Executive (HSE), responsible for enforcing regulations for the safe management of Evaporative Cooling Systems (ECS). They found that 33% of the 625 sites inspected breached regulations, and in 12 sites the breaches were serious enough to prohibit continued use of the ECS until improvement were made. Major problems identified included "lack of training; failure to maintain the cleanliness of cooling towers and the water within them; risk assessments either being absent or not up to date, i.e., no longer representing the risks present; and Written Control Schemes being absent or insufficiently detailed" (Crook, 2020).

In some countries, where states, counties or similar regional divisions are able to set their own health and safety policies the complexity of regulation and enforcement is even more complex e.g. as in the USA and Australia. This can lead to confusion where, for example, companies and water management or treatment contractors, and specialists work across borders with differing regulatory requirements.

Chain of causation

For a system to pose a significant risk of Legionnaires' disease there must be a source which contains virulent *Legionella* and has the potential to be transmitted. Transmission is primarily by inhalation of viable legionellae within aerosols derived from water sprays. In Australasia, potting compost is also a significant source of legionellosis primarily caused by *L. longbeachae* and though relatively rare elsewhere, cases have been identified around the world (Irie et al., 1984; Duchin et al., 2000; Casati et al., 2009a; Lindsay et al., 2011; Currie and Beattie, 2015).

By the time aerosols are small enough to be inhaled deep enough into the lungs (less than 5µm) to cause infection, they are not visible and have a negative falling velocity so remain suspended in air for long periods of time with the rate of survival dependent on temperature, humidity and virulence (Table 1) (Dennis and Lee, 1988). In healthcare however, the risk of Legionnaires' disease caused by aspiration of water, including from sucking ice for patient comfort, is underestimated (Wright et al., 1989; Graman et al., 1997). There are many risk factors for patients in the healthcare setting including those with swallowing difficulties, neurological conditions such as motor neurone diseases, dementia, post-surgical patients particularly of the head and neck, and also patients taking narcotics all increase the risk of aspiration pneumonia (Johnson et al., 1985; Korvick and Yu, 1987; Carratala et al., 1994; Loeb et al., 1999; Sabria and Yu, 2002; Manabe et al., 2015; Feng et al., 2019).

Table 1. Equipment containing water likely to exceed 20°C and which may release aerosols.
Evaporative cooling towers and/or evaporative condensers
Hot and cold-water distribution systems
Natural thermal springs and their distribution systems
Spa pools and/or hot tubs
Humidifiers
Spray irrigation systems
Fountains and water features
Any plant or system which uses water from a non-potable source such as river water for evaporative cooling systems
Industrial plant such as paint spay booths and quench tanks
Clinical equipment including nebulizers and incubators

There must be both a source which contains sufficient *Legionella* and a mode of dissemination for a system to pose a risk of cause infection. When considering the range of systems and or equipment which may pose a risk of Legionnaires' disease, the European Technical Guidelines for the Prevention, Control and Investigation, of Infections Caused by *Legionella* species (ECDC, 2017a) explains what one should consider when assessing a range of systems. As the large obvious sources such as evaporative cooling towers, hot and cold-water systems, and spa pools generally will be recognized by risk assessors, other systems especially those previously unrecognized or novel may be overlooked. Risk assessors need to consider all places containing water which could produce aerosols or be used by someone for drinking, making ice, etc. Vessels containing water should not be overlooked if there is a potential to cause aerosols. For example, when emptying a container a sink, the emptying process will produce aerosols. As an example during an outbreak investigation, a sump from a steam cleaner in a hospital which had been left full in a cupboard contained >10^5 cfu/L of the outbreak strain.

It is important that risk assessments are carried out at the various stages in the life of a water system (Table 2) (Anonymous, 2010). As there are many types of water systems and associated equipment within buildings, the risk assessor requires a methodical approach to determine the types and complexity of all water systems and the numbers and types of additional equipment which may pose a risk. Clearly less complex systems are simpler to manage and therefore likely require fewer controls to be put in place (Pankhurst, 2003; Phin et al., 2014). However, for large buildings quantitative risk assessments are required in addition to a multiple skills matrix that may involve a number of risk assessors, including assessors with specialist knowledge of specific or novel systems with input from the Water Safety Group (WSG) (Pancer, 2013; Pancer et al., 2013; WHO, 2011).

Table 2. Stages in the life of a water system when risk assessments should be carried out.
During each stage of the design and construction including the specification of fittings, components and materials
Prior to and throughout the commissioning, decommissioning and recommissioning processes
After refurbishment
Throughout occupation to normal operation
Following adverse monitoring results
Following any significant change which could affect the validity of the risk assessment including changes in key staff, system usage etc.

Risk assessors need to have experience of the type of water system and equipment to be assessed and to be competent with the skills, knowledge, and resources to undertake the task competently. They must have the relevant training and experience and have knowledge of the water quality required for each type of system, legislative requirements, standards, Water Safety Plans (WSPs) as well as the types of hazards which maybe of a microbiological, chemical and physical nature, applicable to each system to be assessed. The assessor also requires the ability to identify all hazards and hazardous events taking account of the likelihood and severity of harm in the context of exposure (type, extent and frequency), the vulnerability of those exposed together with the measures needed to effectively control risks.

The risk assessment must identify and prioritize risks to develop an effective scheme of control to be implemented for each system. As a consequence the risk assessments need to consider how the water system is operated, if and where the system will allow ingress and microbial proliferation and how water borne microorganisms could be disseminated to those who may succumb to disease, particularly those person who may be more vulnerable such as immunocompromised patients (Li et al., 2015).

A risk assessment involves a practical survey of all the water systems on a site and should consider all aspects of operation of the hot and cold-water systems as well as other equipment on site that involves water. Employees, site personnel, specialist users, and maintenance engineers who operate, maintain and manage the systems should be consulted to determine current operational practices and the effectiveness of control regimes. The commissioning, decommissioning, validation data, periods of operation, maintenance, treatment, monitoring, and subsequent management of each individual aspect of operation will require review and validation to ensure site procedures are effective.

The common requirements when assessing the risk of legionellosis associated with a hot and cold water system are highlighted in Table 3. The assessment should include recommendations for remedial actions for controlling *Legionella* where necessary and for identifying who will be responsible for and undertake the actions. Actions should be prioritised, and a review date set for determining completion of these tasks.

Risk management and risk assessment

While the application of WPSs, as advocated by WHO (WHO, 2004), has resulted in improved public water supplies at the point of entry into buildings, the potential for water quality to deteriorate within the internal water distribution systems and associated systems and equipment, water at the POU being a potential risk to human health particularly for those who are immunocompromised. The WHO's *Water Safety in Buildings*

Table 3. Elements of hot and cold-water systems to be investigated when considering contamination by *Legionella* spp.
Quality of supply water
Examination of hot and cold-water tanks for configuration, flow pattern, contamination, materials, condition, temperature, appropriate size and lids
Areas where there is low or no flow, blind ends, dead legs and little used outlets
Heat gain in heat transfer between hot and cold-water system where heat gain or heat loss could result in temperatures supporting the growth of *Legionella* (Only a small area of a system resulting in temperatures between ≥20 °C and ≤50 °C pose a risk to users of the system. For example, where a cold pipe runs close to a pump, hot water or heating pipe)
Assessment of the potential for the water system to become contaminated with legionella and other material such as low flow or stagnation points in the cold and hot water system
Details of any water pre-treatment processes
Assessment of potential for *Legionella* to grow within system and effectiveness of control measures
Chemical and physical water treatment measures
Disinfection and cleaning regimes
Remedial work and maintenance
Review of the *Legionella* control scheme, including management procedures and site records or logbooks with system maintenance records
Routine data monitoring
Water treatment and service reports, cleaning and disinfection records, *Legionella* and other microbial analysis results e.g. temperature, pH, biocide levels, microbial parameters
Confirmation that consideration was given to preventing risk by elimination or substitution before implementing appropriate control measures
Timely and appropriate remedial action to poor temperature or monitoring results as an indicator of the effectiveness and adequacy of the management controls in place

(Cunliffe et al., 2011), has led to WSPs being used widely to minimise risks from water systems from supply (Ashbolt, 2015) to point of use (WHO, 2011; Anonymous, 2013, 2014a, b). The WSP is a proactive approach to identify waterborne hazards and control their dissemination. The appointment of a multidisciplinary group, the WSG, with the necessary skills to manage all risk from water use is a key component of a WSP. Risk assessments will inform and enable the WSG members to develop a management scheme to control the hazards and mitigate risks by implementing effective remedial measures. Importantly, the WSP should be able to distinguish between high and low risks so that priority actions can be focused on mitigating risks that are more likely to cause harm and should also include supporting programmes such as appropriate documentation and record-keeping, training and communication.

Control strategies

One strategy to reduce the risk of system components providing nutrients to support *Legionella* colonization and growth is to ensure that there is a system for approval e.g. in the UK the (Water Regulations Advisory Scheme (WRAS), 2011). Even an appropriately installed and commissioned water system can experience problems if the design has not considered the type of use and operational aspects of the building. For example, there is a propensity to oversize the water tanks and install hand wash stations in excess of requirements or non-sited to give ease of access in a hospital that could lead to underuse of specific outlets. Currently, patients are in hospital for less time and use less water than previously. Water is a valuable and costly resource but volumes used can exceed 500L per patient per bed day (DETR, 1999). Strategies to reduce water usage and reduce the carbon footprint can result not only in efficiencies but may also have untoward consequences, as the temperature and reduced flow leads to conditions supporting microbial growth.

Management of water systems

Water systems are inherently complex and for microbiological risks to be minimized, the system must be understood, managed and monitored. However, these parameters are rarely undertaken effectively due to a lack of resources. The most significant factors identified following outbreaks of Legionnaires' disease are generally down to the human factor with inappropriate maintenance and control procedures and ineffective communication and systems management. For any control strategy to be effective electronic monitoring systems including temperature, flow, residual disinfectants, microbiology results, asset identification, location and clinical cases should be implemented.

Investigation of outbreaks has shown that they are generally related to poor training and communication, flaws in system design, poor commissioning and risk assessments, defects, and breakdowns. The more information that can be gathered, the more likely that the inherent control strategies will maintain the microbiological content of the water system below a threshold to minimize the risk.

Temperature control

Temperature is the traditional strategy for the control of *Legionella* in water systems and will require monitoring on a frequent basis (Blanc et al., 2005). In the UK, recent guidance has indicated that the temperature of hot water out of the calorifier should be a minimum of 60°C, a minimum of 55°C on flow and returns to all outlets and at the start of the hot water return, with a minimum of 50°C at the final connection to the calorifier to reduce scalding (Anonymous, 2014b). However, as a consequence, thermostatic thermal valves (TMV) have been installed at the majority of outlets in healthcare premises without a scalding risk assessment being carried out (Anonymous, 2016b).

From the location of the TMV the hot water will be reduced from 55°C to ~43°C which is comfortable temperature for hand washing but will enable the proliferation of *Legionella* and other waterborne pathogens and is an example of the unintended consequences of the changes that we make to our water systems (Anonymous, 2014a). Therefore, the use of TMV's should be risk assessed to improve the control strategy (Anonymous, 2016b). However, where a scalding risk is identified which could pose a risk of severe harm e.g where there is whole body immersion, using TMVs on baths and showers for example, should be considered to reduce temperature and the TMV should be placed as close to the POU as possible (Anonymous, 2016c).

Physical methods of control include keeping the water moving, regular flushing and water filtration. Where a recognized risk of waterborne pathogen presence has been identified then point of use filters (bacterial retention filters) are the most commonly recognized way of implementing a rapid control measure until engineering works have been carried out to remediate the problem in the water system (Anonymous, 2016b; Marchesi et al., 2011; Sheffer et al., 2005; Zhou et al., 2013). However, as with any control strategy the misuse of filters, e.g. patients removing the filters to achiever a higher flow rate, may result in the exposure of patients, visitors and staff to *Legionella* (Florentin et al., 2016).

Chemical controls

Biocides themselves are not a panacea, and like temperature control, will not reduce the microbiological risk in a poorly designed, operated and managed water system. Neither will the use of biocides remove the need to ensure that monitoring of the important aspects of the water system is carried out. Different biocides have different requirements, and it is important that before any chemicals are applied that the water system is understood including the type of water to ensure that the appropriate chemical concentrations and contact times are achieved. The effectiveness of disinfectants are determined by the concentration, contact time, pH value, temperature, concentration of organic matter, and the number and types of microorganisms in the water. Where there are areas of low or no flow such as in deadlegs, blind ends, and areas out of use, it does not matter how much biocide is dosed into the system. The risk will not be mitigated unless the biocide reaches all areas. In addition, members of the WSG should understand the use of chemicals in their water system and the impact that this may have on particular patients.

There are a number of chemical treatments systems, typically oxidizing biocides, that have been used for the treatment of hot and cold water systems including chlorine dioxide, monochloramine, , silver stabilized hydrogen peroxide, copper and silver ionization, chlorine, ozone, and UV treatment (Anonymous, 2014b) sometimes in combination (Biurrun et al., 1999; Casari et al., 2007; Pianetti et al., 2008). In healthcare premises the impact on patients in other parts of the hospital needs to be taken in to account to ensure the wellbeing of vulnerable patient such as in renal dialysis units where these chemicals would be toxic to patients (Anonymous, 2016b; Hoenich, 2009).

Chlorine dioxide

Chlorine dioxide is an oxidising disinfectant that is soluble in water and has been effective at controlling both *Legionella* and biofilm growth in hot and cold water systems (Walker et al., 1995; Tesauro et al., 2010). However, reports of its ineffectiveness have also been published (Hosein et al., 2005; Casini et al., 2014). Chlorine dioxide is usually produced on site from a chlorite-based precursor using a chlorine dioxide generator or dosing system by reaction with one or more other chemical precursors or by a catalytic oxidation process. In the UK, use of chlorine dioxide to control *Legionella* is subject to BS EN 12671 (Anonymous, 2016a) and national conditions of use require that the combined concentration of chlorine dioxide, chlorite and chlorate in the drinking water does not exceed 0.5 mg/l as chlorine dioxide. Concentrations (as total oxidant) of 0.1 to 0.5 mg/l at an outlet is usually sufficient, and occasionally higher concentrations may be required to control *Legionella* although alternative drinking water supplies may be needed. The dosage rate and time to reach stable concentrations will depends on the length and complexity of the water distribution system, the water turnover rate, and the extent to which the water system is contaminated with an established biofilm (Loret et al., 2005). Chlorine dioxide is not affected by the pH value or hardness of the water, but it is sometimes difficult to monitor chlorine dioxide samples in domestic hot water systems due to its increased volatility causing the chlorine dioxide reserve to be lost when taking a water sample. As with any chemical where there are infrequently used outlets, a programme of regularly flushing the outlets should be maintained until the chemical residual is detected. When shock dosing is required then offline elevated concentrations of chlorine dioxide (20- 50 mg/l) may also be required, but only following a detailed risk assessment and the system should be flushed thoroughly afterwards to prevent corrosion of the copper or steel pipework. For a combined hot and cold water system, a chlorine dioxide dosing system should be installed on the inlet to the cold water tank supplying water to the system. In the case of hot water distribution systems with calorifiers or water heaters operating at 60 °C, there will be a tendency for chlorine dioxide to be lost by gassing off.

Copper and silver ionisation

Ionisation is the term given to the electrolytic generation of copper and silver ions in water and is generated by passing a low electrical current between two copper and silver or silver allow electrodes. Like many systems, copper and silver ionisation has been shown effective at controlling *Legionella,* particularly in hot water systems where it can penetrate and control established biofilms but fails if the water system is not managed appropriately (Blanc et al., 2005). Controls and regular water testing need to be maintained to ensure that the copper level do not exceed 2.0 mg/l as Cu^{2+} and the silver level does not exceed 0.1 mg/l as Ag^{+}. Concentrations of 0.2 mg/l copper and more than 0.02 mg/l silver are recommended at outlets. It works best with soft water, as in hard water areas the build-up of scale on the electrodes reduces efficiency. The ionisation process is pH-sensitive and dosing levels may need increasing for pH levels greater than 7.6.

Chlorine

Most mains water supplies will contain a low level of chlorine residual (0.1-0.5 mg/l) at the point of supply to a building, but supplementary dosing with the controlled addition of a further chlorine-based product may be required for the control of legionella and biofilm (Biurrun et al., 1999; Muraca et al., 1987). The WHO has set a maximum exposure value of 5.0 mg/l for total chlorine as a residual disinfectant in drinking water. However, it is rarely used continuously in domestic water in buildings at concentrations higher than 1.0 mg/l. Such concentrations would result in unpalatable water and possibly corrosion (Pianetti et al., 2008). When diluted in water chlorine dissociates it forms hypochlorous acid and hypochlorite ions.

Silver stabilised hydrogen peroxide

Silver stabilised hydrogen peroxide has a history of use in the control of *Legionella* in water systems (Marchesi et al., 2016). A silver hydrogen peroxide solution is injected directly into the water system. If applied and maintained according to the manufacturers' instructions, it can be an effective means of control. As with any water treatment programme, it should be validated to ensure that it is effective in controlling *Legionella*. Silver hydrogen peroxide should not be used in water systems supplying dialysis units.

Monochloramine

Water treatment with monochloramine has previously been used for the treatment of supply water and although an uncommon strategy for supplemental disinfection of hospital water systems, it has been used (Kool et al., 1999; Marchesi et al., 2013; Casini et al., 2014; Mancini et al., 2015). Monochloramine is formed when ammonia and free chlorine are mixed in water. Monochloramine's disinfecting action is slower than that of free chlorine, so it is less useful for initial disinfection. On the other hand, it is more stable than free chlorine. The oxidising capacity of monochloramine relatively lower than that of other disinfectants, and an effective residual can be maintained over long distances and for longer time periods which can reduce both cost and systemic effectiveness (Kirmeyer GJ et al., 1993). Studies have evaluated the use of monochloramine in hospital hot water systems (Casini et al., 2014; Mancini et al., 2015) and found that monochloramine significantly reduced levels of *L. pneumophila.* Though, one study found no effect on *P. aeruginosa* (Casini et al., 2014).

Rare, unusual, novel, and special settings to be considered

An increasing number of Legionnaires' disease cases are reported every year, and the majority are still associated with traditional water systems in a range of buildings and hotel complexes (ECDC, 2016). Yet, there are a wide range of other water sources that have been identified as sources of the infection. Some of these sources are unusual, particularly in healthcare settings including bubblers for oxygen-therapy, paediatric incubators, dental chairs, continuous positive airway pressure machines (CPAP) and compost in a hospital plant potting compost (Delia and Lagana, 2002; Dhillon et al., 2009; Schnirman, 2017). Ultimately, *Legionella* may be a fastidious organism in the laboratory where it will only grow on specialised medium, but can be found in a wide range of weird and wonderful environments in our everyday world. Travel associated Legionnaires' disease cases represent a significant cause of travel associated respiratory tract infections and impacts disproportionately on otherwise healthy individuals as a consequence of their travel abroad or within their own country (Decludt et al., 2004; Guyard and Low, 2011). However, it is still surprising that more than 40 years following the discovery of the bacterium that the majority of sources are not yet understood (Knox et al., 2017). This presents a dilemma for healthcare professionals who are trying to protect the public, as health officials are not able to identify and control the source of infection. TALD are frequently reported in travellers who stayed in accommodations sites such as hotels or cruise ships, and as a consequence, many of those sources are difficult to identify (Castellani Pastoris et al., 1999; Regan et al., 2003).

Travelling may be a risk factor when it comes to exposure to *Legionella* either when we are commuting or when we may be working. So whilst we tend to think of outbreaks on ships as being associated with cruise ships (Castellani Pastoris et al., 1999; Regan et al., 2003) there are a number of cases and studies of non-passenger vessels such as naval ships, merchant ships and ferries (Azara et al., 2006; Ahlen et al., 2013). In some instances the growth of *Legionella* may due to particular risk systems such as the ship cabin shower or hospital shower having been identified as being positive for *Legionella* (Collins et al., 2016), or the outbreak may have been related to fresh water taken on board at bunkering stations (Ahlen et al., 2013). Even working on water systems on cargo ships and cleaning hulls with water blasters has resulted in fatalities due to Legionnaires' disease (Cayla et al., 2001).

Risk of vehicles

Travelling in vehicles has not always been recognised as a risk. However, there is one report of a holiday maker having travelled in a camper van for three months who was infected with *Legionella* that matched the water samples from the taps and showers (Euser et al., 2016). Whilst no cases were identified, a survey of recreational vehicles from three different campsites in the USA showed 20% were positive for *Legionella* spp. from water outlets (24%) and water tanks (9%) (Litwin et al., 2013). In 2010 Wallenstein et al., reported that professional drivers were five times more commonly represented among community acquired sporadic cases in England and Wales and that 20% of community acquired sporadic cases could be attributed to such activities (Wallensten et al., 2010). The plausible reasons for this increased exposure were thought to be either driving through industrial areas and driving in vehicles with windscreen wiper fluid not containing added screen-wash that would otherwise have controlled the presence of the *Legionella* spp. (Wallensten et al., 2010; Palmer et al., 2012). Schwake et al., demonstrated survival of *Legionella* spp. in certain windshield washer fluids, as well as growth within school bus washer fluid reservoirs, and the presence of viable cells in bus washer fluid aerosols (Schwake et al., 2015). Car drivers themselves may also be at risk from the air conditioning system, as a study in Scandinavia found that 30% of air filters from vehicles were positive for *L. pneumophila*, and in Japan, 50% of swabs were taken from the aluminium heat-exchanger fins of evaporators were positive, though no cases have been identified in either publication (Sakamoto et al., 2009; Alexandropoulou et al., 2013). A study in Turkey indicated that 19% of long distance bus drivers were seropositive for *Legionella* antibodies indicating that this occupation may be a risk factor for legionellosis (Polat et al., 2007).

Industrial risk

Whilst industrial cooling systems tend to be linked to Legionnaires' disease cases (Hugosson et al., 2007), there are a few other industrial circumstances in which outbreaks have occurred. Visiting and working in the cleaning line of automobile engine manufacturing plant was associated with four cases of Legionnaires' disease (Fry et al., 2003). In 2009, an outbreak of 11 Legionnaires' disease cases was declared in Spain and using spatial-temporal and sequence-based typing analyses indicated that a milling machine used in street repaving as a source of the microorganisms (Coscolla et al., 2010). Unusually, the water tanks of the machine had been filled from a warm natural spring. Aerosols were created by atomisers during the milling process and those nearby were exposed. Both the water tanks and atomisers were positive for *L. pneumophila* serogroup 1. As these milling machines are mobile and move frequently it is surprising that they were identified as a source at all. Waste water treatment plants have also been implicated in causing outbreaks of Legionnaires' disease (Caicedo et al., 2019). Paper mills, which use large quantities of water and have large quantities of organic matter have also been implicated in cases of Legionnaires' disease in paper mill workers, one of which was installing a pump (Kusnetsov et al., 2010). Working on pumps also appears to have been linked to Legionnaires' disease risk as indicated by a case associated with exposure to a pressure test pump in Holland in 2012 (Euser et al., 2014).

Risk of rainwater and water storage

Whilst we are able to put control measures in place for industrial systems, it is more difficult to do for natural water sources. One of the areas that remains overlooked is that of weather and rainfall. Studies have demonstrated that temperature, rainfall and atmospheric pressure are all associated with Legionnaires' disease risk and higher risk associated with simultaneous increase in temperature and rainfall (Beaute et al., 2016). Indeed, rainfall is a significant risk factor for sporadic Legionnaires' disease (Garcia-Vidal et al., 2013). Climate change and adverse weather incidence is something that we have to consider for the future, and it has been demonstrated that flooding is a risk for Legionnaires' disease (Schalk et al., 2012). Water is obviously the common theme for Legionnaires' disease, and in many places water is a valuable product often collected for reuse as grey water. However, collected and reused water can harbour a wide range of microorganisms including *Legionella* spp. and may pose a risk in terms of aerosol transmission (Simmons et al., 2008; Dobrowsky et al., 2014; Bae et al., 2019). Just like any water storage system, grey water harvesting has to be designed to minimise risk (Ward et al., 2010).

Another environment over which we have little control is that of soil and the risk from *Legionella longbeachae*. Whilst *L. longbeachae* is associated with the use of potting soils and responsible for approximately 50% of Legionnaires' disease cases in Australasia (Steele et al., 1990), it was not initially associated with potting composts in Greece, Switzerland, and the United Kingdom examined between March 1989 and May 1990. However, since 1990 outbreaks of *L. longbeachae* have now been associated with other countries including Japan, USA, Switzerland, and Scotland which may indicate improvements in detection technologies (Centers for Disease and Prevention, 2000; Koide et al., 2001; Kubota et al., 2007; Casati et al., 2009b; Pravinkumar et al., 2010; Currie et al., 2013).

Person-to-person risk

While person to person has been a considered route, until recently there had not been sufficient evidence to prove such a link. However, following a large community outbreak of Legionnaires' disease in Portugal, a male patient was exposed to *Legionella* from a cooling tower relocated 300km to his parents' home where his mother was subsequently diagnosed with Legionnaires' disease from the same type of *Legionella* as her son (Correia et al., 2016). Whilst this is an isolated report, a recent study suggested that humans may themselves be vectors where transmission, at least partially via humans, leads to adaptation in man-made water systems selecting for strains with the greatest fit for human infection. Using WGS analysis, the authors used this hypothesis to indicate a global transmission route (David et al., 2016).

References

(WHO), W.H.O. (2017). Drinking Water Parameter Cooperation Project :Support to the revision of Annex I Council Directive 98/83/EC on the Quality of Water Intended for Human Consumption(Drinking Water Directive).

Ahlen, C., Aas, M., Nor, A., Wetteland, P.I., Johansen, H., Sorbo, T., Sommerfelt-Pettersen, J.K., and Iversen, O.J. (2013). Legionella pneumophila in Norwegian naval vessels. Tidsskr Nor Laegeforen *133*, 1445-1448.

Alexandropoulou, I.G., Konstantinidis, T.G., Parasidis, T.A., Nikolaidis, C., Panopoulou, M., and Constantinidis, T.C. (2013). First report of Legionella pneumophila in car cabin air filters. Are these a potential exposure pathway for professional drivers? Scand J Infect Dis *45*, 948-952.

Amemura-Maekawa, J., Kura, F., Chida, K., Ohya, H., Kanatani, J.I., Isobe, J., Tanaka, S., Nakajima, H., Hiratsuka, T., Yoshino, S., *et al.* (2018). Legionella pneumophila and Other Legionella Species Isolated from Legionellosis Patients in Japan between 2008 and 2016. Appl Environ Microbiol *84*.

Anonymous (2010). BS 8580 Water quality. Risk assessments for Legionella control. Code of Practice7: http://shop.bsigroup.com/ProductDetail/?pid=000000000030200235 Accessed March 2017.

Anonymous (2013). Legionnaires' disease. Part 3. The control of legionella bacteria in other risk systems. http://www.hse.gov.uk/pubns/books/hsg274.htm.

Anonymous (2014a). Legionnaires' disease. Part 1. The control of legionella bacteria in evaporative cooling systems. http://www.hse.gov.uk/pubns/priced/hsg274part1.pdf.

Anonymous (2014b). Legionnaires' disease. Part 2. The control of legionella bacteria in hot and cold water systems http://www.hse.gov.uk/pUbns/priced/hsg274part2.pdf.

Anonymous (2016a). BS EN 12671 Chemicals used for treatment of water intended for human consumption. Chlorine dioxide generated in situ British Standards Institution. http://shop.bsigroup.com/ProductDetail/?pid=000000000030300940 Accessed March 2017.

Anonymous (2016b). Health Technical Memorandum 04-01: Safe water in healthcare premises. Department of Health. https://www.gov.uk/government/publications/hot-and-cold-water-supply-storage-and-distribution-systems-for-healthcare-premises [accessed 23.02.17].

Anonymous (2016c). Health Technical Memorandum 04-01: Supplement. Performance specification D 08: thermostatic mixing valves (healthcare premises). https://www.gov.uk/government/publications/hot-and-cold-water-supply-storage-and-distributionsystems-for-healthcare-premises.

Ashbolt, N.J. (2015). Environmental (Saprozoic) Pathogens of Engineered Water Systems: Understanding Their Ecology for Risk Assessment and Management. Pathogens *4*, 390-405.

Azara, A., Piana, A., Sotgiu, G., Dettori, M., Deriu, M.G., Masia, M.D., Are, B.M., and Muresu, E. (2006). Prevalence study of Legionella spp. contamination in ferries and cruise ships. BMC Public Health *6*, 100.

Bae, S., Maestre, J.P., Kinney, K.A., and Kirisits, M.J. (2019). An examination of the microbial community and occurrence of potential human pathogens in rainwater harvested from different roofing materials. Water Res *159*, 406-413.

Beaute, J., Sandin, S., Uldum, S.A., Rota, M.C., Brandsema, P., Giesecke, J., and Sparen, P. (2016). Short-term effects of atmospheric pressure, temperature, and rainfall on notification rate of community-acquired Legionnaires' disease in four European countries. Epidemiol Infect, 1-11.

Biurrun, A., Caballero, L., Pelaz, C., Leon, E., and Gago, A. (1999). Treatment of a Legionella pneumophila-colonized water distribution system using copper-silver ionization and continuous chlorination. Infect Control Hosp Epidemiol *20*, 426-428.

Blanc, D.S., Carrara, P., Zanetti, G., and Francioli, P. (2005). Water disinfection with ozone, copper and silver ions, and temperature increase to control Legionella: seven years of experience in a university teaching hospital. J Hosp Infect *60*, 69-72.

Caicedo, C., Rosenwinkel, K.H., Exner, M., Verstraete, W., Suchenwirth, R., Hartemann, P., and Nogueira, R. (2019). *Legionella* occurrence in municipal and industrial wastewater treatment plants and risks of reclaimed wastewater reuse: Review. Water Res *149*, 21-34.

Carratala, J., Gudiol, F., Pallares, R., Dorca, J., Verdaguer, R., Ariza, J., and Manresa, F. (1994). Risk factors for nosocomial Legionella pneumophila pneumonia. AmJRespirCrit Care Med *149*, 625-629.

Casari, E., Ferrario, A., and Montanelli, A. (2007). Prolonged effect of two combined methods for Legionella disinfection in a hospital water system. Ann Ig *19*, 525-532.

Casati, S., Gioria-Martinoni, A., and Gaia, V. (2009a). Commercial potting soils as an alternative infection source of Legionella pneumophila and other Legionella species in Switzerland. Clin Microbiol Infect *15*, 571-575.

Casati, S., Gioria-Martinoni, A., and Gaia, V. (2009b). Commercial potting soils as an alternative infection source of Legionella pneumophila and other Legionella species in Switzerland. Clin Microbiol Infect *15*, 571-575.

Casini, B., Buzzigoli, A., Cristina, M.L., Spagnolo, A.M., Del Giudice, P., Brusaferro, S., Poscia, A., Moscato, U., Valentini, P., Baggiani, A., *et al.* (2014). Long-term effects of hospital water network disinfection on Legionella and other waterborne bacteria in an Italian university hospital. Infect Control Hosp Epidemiol *35*, 293-299.

Castellani Pastoris, M., Lo Monaco, R., Goldoni, P., Mentore, B., Balestra, G., Ciceroni, L., and Visca, P. (1999). Legionnaires' disease on a cruise ship linked to the water supply system: clinical and public health implications. Clin Infect Dis *28*, 33-38.

Cayla, J.A., Maldonado, R., Gonzalez, J., Pellicer, T., Ferrer, D., Pelaz, C., Gracia, J., Baladron, B., Plasencia, A., and Legionellosis study, g. (2001). A small outbreak of Legionnaires' disease in a cargo ship under repair. Eur Respir J *17*, 1322-1327.

Centers for Disease, C., and Prevention (2000). Legionnaires' Disease associated with potting soil--California, Oregon, and Washington, May-June 2000. MMWR Morb Mortal Wkly Rep *49*, 777-778.

Collins, S.L., Stevenson, D., Mentasti, M., Shaw, A., Johnson, A., Crossley, L., and Willis, C. (2016). High prevalence of Legionella in non-passenger merchant vessels. Epidemiol Infect, 1-9.

Colville, A.C., J.; Dearden, D.; Slack, R.C.B.; Lee, J.V. (1993). Outbreak of Legionnaires' disease at University Hospital, Nottingham: epidemiology,microbiology and control. Epidemiol Infect *110*, :105-116.

Correia, A.M., Ferreira, J.S., Borges, V., Nunes, A., Gomes, B., Capucho, R., Goncalves, J., Antunes, D.M., Almeida, S., Mendes, A., *et al.* (2016). Probable Person-to-Person Transmission of Legionnaires' Disease. N Engl J Med *374*, 497-498.

Coscolla, M., Fenollar, J., Escribano, I., and Gonzalez-Candelas, F. (2010). Legionellosis outbreak associated with asphalt paving machine, Spain, 2009. Emerg Infect Dis *16*, 1381-1387.

Crook, B., Willerton, L., Smith, D., Wilson,L., Poran,V., Helps, J., Mcermott, P., (2020). Legionella risk in evaporative cooling systems and underlying causes of associated breaches in health and safety compliance

nternational Journal of Hygiene and Environmental Health *224*.

Cunliffe, D., Bartram, J., Briand, E., Chartier, Y., Colbourne, J., Drury, D., Lee, J.V., Schaefer, B., and Surman-Lee, S., eds. (2011). Water Safety In Buildings.

Currie, S.L., and Beattie, T.K. (2015). Compost and Legionella longbeachae: an emerging infection? Perspect Public Health *135*, 309-315.

Currie, S.L., Beattie, T.K., Knapp, C.W., and Lindsay, D.S. (2013). Legionella spp. in UK composts-a potential public health issue? Clin Microbiol Infect.

David, S., Rusniok, C., Mentasti, M., Gomez-Valero, L., Harris, S.R., Lechat, P., Lees, J., Ginevra, C., Glaser, P., Ma, L., *et al.* (2016). Multiple major disease-associated clones of Legionella pneumophila have emerged recently and independently. Genome Res *26*, 1555-1564.

De Giglio, O., Fasano, F., Diella, G., Lopuzzo, M., Napoli, C., Apollonio, F., Brigida, S., Calia, C., Campanale, C., Marzella, A., *et al.* (2019). Legionella and legionellosis in touristic-recreational facilities: Influence of climate factors and geostatistical analysis in Southern Italy (2001-2017). Environ Res *178*, 108721.

Decludt, B., Campese, C., Lacoste, M., Che, D., Jarraud, S., and Etienne, J. (2004). Clusters of travel associated legionnaires' disease in France, September 2001- August 2003. Euro Surveill *9*, 12-13.

Delia, S., and Lagana, P. (2002). [Unusual sources of L. pneumophila in hospital environment]. Ann Ig *14*, 443-446.

Dennis, P.J., and Lee, J.V. (1988). Differences in aerosol survival between pathogenic and non-pathogenic strains of Legionella pneumophila serogroup 1. J Appl Bacteriol *65*, 135-141.

DETR (1999). Energy consumption in hospitals http://www.cibse.org/getmedia/a9ab0fc1-97ed-4048-b6b5-936116334bc4/ECG72-Energy-Consumption-in-Hospitals-1999.pdf.aspx.

Dhillon, R., Bastiampillai, T., and Hong, S. (2009). An unusual case of hospital-acquired infection: Legionella Longbeachae. Australas Psychiatry *17*, 337-338.

Dobrowsky, P.H., De Kwaadsteniet, M., Cloete, T.E., and Khan, W. (2014). Distribution of indigenous bacterial pathogens and potential pathogens associated with roof-harvested rainwater. Appl Environ Microbiol *80*, 2307-2316.

Duchin, J.S., Koehler, J., Kobayashi, J.M., Rakita, R.M., Olson, K., Hampson, N.B., Gilbert, D.N., Jackson, J.M., Stefonek, K.R., Kohn, M.A., *et al.* (2000). Legionnaires' disease associated with potting soil - California, Oregon, and Washington, May-June 2000. Morbidity and Mortality Weekly Report *49*, 777-778.

ECDC (2016). Annual epidemiological report Legionnaires' disease http://ecdc.europa.eu/en/healthtopics/legionnaires_disease/surveillance/

Pages/annual-epidemiological-report-2016.aspx#sthash.B7S3Hw20.dpuf Accessed March 2017.

European Centre for Disease Prevention and Control (ECDC) (2017a). European technical guidelines for the prevention, control and investigation, of infections caused by *Legionella* species, pp. 125.

European Centre for Disease Prevention and Control (ECDC) (2017b). Surveillance report. Annual epidemiological report for 2015 legionnaires' disease (Stockholm), pp. 7.

European Centre for Disease Prevention and Control (ECDC) (2019). Legionnaires' disease. Annual epidemiological report for 2017 (Stockholm, Sweden: ECDC), pp. 6.

Euser, S.M., Boogmans, B., Brandsema, P., Wouters, M., and Den Boer, J.W. (2014). Legionnaires' disease after using an industrial pressure test pump: a case report. J Med Case Rep *8*, 31.

Euser, S.M., Diederen, B.M., Bakker, M., Honing, M.L., Bruin, J.P., Brandsema, P.S., Reijnen, L., and Den Boer, J.W. (2016). Legionnaires' disease after a campervan holiday: a case report. J Travel Med *23*.

Feng, M.C., Lin, Y.C., Chang, Y.H., Chen, C.H., Chiang, H.C., Huang, L.C., Yang, Y.H., and Hung, C.H. (2019). The Mortality and the Risk of Aspiration Pneumonia Related with Dysphagia in Stroke Patients. J Stroke Cerebrovasc Dis *28*, 1381-1387.

Flemming, H.C., and Bendinger, B. (2014). The last meters before the tap: where drinking water quality is at risk (chapter 8). In Microbial growth in drinking-water distribution systems Problems, causes, prevention and research needs, D. van der Kooij, and P.W. van der Wielen, eds. (London, UK: IWA Publishing), pp. 207-238.

Florentin, A., Lizon, J., Asensio, E., Forin, J., and Rivier, A. (2016). Water and surface microbiologic quality of point-of-use water filters: A comparative study. Am J Infect Control.

Fry, A.M., Rutman, M., Allan, T., Scaife, H., Salehi, E., Benson, R., Fields, B., Nowicki, S., Parrish, M.K., Carpenter, J., *et al.* (2003). Legionnaires' disease outbreak in an automobile engine manufacturing plant. J Infect Dis *187*, 1015-1018.

Garcia-Vidal, C., Labori, M., Viasus, D., Simonetti, A., Garcia-Somoza, D., Dorca, J., Gudiol, F., and Carratala, J. (2013). Rainfall is a risk factor for sporadic cases of Legionella pneumophila pneumonia. PLoS One *8*, e61036.

Graman, P.S., Quinlan, G.A., and Rank, J.A. (1997). Nosocomial legionellosis traced to a contaminated ice machine. InfectControl HospEpidemiol *18*, 637-640.

Guyard, C., and Low, D.E. (2011). Legionella infections and travel associated legionellosis. Travel Med Infect Dis *9*, 176-186.

Haley, C.E., Cohen, M.L., Halter, J., and Meyer, R.D. (1979). Nosocomial Legionnaires' disease: a continuing common-source epidemic at Wadsworth Medical Center. Ann Intern Med *90*, 583-586.

Health and Safety Executive (HSE) (2013a). Legionnaires' disease: Technical guidance. Part 2: The control of *Legionella* bacteria in hot and cold water systems (United Kingdom: HSE Books).

Health and Safety Executive (HSE) (2013b). Legionnaires'disease: Technical guidance. Part 1: The control of *Legionella* bacteria in evoporative cooling systems.

Health and Safety Executive (HSE) (2013c). Legionnaires' disease: Technical guidance. Part 3: The control of *Legionella* bacteria in other risk systems.

Health Safety and Executive (HSE) (2013). Legionnaires'disease: The control of *Legionell*a bacteria in water systems. Approved code of practice and guidance on regulations.

Hoenich, N.A. (2009). Disinfection of the hospital water supply: a hidden risk to dialysis patients. Crit Care *13*, 1007.

Hosein, I.K., Hill, D.W., Tan, T.Y., Butchart, E.G., Wilson, K., Finlay, G., Burge, S., and Ribeiro, C.D. (2005). Point-of-care controls for nosocomial legionellosis combined with chlorine dioxide potable water decontamination: a two-year survey at a Welsh teaching hospital. J Hosp Infect *61*, 100-106.

Hugosson, A., Hjorth, M., Bernander, S., Claesson, B.E., Johansson, A., Larsson, H., Nolskog, P., Pap, J., Svensson, N., and Ulleryd, P. (2007). A community outbreak of Legionnaires' disease from an industrial cooling tower: assessment of clinical features and diagnostic procedures. Scand J Infect Dis *39*, 217-224.

Irie, S., Akagi, K., Hiraga, T., Tada, S., Kimura, I., and Saito, A. (1984). [The first case of Legionella longbeachae pneumonia in Japan]. Nihon Kyobu Shikkan Gakkai Zasshi *22*, 518-522.

Johnson, J.T., Yu, V.L., Best, M.G., Vickers, R.M., Goetz, A., Wagner, R., Wicker, H., and Woo, A. (1985). Nosocomial legionellosis in surgical patients with head-and-neck cancer: implications for epidemiological reservoir and mode of transmission. Lancet *2*, 298-300.

Kirmeyer GJ, Foust GW, Pierson GL, Simmler JJ, and MW.., L. (1993). Optimizing chloramine treatment. Denver: American Water Works. American Water Works Research Foundation, .

Knox, N.C., Weedmark, K.A., Conly, J., Ensminger, A.W., Hosein, F.S., Drews, S.J., and and the Legionella Outbreak Investigative, T. (2017). Unusual Legionnaires' outbreak in cool, dry Western Canada: an investigation using genomic epidemiology. Epidemiol Infect *145*, 254-265.

Koide, M., Arakaki, N., and Saito, A. (2001). Distribution of Legionella longbeachae and other legionellae in Japanese potting soils. J Infect Chemother *7*, 224-227.

Kool, J.L., Carpenter, J.C., and Fields, B.S. (1999). Effect of monochloramine disinfection of municipal drinking water on risk of nosocomial Legionnaires' disease. Lancet *353*, 272-277.

Kool, J.L., Fiore, A. E., Kioski, C. M., Brown, E. W., Benson, R. F., Pruckler, J. M., Glasby, C., Butler, J. C., Cage, G. D., Carpenter, J. C., Mandel, R. M., England, B. and Breiman, R. F. (1998). More than 10 Years of Unrecognized Nosocomial Transmission of Legionnaires'

Disease Among Transplant Patients. Infection Control and Hospital Epidemiology, 898-904.

Korvick, J.A., and Yu, V.L. (1987). Legionnaires' disease: an emerging surgical problem. AnnThoracSurg *43*, 341-347.

Kubota, M., Tomii, K., Tachikawa, R., Harada, Y., Seo, R., Kaji, R., Takeshima, Y., Hayashi, M., Nishimura, T., and Ishihara, K. (2007). [Legionella longbeachae pneumonia infection from home garden soil]. Nihon Kokyuki Gakkai Zasshi *45*, 698-703.

Kusnetsov, J., Neuvonen, L.-K., Korpio, T., Uldum, S.A., Mentula, S., Putus, T., Tran Minh, N.N., and Martimo, K.-P. (2010). Two Legionnaires' disease cases associated with industrial waste water treatment plants: a case report. BMC Infect Dis *10*, 343.

Lepine, L.A., Jernigan, D. B., Butler, J. C., Pruckler, J. M., Benson, R. F., Kim, G., Hadler, J. L., Cartter, M. L. and Fields, B. S. (1998). A Recurrent Outbreak of Nosocomial Legionnaires' Disease Detected by Urinary Antigen Testing: Evidence for Long-Term Colonization of a Hospital Plumbing System. Infection Control and Hospital Epidemiology *19*, 905-910.

Li, Y., Zhang, Y., Fan, H., and Wei, L. (2015). Outbreak of six cases of nosocomial *Legionella pneumophila* pneumonia. Chinese journal of tuberculosis and respiratory diseases *38*, 294-297.

Lindsay, D.S., Brown, A.W., Brown, D.J., Pravinkumar, J., Anderson, E., and Edwards, G.F. (2011). Legionella longbeachae serogroup 1 infections linked to Potting Compost. JMedMicrobiol.

Litwin, C.M., Asebiomo, B., Wilson, K., Hafez, M., Stevens, V., Fliermans, C.B., Fields, B.S., and Fisher, J.F. (2013). Recreational vehicle water tanks as a possible source for legionella infections. Case Rep Infect Dis *2013*, 286347.

Loeb, M., McGeer, A., McArthur, M., Walter, S., and Simor, A.E. (1999). Risk factors for pneumonia and other lower respiratory tract infections in elderly residents of long-term care facilities. Arch Intern Med *159*, 2058-2064.

Loret, J.F., Robert, S., Thomas, V., Cooper, A.J., McCoy, W.F., and Levi, Y. (2005). Comparison of disinfectants for biofilm, protozoa and Legionella control. J Water Health *3*, 423-433.

Manabe, T., Teramoto, S., Tamiya, N., Okochi, J., and Hizawa, N. (2015). Risk Factors for Aspiration Pneumonia in Older Adults. PLoS One *10*, e0140060.

Mancini, B., Scurti, M., Dormi, A., Grottola, A., Zanotti, A., and Cristino, S. (2015). Effect of monochloramine treatment on colonization of a hospital water distribution system by Legionella spp.: a 1 year experience study. Environ Sci Technol *49*, 4551-4558.

Marchesi, I., Ferranti, G., Bargellini, A., Marchegiano, P., Predieri, G., Stout, J.E., and Borella, P. (2013). Monochloramine and chlorine dioxide for controlling Legionella pneumophila contamination: biocide levels and disinfection by-product formation in hospital water networks. J Water Health *11*, 738-747.

Marchesi, I., Ferranti, G., Mansi, A., Marcelloni, A.M., Proietto, A.R., Saini, N., Borella, P., and Bargellini, A. (2016). Control of Legionella Contamination and Risk of Corrosion in Hospital Water Networks following Various Disinfection Procedures. Appl Environ Microbiol *82*, 2959-2965.

Marchesi, I., Marchegiano, P., Bargellini, A., Cencetti, S., Frezza, G., Miselli, M., and Borella, P. (2011). Effectiveness of different methods to control legionella in the water supply: ten-year experience in an Italian university hospital. J Hosp Infect *77*, 47-51.

Muraca, P., Stout, J.E., and Yu, V.L. (1987). Comparative assessment of chlorine, heat, ozone, and UV light for killing Legionella pneumophila within a model plumbing system. Appl Environ Microbiol *53*, 447-453.

O'Mahony, J., Stanwell-Smith, R.E., Tillett, H.E., Harper, D., Hutchison, J.G.P., Farrell, I.D., Hutchinson, D.N., Lee, J.V., Dennis, P.J., Duggal, H.V., *et al.* (1990). The Stafford outbreak of legionnaires'disease. Epidemiology and Infection *104*, 361-380.

Official Journal of the European Communities (1998). Council directive 98/83/EC of 3 November 1998 on the quality of water intended for human consumption (the drinking water directive) (The Council of the European Union), pp. 23.

Palmer, M.E., Longmaid, K., Lamph, D., Willis, C., Heaslip, V., and Khattab, A. (2012). Legionella pneumophila found in windscreen washer fluid without added screenwash. Eur J Epidemiol *27*, 667.

Pancer, K. (2013). Sequence-based typing of *Legionella pneumophila* strains isolated from hospital water distribution systems as a complementary element of risk assessment of legionellosis in Poland. Ann Agric Environ Med *20*, 436-440.

Pancer, K., Matuszewska, R., Bartosik, M., Kacperski, K., and Krogulska, B. (2013). Persistent colonization of 2 hospital water supplies by L. pneumophila strains through 7 years - Sequence-based typing and serotyping as useful tools for a complex risk analysis. Ann Agric Environ Med *20*, 687-694.

Pankhurst, C.L. (2003). Risk assessment of dental unit waterline contamination. Prim Dent Care *10*, 5-10.

Parr, A., Whitney, E.A., and Berkelman, R.L. (2015). Legionellosis on the Rise: A Review of Guidelines for Prevention in the United States. J Public Health Manag Pract *21*, E17-26.

Phin, N., Cresswell, T., Parry-Ford, F., and Incident Control, T. (2014). Case of Legionnaires' disease in a neonate following a home birth in a heated birthing pool, England, June 2014. Euro Surveill *19*.

Pianetti, A., Sabatini, L., Citterio, B., Sisti, E., Pierfelici, L., and Bruscolini, F. (2008). Inactivation of Legionella pneumophila by combined systems of copper and silver ions and free chlorine. Ig Sanita Pubbl *64*, 27-40.

Polat, Y., Ergin, C., Kaleli, I., and Pinar, A. (2007). [Investigation of Legionella pneumophila seropositivity in the professional long distance drivers as a risky occupation]. Mikrobiyol Bul *41*, 211-217.

Pravinkumar, S.J., Edwards, G., Lindsay, D., Redmond, S., Stirling, J., House, R., Kerr, J., Anderson, E., Breen, D., Blatchford, O., *et al.* (2010).

A cluster of Legionnaires' disease caused by Legionella longbeachae linked to potting compost in Scotland, 2008-2009. Euro Surveill *15*, 19496.

Regan, C.M., McCann, B., Syed, Q., Christie, P., Joseph, C., Colligan, J., and McGaffin, A. (2003). Outbreak of Legionnaires' disease on a cruise ship: lessons for international surveillance and control. Commun Dis Public Health *6*, 152-156.

Sabria, M., and Yu, V.L. (2002). Hospital-acquired legionellosis: solutions for a preventable infection. Lancet Infect Dis 2, 368-373.

Sakamoto, R., Ohno, A., Nakahara, T., Satomura, K., Iwanaga, S., Kouyama, Y., Kura, F., Noami, M., Kusaka, K., Funato, T.*, et al.* (2009). Is driving a car a risk for Legionnaires' disease? Epidemiol Infect *137*, 1615-1622.

Schalk, J.A., Docters van Leeuwen, A.E., Lodder, W.J., de Man, H., Euser, S., den Boer, J.W., and de Roda Husman, A.M. (2012). Isolation of Legionella pneumophila from pluvial floods by amoebal coculture. Appl Environ Microbiol *78*, 4519-4521.

Schnirman, e.a. (2017). A case of legionella pneumonia caused by home use of continuous positive airway pressure. In SAGE Open Med Case RepeCollection 2017.

Schwake, D.O., Alum, A., and Abbaszadegan, M. (2015). Automobile windshield washer fluid: A potential source of transmission for Legionella. Sci Total Environ *526*, 271-277.

Sheffer, P.J., Stout, J.E., Wagener, M.M., and Muder, R.R. (2005). Efficacy of new point-of-use water filter for preventing exposure to Legionella and waterborne bacteria. Am J Infect Control *33*, S20-25.

Simmons, G., Jury, S., Thornley, C., Harte, D., Mohiuddin, J., and Taylor, M. (2008). A Legionnaires' disease outbreak: a water blaster and roof-collected rainwater systems. Water Res *42*, 1449-1458.

Sonder, G.J., van den Hoek, J.A., Bovée, L.P., Aanhane, F.E., Worp, J., Du Ry van Beest Holle, M., van Steenbergen, J.E., den Boer, J.W., Ijzerman, E.P., and Coutinho, R.A. (2008). Changes in prevention and outbreak management of Legionnaires' disease in the Netherlands between two large outbreaks in 1999 and 2006. Euro surveillance *13*, pii 18983.

Steele, T.W., Moore, C.V., and Sangster, N. (1990). Distribution of Legionella longbeachae serogroup 1 and other legionellae in potting soils in Australia. Appl Environ Microbiol *56*, 2984-2988.

Tesauro, M., Bianchi, A., Consonni, M., Pregliasco, F., and Galli, M.G. (2010). Environmental surveillance of *Legionella pneumophila* in two Italian hospitals. Ann Ist Super Sanita *46*, 274-278.

van der Lugt, W., Euser, S.M., Bruin, J.P., Den Boer, J.W., Walker, J.T., and Crespi, S. (2017). Growth of *Legionella anisa* in a model drinking water system to evaluate different shower outlets and the impact of cast iron rust. Int J Hyg Environ Health *220*, 1295-1308.

Van Kenhove, E., Dinne, K., Janssens, A., and Laverge, J. (2019). Overview and comparison of *Legionella* regulations worldwide (in press). American Journal of Infection Control.

Walker, J., and Moore, G. (2015). *Pseudomonas aeruginosa* in hospital water systems: Biofilms, guidelines, and practicalities. Journal of Hospital Infection *89*, 324-327.

Walker, J.T. (2018). The influence of climate change on waterborne disease and *Legionella*: A review. Perspectives Public Health *138*, 282-286.

Walker, J.T., Mackerness, C.W., Mallon, D., Makin, T., Williets, T., and Keevil, C.W. (1995). Control of Legionella pneumophila in a hospital water system by chlorine dioxide. J Ind Microbiol *15*, 384-390.

Wallensten, A., Oliver, I., Ricketts, K., Kafatos, G., Stuart, J.M., and Joseph, C. (2010). Windscreen wiper fluid without added screenwash in motor vehicles: a newly identified risk factor for Legionnaires' disease. Eur J Epidemiol *25*, 661-665.

Wang, H., Bedard, E., Prevost, M., Camper, A.K., Hill, V.R., and Pruden, A. (2017). Methodological approaches for monitoring opportunistic pathogens in premise plumbing: A review. Water Res *117*, 68-86.

Ward, S., Memon, F.A., and Butler, D. (2010). Harvested rainwater quality: the importance of appropriate design. Water Sci Technol *61*, 1707-1714.

Water Regulations Advisory Scheme (WRAS) (2011). Water quality effects of non-metallic materials in flexible hoses and water fitting components, pp. 3.

World Health Organization (WHO) (2004). Guidelines for drinking water-quality. Third Edition (Volume 1). Recommendations (Geneva, Switzerland), pp. 540.

World Health Organization (WHO) (2011). Water safety in buildings. Edited by: David Cunliffe, Jamie Bartram, Emmanuel Briand, Yves Chartier, Jeni Colbourne, David Drury, John Lee, Benedikt Schaefer and Susanne Surman-Lee. http://apps.who.int/iris/bitstream/10665/76145/1/9789241548106_eng.pdf.

World Health Organization (WHO) (2017). Drinking Water Parameter Cooperation Project :Support to the revision of Annex I Council Directive 98/83/EC on the Quality of Water Intended for Human Consumption (Drinking Water Directive).

Wright, J.B., Athar, M.A., van Olm, T.M., Wootliff, J.S., and Costerton, J.W. (1989). Atypical legionellosis: isolation of Legionella pneumophila serogroup 1 from a patient with aspiration pneumonia. JHospInfect *13*, 187-190.

Zhou, Z.Y., Hu, B.J., Qin, L., Lin, Y.E., Watanabe, H., Zhou, Q., and Gao, X.D. (2013). Removal of waterborne pathogens from liver transplant unit water taps in prevention of healthcare-associated infections: a proposal for a cost-effective, proactive infection control strategy. Clin Microbiol Infect.

Chapter 9

European Surveillance of Legionnaires' Disease

Birgitta de Jong and Lara Payne Hallström*

European Centre for Disease Prevention and Control (ECDC), Solna, Sweden

*lara.payne@ecdc.europa.eu

DOI: https://doi.org/10.21775/9781913652531.09

Abstract

Legionnaires' disease and Pontiac fever are both diseases with important public health implications and require prompt and thorough responses to outbreaks for future prevention. Effective methods for defining, diagnosing, reporting and responding to legionellosis outbreaks ideally should be standardized across countries. Therefore, the European Union surveillance methods for countering Legionnaires' disease is a useful model especially for travel-associated Legionnaires' disease (TALD) cases which are on the rise. Multi-country surveillance in the EU and European Economic Area (EEA) has evolved since the first organizational efforts in the 1980's to the currently responsible, European Legionnaires' Disease Surveillance Network (ELDSNet). This chapter outlines the practices of the EU surveillance of Legionnaires' disease including their schemata, definitions, responsibilities of participating members, methods and the results of the data collected since the program's inception. Lastly, improvements must still be made as the incidence of Legionnaires' disease in the EU is likely underestimated due to underreporting and/or underdiagnosis. Nearly 70% of cases are reported from only four countries, France, Germany, Italy and Spain, which represent 50% of the EU population.

Introduction

Legionnaires' disease is a statutorily notifiable disease in all EU and EEA countries. However, the case definitions across different countries varies. In some countries legionellosis, in general, is a notifiable disease which includes cases of both Legionnaires' disease and Pontiac fever. However, in other countries only Legionnaires' disease is a notifiable disease. There are also discrepancies between countries in the laboratory methods used. A case definition that includes both epidemiological and microbiological criteria is therefore essential when performing surveillance. Such a definition should accomplish the following:

- Set standards for defining cases;
- Compare data between countries;
- Verify the diagnosis and share laboratory results;
- Facilitate international investigations and collaborations;
- Develop best practice within and between countries.

For EU surveillance, a case definition for Legionnaires' disease has been agreed on [See Box]. Since this case definition only deals with cases that have clinical criteria of pneumonia, all cases of Pontiac fever are excluded from the EU surveillance of *Legionella* infections.

Multi-country surveillance of Legionnaires' disease within Europe has existed since the mid-1980s. Since the establishment of the European Centre for Disease Prevention and Control (ECDC) in 2005, surveillance of Legionnaires' disease at the EU-level is carried out by ECDC. One of the ECDC's objectives is to strengthen the surveillance of communicable

2012 EU case definition of Legionnaires' disease [#Commission Implementing Decision 2012/506/EU of 8 August 2012 amending Decision 2002/253/EC laying down case definitions for reporting communicable diseases to the Community network under Decision No 2119/98/EC of the European Parliament and of the Council (2012).

Clinical criteria

Any person with pneumonia

Laboratory criteria for case confirmation

At least one of the following three:

☐ Isolation of *Legionella* spp. from respiratory secretions or any normally sterile site

☐ Detection of *Legionella pneumophila* antigen in urine

☐ Significant rise in specific antibody level to *Legionella pneumophila* serogroup 1 in paired serum samples.

Laboratory criteria for a probable case

At least one of the following four:

☐ Detection of *Legionella pneumophila* antigen in respiratory secretions or lung tissue, e.g. by DFA staining using monoclonal-antibody-derived reagents

☐ Detection of *Legionella* spp. nucleic acid in respiratory secretions, lung tissue or any normally sterile site;

☐ Significant rise in specific antibody level to *Legionella pneumophila* other than serogroup 1 or other *Legionella* spp. in paired serum samples

☐ Single high level of specific antibody to *Legionella pneumophila* serogroup 1 in serum.

Case classification

Probable case: Any person meeting the clinical criterion AND at least one laboratory criterion for a probable case

Confirmed case: Any person meeting the clinical criterion AND at least one laboratory criterion for a confirmed case diseases in the EU and another is to provide guiding principles for response to public health threats related to communicable diseases at the EU level. For example, ECDC supports Member States by coordinating investigations of a large outbreaks that involve individuals from several different Member States or outbreaks occurring outside of the EU but affecting EU citizens. ECDC also collects annual surveillance data on Legionnaires' disease from the EU/EEA countries to analyse disease trends. To achieve these objectives of surveillance, a disease-specific network called the European Legionnaires' Disease Surveillance Network (ELDSNet) was established in 2010.

Representatives from the national authorities with responsibility for public health in each EU/EEA country nominate the official participatory members for the network. These members generally include one public health epidemiologist from the national public health institute or ministry of health and one microbiologist from the national or regional *Legionella* reference laboratory.

Prior to ECDC, a network called the European Working Group for *Legionella* Infections (EWGLI) was already established in 1986. The name, EWGLINET, was adopted in May 2002 in order to distinguish the surveillance from other activities carried out by EWGLI. From 1993 to the end of March 2010, EWGLI and EWGLINET was managed by the coordinating centre at the Health Protection Agency's Centre for Infections in London (formally the Communicable Disease Surveillance Centre of the Public Health Laboratory Service). From April 2010, running the network became the responsibility of ECDC and the network was renamed, ELDSNet. The prime aim of ELDSNet is to detect, control and prevent cases, clusters and outbreaks of Legionnaires' disease within EU/EEA countries. In addition, ELDSNet assists, where possible, with detection and response outside these countries. The network provides the means within the EU to share information and collaborate on Member State actions so that residents of European countries are increasingly protected from acquiring Legionnaires' disease linked to travel within their own countries or abroad.

EU Surveillance

The surveillance data are from two different schemes: the first scheme covers all cases reported annually from EU Member States and EEA countries with the following objectives:

- Monitor trends over time and to compare them across Member States
- Monitor the morbidity and mortality due to Legionnaires' disease and identify population groups at risk and in need of targeted preventive measures.
- Provide evidence-based data for public health decisions and actions by the EU and/or Member States level

The second near real-time reporting scheme covers travel-associated cases of Legionnaires' disease (TALD), including some reports from countries outside the EU/EEA. This second scheme primarily aims at identifying clusters of cases associated with accommodation sites that may otherwise not have been detected at the national level and enabling timely investigation and control measures at the implicated accommodation sites in order to prevent further cases of Legionnaires' disease.

Data collection methods

The collection of national data covering all reported cases annually is done through a dataset call in the beginning of the year. The appointed ELDSNet members in each European country are asked to electronically report to the European Surveillance System (TESSy) database all cases reported nationally in their country during the previous year following a strict protocol. About 20% of reported cases to the EU level since 2005 are travel-associated (domestic or international travel).

The travel-associated Legionnaires' disease surveillance scheme has distinct objectives. Particularly, ECDC receives reports of individual cases of TALD from TESSy on a daily basis in near real-time. The reporting country is generally the country where the case is diagnosed. Therefore, the reporting country can differ from the case's country of residence depending on the travel itinerary of the case. Case reports include age, sex, date of onset of disease, method of diagnosis and travel information for the places where the individual stayed within two to ten days prior to the onset of illness. Only cases who stayed at a commercial or public accommodation sites are reported as opposed to cases of Legionnaires' disease who stayed with relatives or friends. All network members transmit case information via a secure part of the ECDC web portal. Case reports received are reviewed, and the location is searched for other cases associated with the location at any time since 1987 when recording began. With complete and rapid reporting, the surveillance network can detect clusters of TALD in residents from two or more countries travelling to a single holiday destination or staying in the same commercial accommodation site. Receipt of the information leads to specific and timely action by the network countries in order to protect travellers staying at the indicated sites within EU. After receiving the report, each new case is classified as a single case or as part of a cluster in accordance with the definitions agreed upon by the network.

If the travel-associated infection is linked to countries outside the EU/EEA area, ECDC liaises with the national authorities either directly or through collaboration with the World Health Organization (WHO). This way, prevention of TALD is promoted outside the EU/EEA area. As at April 2018, TALD surveillance is the only indicator-based surveillance scheme operating at ECDC year-round as near real-time surveillance reporting.

The daily surveillance of TALD at a European level supplements national level surveillance by identifying clusters of cases which may originate from different countries. These clusters otherwise may not have been detected at the national level. The rapid reporting of cases and clusters enable timely investigations and implementation of control measures at the implicated accommodation sites in order to prevent further infections.

ELDSNet definitions of travel-associated Legionnaires' disease cases

Single case

Case of Legionnaires' disease who in the two to ten days before onset of illness stayed at or visited a commercial accommodation site that has not been associated with any other cases of Legionnaires' disease, or case who stayed at an accommodation site linked to other cases with date of onset more than two years apart.

Cluster

Two or more cases who stayed at or visited the same commercial accommodation site in the two to ten days before onset of illness and whose onset is within the same two-year period.

A cluster can be additionally categorised as a:

* Rapidly evolving cluster: when three or more cases have onset of disease within a three-month period occurring in the six months preceding or following a cluster notification or update. This categorisation indicates a possible recent increase of exposure risk.

* Complex cluster: if a cluster involves cases that are associated with staying at one or more accommodation sites also being part of clusters, it is handled as a 'complex cluster'.

If any further cases associated with the cluster site occur more than two years after the last case, they will be reported as new single cases, although the country of infection will receive information on all previous cases linked to the accommodation site regardless of the time period elapsed.

As at April 2018, 28 EU Member States, two EEA countries and two countries outside the EU/EEA area were contributing or receiving data on TALD cases. ELDSNet has contacts in 30 countries outside the EU/EEA area who are receiving data directly from ELDSNet.

TALD notifications and investigations

All notifications regarding single TALD cases and clusters aim to be sent out within one business day from the reception of the case report. Single cases are reported by ELDSNet to the country where the accommodation site is situated and to the reporting country. The hosting country is then expected to ensure that the notified site receives a checklist outlining good practices for minimising risk of *Legionella* infection. This checklist is found within a leaflet for managers running a tourist accommodation (ECDC Health Information). Many countries carry out environmental investigations for single cases, and the investigation assessment of these can be voluntarily submitted to ELDSNet.

Clusters are reported to the country where the accommodation site is located, the reporting country and all ELDSNet network members. The hosting country is expected to arrange for the accommodation site to be inspected by a competent institution authorised by the national authority such as a local or regional public health authority. This must be accomplished promptly as well as a risk assessment according to guidelines of the country or the European Technical guidelines (2017). This should be followed by an environmental investigation such as sampling of accommodation site water systems. The investigation informs the implementation of control measures to minimise any possible risk for Legionnaires' disease and support prevention. These actions should be reported to ELDSNet using the specific forms.

Form A should have information on whether or not an inspection and risk assessment have been carried out at the accommodation site. It should be returned to ECDC within two weeks of receipt of the cluster notification. Form A should state whether control measures are in progress and whether the hotel remains open or not. Form B should have information on the investigations and control measures that have been implemented at the accommodation site, including the results of sampling that took place. It should be returned to ECDC within six weeks of receipt of the cluster notification.

If these forms are not received from an EU/EEA country, or if it contains information that the accommodation site is not implementing any recommended control measures, then the site's name is published on the accommodations list on the public section of the ECDC website. The accommodation site list on the ECDC public website shows the current accommodation sites in the EU/EEA countries in which clusters of Legionnaires' disease have been identified, where there has been no complete assessment of the risk for *Legionella* infection and where ELDSNet believes there may be an increased risk to travellers.

Accommodation sites associated with clusters situated outside of the EU/EEA area do not appear on the accommodation site list. Rather, these clusters are reported to those tour operators that are subscribed to receive specific summary reports from ELDSNet concerning the clusters. Subscribed tour operators are routinely informed of all clusters located in countries outside the EU/EEA area, because the ECDC does not require timely information on the control measures being implemented at these accommodation sites. In the event of a rapidly evolving cluster occurring inside EU/EEA area or worldwide, subscribed tour operators are informed. Tour operators on the subscriber list will be informed when an accommodation in an EU/EEA country is intended to be published or removed from the ECDC public website. Tour operators are also informed of any voluntarily completed form B for other sites outside the EU/EEA area submitted by the national public health authorities.

Reported disease in the EU/EEA

All reported cases

The number of reported cases in the EU/EEA countries has fluctuated during the last ten years (Figure 1). A sharp increase in the number of reported cases was seen in 2014 when a large outbreak affecting almost 400 people occurred in Portugal. However, even if this outbreak was not included in the figures, the year of 2014 had the highest reported number of cases (7022 cases). Since there were no changes in laboratory methods or reporting systems, this increase of cases remains unexplained. The high level of reported cases continued into 2015 with 7034 cases, and no major outbreaks were reported.

Figure 1. Notification rate of Legionnaires' disease in the EU/EEA* by year of reporting, 1995-2015. *EWGLINET member countries not belonging to the EU/EEA were excluded for 1995-2008. *Source:* https://ecdc.europa.eu/sites/portal/files/documents/Legionnares-disease-europe-2015.pdf

The majority of cases are reported from countries in the Western part of Europe (Figure 2). Four countries, France, Germany, Italy and Spain usually report the vast majority of cases in the EU/EEA area. In 2015, these four countries they accounted for 69% of all cases. On the other hand, the 15 lowest reporting countries merely reported 3% of all cases. This shows how unevenly distributed the reported cases are among the EU/EEA countries.

There is a seasonal variation for Legionnaires' disease, and August is the month when most cases are reported to have onset of disease. Notification rates also increase with age. People older than 50 years of age accounted for 81% of cases with known age and sex in 2015. Legionnaires' disease is more common in males, with an overall male-to-

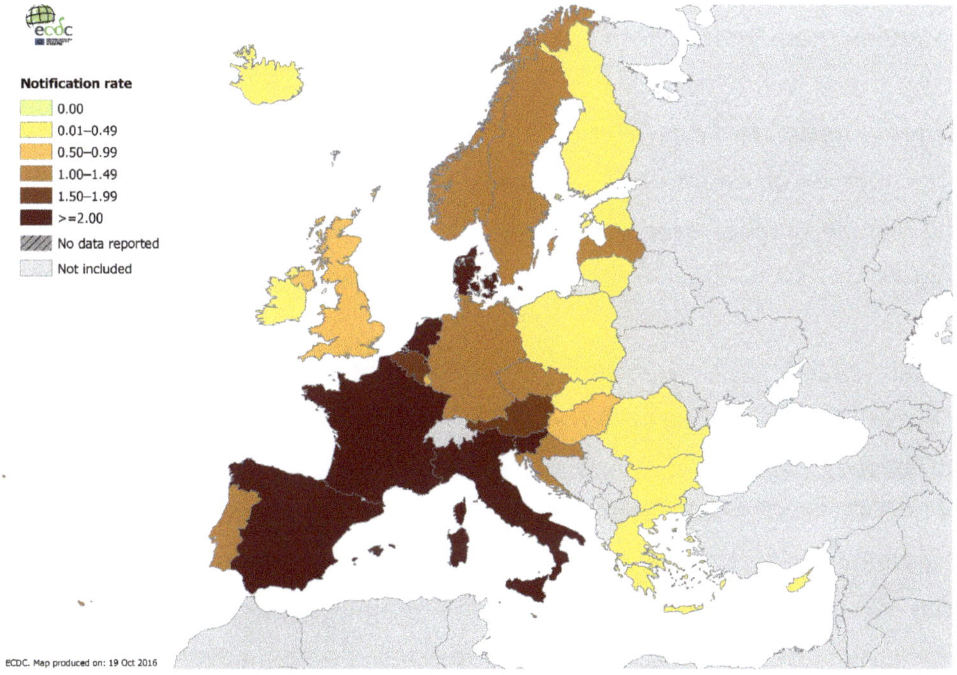

Figure 2. Notifications of Legionnaires' disease per 100 000, by reporting country, EU/EEA, 2015. *Source:* https://ecdc.europa.eu/sites/portal/files/documents/Legionnares-disease-europe-2015.pdf

female ratio of 2.5 - 3.0 to 1. About 75% of the cases are reported as community-acquired and roughly 20% are travel associated. Healthcare-associated cases represents a considerable proportion of cases in older age groups, but in the total figures this setting represents less than 10%.

The reported mortality rate of Legionnaires' disease in 2015 was 0.8 per million inhabitants, which is consistent with the rates recorded since 2008 which ranged from 0.7 to 0.9 per million inhabitants. Of 4893 cases with a known outcome, 396 were reported to have died giving a case-fatality ratio (CFR) of 8%. Case fatality is higher for older age groups, both in males and females. In individuals above 50 years of age, case fatality is higher. A similar pattern is demonstrated for both males and females.

Almost 90% of cases are reported to have been diagnosed with a urinary antigen test, but there is a large variety in diagnostic tests used in the different EU countries. PCR tests are increasingly used, and in some countries more than 20% of cases are diagnosed with PCR.

Travel-associated cases

The number of reported TALD cases with onset of disease in 2015 was the highest number ever reported to the network with 1141 reported cases (Figure 3).

These cases were reported from 25 countries including 22 EU/EEA Member States and three non-EU/EEA countries which were Switzerland (30 cases), the USA (6 cases) and Australia (2 cases). Two-thirds (66.2%) of all TALD cases were reported by the United Kingdom, Italy, France and the Netherlands. For the TALD cases, there was also a seasonal variation with two-thirds of the cases in 2015 emerging between June and October. There is also a similarity to the overall Legionnaires' disease sex distribution, as over two-thirds (69%) of the reported TALD cases are male. Cases had a median age of 62 years and 82% of cases were aged

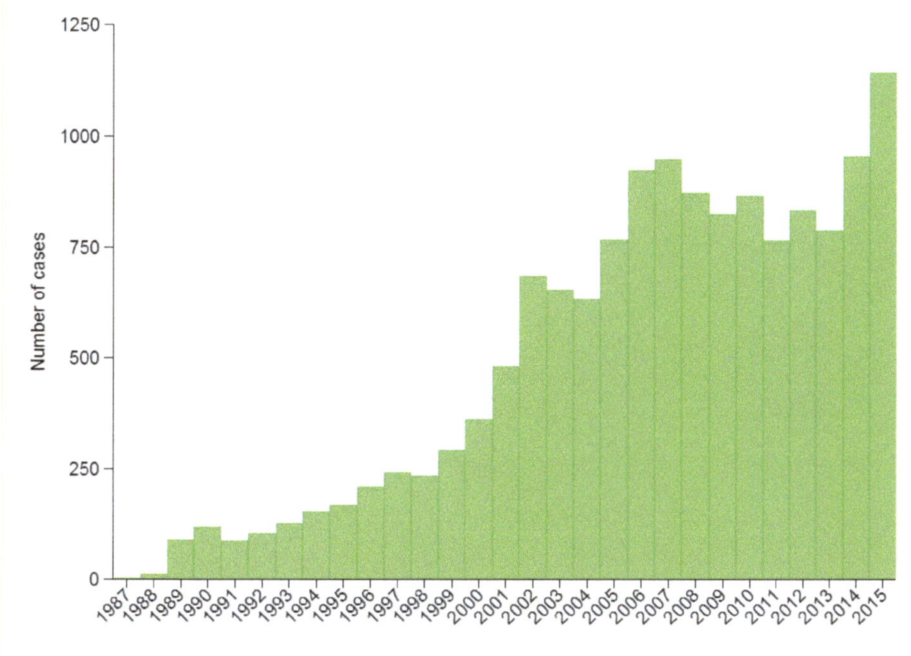

Figure 3. Number of travel-associated cases of Legionnaires' disease reported to ELDSNet, by year, 1987-2015. *Source:* https://ecdc.europa.eu/sites/portal/files/documents/Legionnares-disease-europe-2015.pdf

50 years or older. Around 90% of the TALD cases are diagnosed using a urinary antigen test. Since the TALD cases are reported in a very timely manner, the clinical outcome of infection is often known for only about half of the cases, and the case-fatality rate for those cases with a reported outcome is around 4%. The TALD cases reported in 2015 were associated with a cumulative total of 1606 stays in various destinations around the world, and 76% were within the EU/EEA area (Figure 4).

In total, 167 new clusters were detected in 2015, and the majority were comprised of two cases. In 60% of the clusters, the first two reported cases were from different countries. Thus, most of these clusters would likely not have been detected without the ELDSNet network and EU

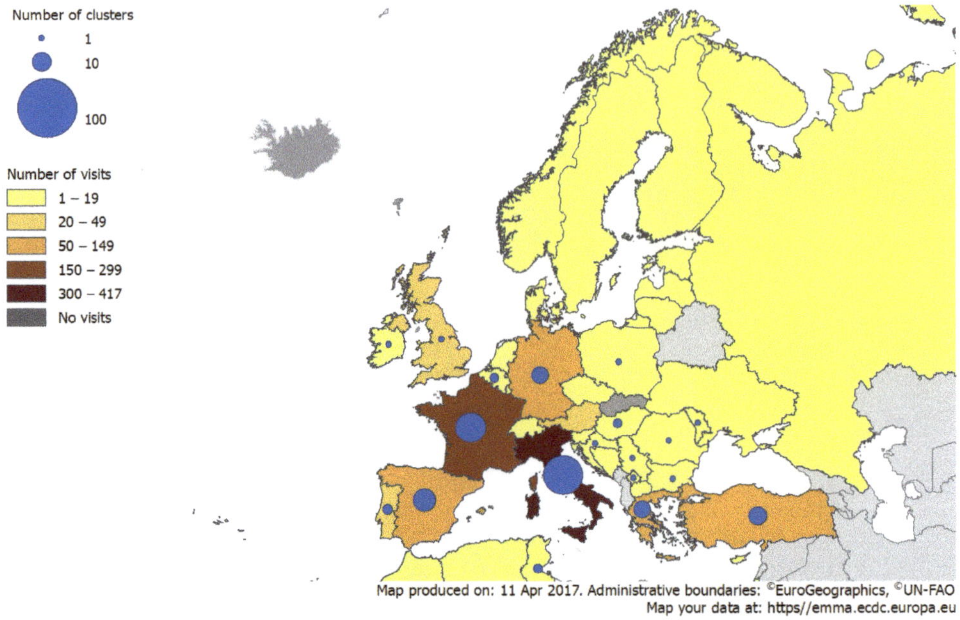

Figure 4. Number of accommodation site visits and clusters of travel-associated cases of Legionnaires' disease per destination country, EU/EEA and neighbouring countries, 2015. *Source: https://ecdc.europa.eu/sites/portal/files/documents/Legionnares-disease-europe-2015.pdf*

surveillance scheme. The number of TALD clusters at the subnational (NUTS2) is shown in Figure 5.

For all 167 clusters, the Forms A and B were received. *Legionella* bacteria were found in 60% of all cluster sites where environmental water sampling results were reported. Eight accommodation site names were published on the ECDC website, because the form B stated that satisfactory control measures had not been implemented in a timely manner.

Conclusions

Notification rates of Legionnaires' disease vary from 0.1 to 30.0 cases per million population in the EU/EEA countries. Four countries (France, Germany, Italy, and Spain) accounted for 69% of all notified cases in 2015, although their combined populations only represent approximately

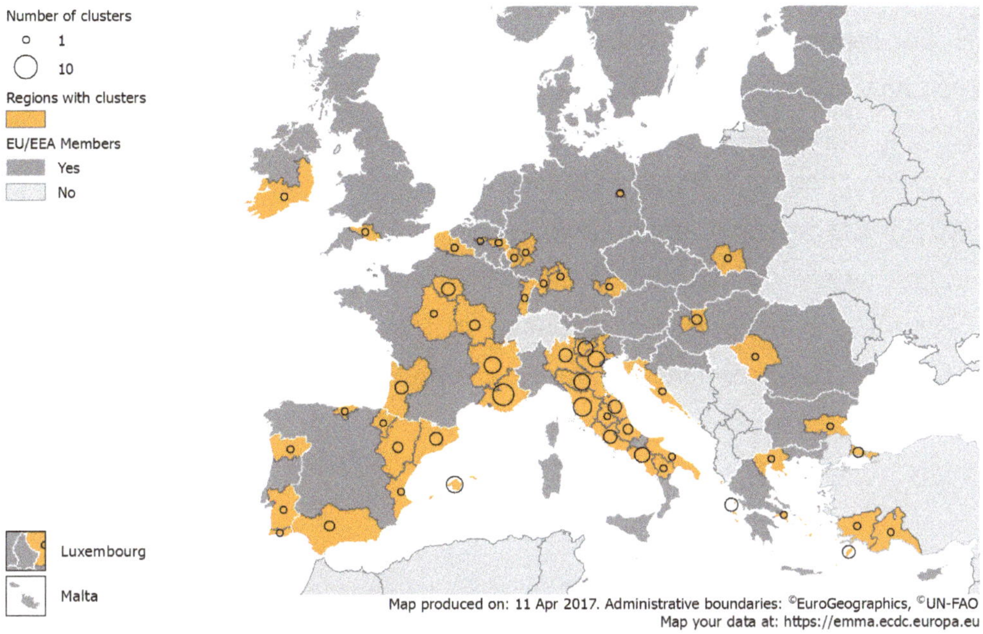

Figure 5. Number of clusters* of travel-associated Legionnaires' disease per destination area (NUTS2), EU/EEA, 2015 excluding complex clusters. *Source: https://ecdc.europa.eu/sites/portal/files/documents/Legionnares-disease-europe-2015.pdf*

50% of the EU/EEA population. Many countries have a notification rate below 0.5 cases per 100 000 population which has been unchanged for the past five years and unlikely to reflect the true incidence of Legionnaires' disease. It is likely that the lower rates represent a considerable underestimation of the incidence which could be due to both underdiagnosis and underreporting. The European surveillance of the travel-associated cases as performed by ELDSNet is a very good example of how multi-country joint surveillance schemes can facilitate the implementation of preventive measures, and thereby reduce potential exposure sources for travellers at risk for Legionnaires' disease.

References

ECDC Health Information Leaflet for managers of tourist accommodation on how to reduce the risk of Legionnaires' disease.

(2017). European Technical Guidelines for the Prevention, Control and Investigation, of Infections Caused by *Legionella* species.

(2012) Commission Implementing Decision 2012/506/EU of 8 August 2012 amending Decision 2002/253/EC laying down case definitions for reporting communicable diseases to the Community network under Decision No 2119/98/EC of the European Parliament and of the Council.

Chapter 10

Epidemiological Genotyping of *Legionella pneumophila*: from Plasmids to Sequence-Based Typing

Norman K. Fry[1] and Sophie Jarraud*[2,3,4]

[1]Respiratory and Vaccine Preventable Bacteria Reference Unit, Public Health England - National Infection Service, Colindale, London, UK
[2]Inserm, U1111, Université Lyon 1, CNRS, UMR5308, École Normale Supérieure de Lyon, University Lyon, F-69007, France
[3]Hospices Civils de Lyon, Groupement Hospitalier Nord, National Reference Centre of Legionella, Institute for Infectious Agents, 103 Grande rue de la Croix Rousse, 69004 Lyon, France
[4]CIRI, Centre International de Recherche en Infectiologie, Equipe Pathogénèse des Légionelles, Lyon, France

*sophie.jarraud@univ-lyon1.fr

DOI: https://doi.org/10.21775/9781913652531.10

Abstract
Genotyping is used for the identification and differentiation of *Legionella* types within a species and most often for the species, *L. pneumophila,* which causes most cases of human infection. Due to the complex nature of *Legionella* outbreaks and sporadic infections which are often associated with travel, effective genotyping methods must be transportable and accessible with enough resolution to identify closely related strains. This chapter describes each genotyping method developed for the discernment of *L.*

pneumophila including the discovery of diversity within plasmid genomes, restriction endonuclease analysis, pulsed-field gel electrophoresis, restriction fragment length polymorphism analysis and ribotyping, polymerase chain reaction (PCR) typing, amplified fragment length polymorphism analysis and finally sequence-based typing (SBT) which is the recent gold-standard. The shortfalls of each typing method will be discussed and how these flaws lead us into the era of next generation sequencing (NGS) which has already begun to revolutionize typing methodology.

Introduction

A wide range of typing methods have been applied to the genus *Legionella*. The majority of these methods have been applied to the species most commonly associated with clinical disease, *Legionella pneumophila*. This chapter will address the application of genotypic typing methods to *L. pneumophila* from the early 1980's starting with the discovery of plasmids within the genus and until the international consensus *L. pneumophila* seven allele sequence-based typing (SBT) scheme.

Broadly speaking microbial typing methods may be divided into phenotypic and genotypic methods. Phenotypic methods can depend on observable physical and morphological or biochemical characteristics of the organism as well as the expression of specific traits. Genotypic methods involve direct or indirect DNA-based analysis of the genome of the organism. Although this chapter will focus on genotyping, typically some phenotypic testing is performed in order to selectively screen isolates prior to genotypic testing. For *L. pneumophila* and several other species, initial identification to species level is normally followed by serogrouping.

To date, sixteen serogroups (sgs) of *L. pneumophila* have been described (Lück *et al.*, 1995). It is likely that more sgs exist, though they may occur rarely in human disease. Although there are commercial latex agglutination kits available allowing differentiation of isolates belonging to sg 1 from sg 2 to sg 14 (or sg 2 to sg 15), typically individual serogrouping from sg 1 to sg 16 using monoclonal antibodies (mAbs) and/or rabbit antisera is restricted to reference laboratories. Further differentiation using mAb subgrouping can be performed for isolates belonging to some of these serogroups. Strains of *L. pneumophila* belonging to sg 1 may also be subdivided as possessing, or not possessing, the virulence-associated epitope recognized by the mAb 3/1 (using the 'Dresden mAb panel') previously designated mAb 2 (Helbig *et al.*, 2002). Monoclonal antibody subgrouping is typically performed by indirect immunofluorescence or indirect ELISA methods (Lück *et al.*, 2013). As the majority of clinical infections are caused by *L. pneumophila* sg 1, most genotypic typing methods have focussed on the differentiation of strains belonging to this serogroup.

Genotypic typing methods

Plasmid analysis

The presence of plasmids in *L. pneumophila* was first reported by Knudson and Mikesell (1980). This subsequently led to the investigation of both clinical and environmental *L. pneumophila* isolates for plasmids in order to study the epidemiology of Legionnaires' Disease (Brown *et al.*, 1982, 1985; Maher *et al.*, 1983, 1987; Nolte *et al.*, 1984). A variety of extraction methods was used, including alkaline lysis followed by horizontal agarose electrophoresis and ethidium bromide (EtBr) staining. It was postulated that *Legionella* plasmids carry antibiotic resistance genes and virulence factors, and a correlation between a higher persistence of plasmids found in environmental strains of *L. pneumophila* was reported. However, clinical strains may also possess plasmids. The size of legionella plasmids described ranges from ca. 50.5 kb to ca. 95 MD (López de Felipe and Martínez-Suárez, 1991; Lévesque *et al.*, 2014). Methods for evaluating plasmids are generally precluded from wide application in the epidemiological typing of *Legionella.* This is due to the potential loss of plasmids on subculture on complex artificial media, difficulties in the intact isolation of those of high molecular size, the presence of similar size plasmids with indistinguishable restriction endonuclease analysis profiles from unrelated sources, and the fact that many clinical isolates appeared not to possess plasmids. Currently, whole genome sequencing (WGS) can offer many solutions to the further characterisation of plasmid genomes.

Restriction endonuclease analysis (REA)

The application of restriction endonuclease analysis (REA) to the epidemiological typing of *L. pneumophila* was first described by van Ketel *et al.* (1984) using *Eco*RI and *Hin*dIII, two enzymes with six-base recognition sequences. The products of digestion were separated following overnight electrophoresis using agarose gels, and the resulting patterns compared visually following image capture. Other enzymes used to perform either single- or double-digestions included, *Bam*HI, *Hpa*I, and *Hpa*II (Tompkins *et al.* 1987, van Ketel, 1988) with the double-digestion combinations of *Eco*RI-*Hind*III and *Hpa*I-*Hpa*II giving optimal results. In an attempt to achieve greater resolution of the DNA fragments of lower molecular size, a further variation was employed using vertical polyacrylamide gel electrophoresis (PAGE) followed by silver staining (Haertl and Bandlow, 1991). The analyses of these patterns tended to be visual. Whilst comparison of strains within a gel could prove useful, inter-gel comparison was difficult, and somewhat subjective.

Pulsed-field gel electrophoresis

Pulsed-Field Gel Electrophoresis (PFGE) also referred to as macrorestriction analysis (MRA) is a variation of horizontal electrophoresis allowing separation of large DNA restriction fragments using restriction endonucleases which cut at rare sites. The resulting patterns, revealed after conventional ethidium bromide staining, are based upon the variable migration of these fragments in an electrical field of alternating polarity. One of the main endonuclease selected for analysis of *L. pneumophila* has been *Sfi*I, which cuts DNA at the recognition sequence GGCCNNNN↓NGGCC, where N is any base and ↓ is the point of cleavage (Schoonmaker *et al.*,

1992). This method has a high index of discrimination but can show both intra- and inter-laboratory variation. Therefore, there have been confounding attempts to achieve international standardisation. However, PFGE/MRA has been widely used by several reference laboratories to reveal genotypic variation between isolates belonging to the same serogroup including sg 1 and non-sg 1 strains (Decludt *et al.*, 2004; Kura *et al.*, 2006). This method has also been applied to non-*L. pneumophila* species to demonstrate indistinguishable patterns from epidemiologically related clinical and environmental isolates (Lück *et al.*, 1995; Montanaro-Punzengruber et al., 1999; Akermi *et al.*, 2006; Matsui *et al.*, 2010).

Restriction Fragment Length Polymorphism analysis and ribotyping

In order to simplify the analysis of multiple REA products, Southern blotting and hybridisation with DNA probes have been applied to *Legionella* typing. Southern blotting is the transfer of separated DNA fragments following electrophoresis onto a nitrocellulose or nylon membrane which are then fixed by heat or UV treatment. Next, a labelled probe, which was originally radio-isotopic and later biotin-streptavidin or variants, is applied and the membrane hybridised. After being washed at selected stringency the fragments are developed to reveal where the probe has bound. Restriction fragment length polymorphism (RFLP) analysis using labelled ribosomal RNA, typically 16S and 23S rRNA, or ribosomal DNA as probe is called ribotyping. A number of researchers utilised ribotyping (Gaia *et al.*, 1994; Bangsborg *et al.*, 1995) or RFLP analysis with specific DNA probes to characterise *L. pneumophila,* and for many years RFLP analysis was the standard method used for epidemiological typing in the UK (Saunders *et al.*, 1990; 1991; Harrison *et al.*, 1990).

PCR typing: Random Amplified Polymorphic DNA, repetitive extragenic palindromic PCR and arbitrarily-primed PCR

The PCR-based typing techniques, Random Amplified Polymorphic DNA (RAPD), repetitive extragenic palindromic (rep)-PCR and arbitrarily-primed

(AP)-PCR all generate amplified products from random segments of genomic DNA using one or more oligonucleotide primers. In the case of AP-PCR or RAPD these are arbitrary/random sequences. While in the case of rep-PCR, these may be based on repetitive intergenic consensus motifs such as ERIC2 (Williams *et al.*, 1990; Gomez-Lus *et al.*, 1993; van Belkum *et al.*, 1993; Ledesma *et al.*, 1995; Georghiou *et al.*, 1994). The PCR products are typically resolved by standard electrophoresis using agarose gels followed by visualisation by staining with ethidium bromide. This technique does not require prior knowledge of the DNA sequence of the target organism. Gel to gel comparison is difficult, however it is also relatively inexpensive, rapid and easy to perform.

Amplified fragment length polymorphism analysis

Amplified fragment length polymorphism (AFLP) analysis consists of a simple restriction-ligation reaction followed by PCR amplification. Genomic DNA is digested with a specific restriction endonuclease, and specially constructed adapters are ligated to the restriction fragments in a one-step reaction. The resulting restriction fragments, referred to as "tagged fragments", then act as the DNA template in a standard PCR reaction with primers complementary to the adapters. The resulting PCR products are then typically resolved on agarose gels followed by staining with ethidium bromide (Valsangiacomo *et al.*, 1995). Two other variants have also been described with one of the PCR primers labelled either radio-isotopically, e.g., ^{32}P (Vos *et al.*, 1995) or with a fluorescent dye, e.g., 5-carboxyfluorescein (Desai *et al.*, 1998) designated fAFLP with the products resolved by PAGE or capillary electrophoresis (CE). The advantages of AFLP analysis include universal application without prior knowledge of genomic DNA sequence, the ability to use differing combinations of restriction endonucleases and addition of selective nucleotides to PCR primers to help optimise results, relatively quickly methodology and potential for high-throughput with local electronic database pattern storage at relatively low cost. Disadvantages include the variation in fragment sizing precision.

Following its successful performance in a multicentre comparison of genotypic typing methods, non-radioisotopic AFLP analysis was adopted as a standard by the European Working Group for Legionella Infections (EWGLI) (Fry *et al.*, 1999; 2002) later the ESCMID Study Group for Legionella Infections (ESGLI). AFLP was also widely used to allocate *L. pneumophila* strains into designated types by interrogation of a reference database through an internet-based protocol. However, the reliance on analysis of multibanded patterns and sizing variation served to complicate unambiguous type assignation. The fluorescent alternative was also explored (Jonas *et al.*, 2000), but difficulties in comparison across different platforms was reported (Fry *et al.*, 2005).

Sequence-based typing

Sequence based typing (SBT) remains the mainstay identification methods used in outbreak scenarios. Whilst progress toward standardisation of a genotyping method for *L. pneumophila* was made using AFLP, proficiency testing demonstrated that a significant proportion of laboratories could not correctly identify isolates 100% of the time (Fry *et al.*, 2005). This was mainly due to issues with data analysis such as scoring and comparison of banding patterns. Therefore, a universal typing method which would be simple, rapid, discriminatory and truly 'portable' was still required. The many advantages of DNA sequence data are now well-known and include portability between laboratories and the relative ease with which global databases can be established. This was exemplified by the multi-locus sequence typing (MLST) schemes initially developed by Maiden *et al.* (1988) for *Neisseria meningitidis* and Enright and Spratt (1988) for *Streptococcus pneumonia.* These schemes now encompass more than 50 genera and 90 species (see MLST.net; PubMLST.org).

Preliminary studies on *L. pneumophila* investigated a number of gene targets (Gaia *et al.*, 2003). Initially, three of these were selected including the *flaA* gene, which encodes the flagellum subunit protein of *L. pneumophila* sg 1, *proA*, which encodes a zinc metalloprotease found in *L. pneumophila*, and *mompS*, which encodes an outer membrane protein of ca. 29 kDa. Later, three additional genes, *asd*, *mip*, and *pilE*, which encode the *L. pneumophila* aspartate-β-semialdehyde dehydrogenase, macrophage infectivity potentiator protein and type IV pilin, respectively, were added to a consensus typing scheme. Finally, a seventh gene, *neuA* which encodes N-acylneuraminate cytidyltransferase was added (Ratzow *et al.*, 2007). Later the primers were modified following the discovery of a *neuA* homologue (designated *neuAh*) that was refractory to amplification of some strains using the standard *neuA* primers (Farhat *et al.*, 2011; Mentasti *et al.*, 2014).

The consensus 7-allele SBT epidemiological typing scheme for clinical and environmental isolates of *L. pneumophila* was developed by members of ESGLI, and has been evaluated extensively for implementation in the investigation of outbreaks of legionellosis caused by *L. pneumophila*. Using the SBT protocol, the SBT database http://www.hpa-bioinformatics.org.uk/legionella/legionella_sbt/php/sbt_homepage.php allows assignment of the seven ordered alleles, *flaA*, *pilE*, *asd*, *mip*, *mompS*, *proA* and *neuA* represented as a Sequence Type (ST), or allelic profile, of the ordered string of allele numbers separated by commas e.g. ST1, which has the allelic profile 1,4,3,1,1,1,1. This database now contains SBT data from more than 11460 strains, belonging to 2300 Sequence Types and from 61 countries (accessed 11/01/2017; 12:15).

Further developments of the scheme include the direct application to clinical specimens without cultivation with the use of 'nested' PCR and modifications of the primers to including M13-tails (Fry *et al.*, 2006; Ginevra *et al.*, 2009; Mentasti and Fry, 2012; Mentasti *et al.*, 2016). The use of two separate rounds of amplification and the M13-modifications in the second round increases the sensitivity of the PCR and thus the likelihood of successfully obtaining typing data. These tactics help to accelerate the sequencing process by greatly reducing the number of primers required.

Integral to the success of the SBT scheme was the development of novel web-based tools to facilitate database curation. These tools accomplish designation of novel allele types and allelic profiles and automated quality assessment of sequence chromatograms (Underwood *et al.*, 2006). The online tools allow users to assemble forward and reverse trace files to identify matching pre-existing alleles. Putative novel alleles not matching existing alleles are submitted to the database curators for verification.

Therefore, SBT became the gold standard for the epidemiological typing of *L. pneumophila,* and it has been adopted by many countries for national surveillance and outbreak investigation. The methodology has also been applied to the study of genetic diversity in both clinical and environmental settings. New insights into the limited number of STs causing disease and the wide diversity in the environment have been revealed. The widespread nature of some STs such as ST1 appear to have universal distribution, and those having an apparently restricted geographical area such as ST182 have been documented (Borchardt *et al.*, 2008). SBT has revealed that a low number of STs accounts for a disproportionately high amount of clinical cases. Specifically, five STs (ST1, ST23, ST37, ST47, and ST62) were shown to account for approximately half of all epidemiologically unrelated Legionnaires' disease cases reported to the SBT database in northwest Europe (David *et al.*, 2016). The use of a standard methodology by numerous countries, and the curation of a central database has thus provided global representation of the *L. pneumophila* strain distribution and provided information concerning clinical significance.

Multiple-locus variable number tandem repeat analysis

Analysis of polymorphic tandem repeats has been successfully applied to the epidemiological typing of many bacterial species (Kremer *et al.*, 1999; Keim *et al.*, 2000; Lindstedt, 2005). These multiple-locus variable number of tandem repeats (VNTR) assays (MLVA) are based on analysis of tandemly repeated sequences as well as designated microsatellites of less than 9 bp and minisatellites of more than 9 bp in length. Modifications to the initial work by Pourcel *et al.* (2003), led to description of a standard MLVA protocol for *L. pneumophila* sg 1 using standard PCR and gel-based image analysis software. The first MLVA database for *L. pneumophila* was also established which was available for direct query via the internet (Pourcel *et al.*, 2007). The MLVA assay consisted of 8 VNTR loci (MLVA-8), and the number of repeats was determined based on estimated sizes and these repeat numbers were used as alleles. As for AFLP, further modifications such as fluorescent-labelling and use of CE were also included to assist the detection and sizing of MLVA fragments (Nederbragt *et al.*, 2008). This technique was even applied directly to environmental specimens without prior cultivation. With this direct method, problems including PCR inhibition and high background signals were encountered. Thus, determining a full allelic profile was not always possible using CE and therefore, single-strand conformation polymorphism (SSCP) and PAGE were used (Kahlisch *et al.*, 2010). Sequence data from individual amplicons can also be recovered (Kahlisch *et al.*, 2010). Additional and even sometimes surprising variation was demonstrated in particular loci such as Lpms04 (Visca *et al.*, 2011). An extended fluorescent CE twelve loci assay (MLVA-12) was described by Sobral *et al.* (2011) revealing a large number of closely related environmental *L. pneumophila* isolates associated with hot water supply systems but not with disease. As in other MLVA studies, not all loci were amplified in all strains due to presumed sequence variation at the PCR primer binding sites.

Spoligotyping

The high prevalence of particular sequence types found using the SBT scheme such as ST1 and pulsotypes using PFGE such as 'Paris' in both clinical and environmental isolates of *L. pneumophila* have hampered identification of sources of infection during epidemiological investigations when identified in clinical isolates. Therefore, several techniques have been applied for subtyping of the ST1/Paris clonal types. Clustered regularly interspaced short palindromic repeats (CRISPR) have been used as genotyping markers for several microorganisms (Kamerbeek *et al.*, 1997; Schouls *et al.*, 2003). CRISPR array is a succession of direct repeats (DRs) of 30-40 bp interspersed by unique spacer sequences of approximately the same size. Spoligotyping is a reverse dot-blot using spacers as probes and primers designed in the repeat sequence for amplification of spacers. Ginevra *et al.* (2012) developed a spoligotyping tool based on the diversity of a CRISPR/cas locus for subtyping of strains belonging to *L. pneumophila* sg 1 ST1/Paris pulsotype. Successful discrimination of strains previously indistinguishable by ST/PFGE genotypes was achieved which could assist in the identification of environmental sources. A microbead-based multiplex format using flow cytometry has also been developed allowing high-throughput and a better standardization of the assay (Gomgnimbou *et al.*, 2014). The general application of this typing method to all *L. pneumophila* strains is not feasible since not all *L. pneumophila* strains contain CRISPRs. However, the spoligotyping method has also been developed for ST62 isolates and other worldwide clinically predominant *L. pneumophila* STs harbouring one CRISPR locus (Lück *et al.* 2015).

Summary

The relatively recent discovery of the causative agent of Legionnaires' disease and its description in the late 1970s implied that many of these genotypic methods had been previously applied to other microorganisms than *Legionella* (McDade *et al.*, 1977). However, the unique nature of the problem presented by the tracking of *Legionella*, particularly with travel-associated disease, ultimately requires a portable typing methodology capable of fulfilling all of the requirements of a good typing scheme (Struelens *et al.*, 1996, Van Belkum *et al.*, 2007). Of all the genotypic methodologies described here, only SBT has successfully achieved this as demonstrated by its evaluation in coded trials and extensive adoption by multiple countries. The optimal typing system must combine sufficient discrimination with reproducibility, in addition to other essential criteria. No system is perfect, however, and the choice of gene targets for *L. pneumophila* was initially made prior to the availability of multiple whole genomes in the public sequence databases. This resulted in the unintentional inclusion of one gene locus subsequently found to present in more than one copy (*mompS*) and one locus which was later found to be heterogenous (*neuA*). Successful selective amplification of *mompS* was achieved by use of specific asymmetrical primers in the standard protocols and novel primers specific for the *neuA* homologue designed and internationally validated.

There is no doubt that the increasing availability of sequence data derived from WGS will lead to the development of new and improved typing methodologies for *L. pneumophila* and other members of the genus, and this will be explored in other chapters of this book.

References

Akermi M, Doleans A, Forey F, Reyrolle M, Meugnier H, Freney J, Vandenesch F, Etienne J, Jarraud S. (2006). Characterization of the *Legionella anisa* population structure by pulsed-field gel electrophoresis. FEMS Microbiol Lett 258: 204-207.

Bangsborg JM, Gerner-Smidt P, Colding H, Fiehn NE, Bruun B, Høiby N. (1995). Restriction fragment length polymorphism of rRNA genes for molecular typing of members of the family Legionellaceae. J Clin Microbiol 33: 402-406.

Borchardt J, Helbig JH, Lück PC. (2008). Occurrence and distribution of sequence types among Legionella pneumophila strains isolated from patients in Germany: common features and differences to other regions of the world. Eur J Clin Microbiol Infect Dis 27: 29-36.

Brown A, Lema M, Ciesielski CA, Blaser MJ. (1985). Combined plasmid and peptide analysis of clinical and environmental *Legionella pneumophila* strains associated with a small cluster of Legionnaires' disease cases. Infection 13: 163-166.

Brown A, Vickers RM, Elder EM, Lema M, Garrity GM. (1982). Plasmid and surface antigen markers of endemic and epidemic *Legionella pneumophila* strains. Clin Microbiol 16: 230-235.

David S, Rusniok C, Mentasti M, Gomez-Valero L, Harris SR, Lechat P, Lees J, Ginevra C, Glaser P, Ma L, Bouchier C, Underwood A, Jarraud S, Harrison TG, Parkhill J, Buchrieser C. (2016). Multiple major disease-associated clones of *Legionella pneumophila* have emerged recently and independently. Genome Res 26: 1555-1564.

Decludt B, Campese C, Lacoste M, Che D, Jarraud S, Etienne J. (2004). Clusters of travel associated legionnaires' disease in France, September 2001- August 2003. Euro Surveill 9: 12-13.

Desai M, Tanna A, Wall R, Efstratiou A, George R, Stanley J. (1998). Fluorescent amplified-fragment length polymorphism analysis of an outbreak of group A streptococcal invasive disease. J Clin Microbiol 36: 3133-3137.

Enright MC, Spratt BG. (1998). A multilocus sequence typing scheme for *Streptococcus pneumoniae*: identification of clones associated with serious invasive disease. Microbiology. 144: 3049-3060.

Farhat C, Mentasti M, Jacobs E, Fry NK, Lück C. (2011). The N-acylneuraminate cytidyltransferase gene, *neuA*, is heterogenous in *Legionella pneumophila* strains but can be used as a marker for epidemiological typing in the consensus sequence-based typing scheme. J Clin Microbiol 49: 4052-4058.

Fry NK, Afshar B, Visca P, Jonas D, Duncan J, Nebuloso E, Underwood A, Harrison TG. (2005). Assessment of fluorescent amplified fragment length polymorphism analysis for epidemiological genotyping of *Legionella pneumophila* serogroup 1. Clin Microbiol Infect 11: 704-712.

Fry NK, Afshar B, Wewalka G, Harrison TG. (2006). Chapter 41. Epidemiological typing of *Legionella pneumophila* in the absence of isolates, pp. 152-155. *In* NP Cianciotto, KY Abu, PH Edelstein, BS Fields,

DF Geary, TG Harrison, CA Joseph, RM Ratcliff, JE Stout, MS Swanson (eds.), Legionella: State of the Art 30 Years after Its Recognition. ASM Press, Washington, DC.

Fry NK, Alexion-Daniel S, Bangsborg JM, Bernander S, Castellani-Pastoris M, Etienne J, Forsblom B, Gaia V, Helbig JH, Lindsay D, Lück PC, Pelaz C, Uldum SA, Harrison TG. (1999). A multicentre evaluation of genotypic methods for epidemiologic typing of *Legionella pneumophila* serogroup 1: results of a pan-European study. Clin. Microbiol. Infect. 5: 462-477.

Gaia V, Fry NK, Harrison TG, Peduzzi R. (2003). Sequence-based typing of Legionella pneumophila serogroup 1 offers the potential for true portability in legionellosis outbreak investigation. J Clin Microbiol 41: 2932-2939.

Gaia V, Poloni C, Peduzzi R. (1994). Epidemiological typing of *Legionella pneumophila* with ribotyping. Report of two clinical cases. Eur J Epidemiol 10: 303-306.

Georghiou PR, Doggett AM, Kielhofner MA, Stout JE, Watson DA, Lupski JR, Hamill RJ. (1994). Molecular fingerprinting of Legionella species by repetitive element PCR. J Clin Microbiol 32: 2989-2994.

Ginevra C, Jacotin N, Diancourt L, Guigon G, Arquilliere R, Meugnier H, Descours G, Vandenesch F, Etienne J, Lina G, Caro V, Jarraud S. (2012). *Legionella pneumophila* sequence type 1/Paris pulsotype subtyping by spoligotyping. J Clin Microbiol 50: 696-701.

Ginevra C, Lopez M, Forey F, Reyrolle M, Meugnier H, Vandenesch F, Etienne J, Jarraud S, Molmeret M. (2009). Evaluation of a nested-PCR-derived sequence-based typing method applied directly to respiratory samples from patients with Legionnaires' disease. J Clin Microbiol 47: 981-987.

Gomez-Lus P, Fields BS, Benson RF, Martin WT, O'Connor SP, Black CM. (1993). Comparison of arbitrarily primed polymerase chain reaction, ribotyping, and monoclonal antibody analysis for subtyping *Legionella pneumophila* serogroup 1. J Clin Microbiol 31: 1940-1942.

Gomgnimbou MK, Ginevra C, Peron-Cane C, Versapuech M, Refrégier G, Jacotin N, Sola C, Jarraud S. (2014). Validation of a microbead-based format for spoligotyping of *Legionella pneumophila*. J Clin Microbiol 52: 2410-2415.

Haertl R, Bandlow G. (1991). Subtyping of *Legionella pneumophila* serogroup 1 isolates by small-fragment restriction endonuclease analysis. Eur J Clin Microbiol Infect Dis 10: 630-635.

Harrison TG, Saunders NA, Haththotuwa A, Doshi N, Taylor AG. (1992). Further evidence that genotypically closely related strains of *Legionella pneumophila* can express different serogroup specific antigens. J Med Microbiol 37: 155-161.

Helbig JH, Bernander S, Castellani Pastoris M, Etienne J, Gaia V, Lauwers S, Lindsay D, Lück PC, Marques T, Mentula S, Peeters MF, Pelaz C, Struelens M, Uldum SA, Wewalka G, Harrison TG. (2002). Pan-European study on culture-proven Legionnaires' disease: distribution of *Legionella pneumophila* serogroups and monoclonal subgroups. Eur J Clin Microbiol Infect Dis. 21: 710-716.

Jonas D, Meyer HG, Matthes P, Hartung D, Jahn B, Daschner FD, Jansen B. (2000). Comparative evaluation of three different genotyping methods for investigation of nosocomial outbreaks of Legionnaires' disease in hospitals. J Clin Microbiol 38: 2284-2291.

Kahlisch L, Henne K, Draheim J, Brettar I, Höfle MG. (2010). High-resolution *in situ* genotyping of *Legionella pneumophila* populations in drinking water by multiple-locus variable-number tandem-repeat analysis using environmental DNA. Appl Environ Microbiol 76: 6186-6195.

Kamerbeek J, Schouls L, Kolk A, van Agterveld M, van Soolingen D, Kuijper S, Bunschoten A, Molhuizen H, Shaw R, Goyal M, van Embden J. (1997). Simultaneous detection and strain differentiation of *Mycobacterium tuberculosis* for diagnosis and epidemiology. J Clin Microbiol 35: 907-914.

Keim P, Price LB, Klevytska AM, Smith KL, Schupp JM, Okinaka R, Jackson PJ, Hugh-Jones ME. (2000). Multiple-locus variable-number tandem repeat analysis reveals genetic relationships within *Bacillus anthracis*. J Bacteriol 182: 2928-2936.

Knudson GB, Mikesell P. (1980). A plasmid in *Legionella pneumophila*. Infect Immun. 29: 1092-1095.

Kremer K, van Soolingen D, Frothingham R, Haas WH, Hermans PW, Martín C, Palittapongarnpim P, Plikaytis BB, Riley LW, Yakrus MA, Musser JM, van Embden JD. (1999). Comparison of methods based on different molecular epidemiological markers for typing of *Mycobacterium tuberculosis* complex strains: interlaboratory study of discriminatory power and reproducibility. J Clin Microbiol 37: 2607-2618.

Kura F, Amemura-Maekawa J, Yagita K, Endo T, Ikeno M, Tsuji H, Taguchi M, Kobayashi K, Ishii E, Watanabe H. (2006). Outbreak of Legionnaires' disease on a cruise ship linked to spa-bath filter stones contaminated with *Legionella pneumophila* serogroup 5. Epidemiol Infect 134: 385-391.

Ledesma E, Camaró ML, Carbonell E, Sacristán T, Martí A, Pellicer S, Llorca J, Herrero P, Dasí MA. (1995). Subtyping of *Legionella pneumophila* isolates by arbitrarily primed polymerase chain reaction. Can J Microbiol 41: 846-848.

Lévesque S, Plante PL, Mendis N, Cantin P, Marchand G, Charest H, Raymond F, Huot C, Goupil-Sormany I, Desbiens F, Faucher SP, Corbeil J, Tremblay C. (2014). Genomic characterization of a large outbreak of *Legionella pneumophila* serogroup 1 strains in Quebec City, 2012. PLoS One 9: e103852.

Lindstedt BA. (2005). Multiple-locus variable number tandem repeats analysis for genetic fingerprinting of pathogenic bacteria. Electrophoresis. 26: 2567-2582.

López de Felipe F, Martínez-Suárez, JV. (1991). Wide distribution of a 36-megadalton plasmid among clinical and environmental Spanish isolates of Legionella pneumophila serogroup 1. Current Microbiology 23: 233-236.

Lück C, Brzuszkiewicz E, Rydzewski K, Koshkolda T, Sarnow K, Essig A, Heuner K. (2015). Subtyping of the *Legionella pneumophila* "Ulm" outbreak strain using the CRISPR-Cas system. Int J Med Microbiol 305: 828-837.

Lück C, Fry NK, Helbig JH, Jarraud S, Harrison TG. (2013). Typing methods for legionella. Methods Mol Biol 954: 119-148.

Lück PC, Helbig JH, Ehret W, Ott M. (1995). Isolation of a *Legionella pneumophila* strain serologically distinguishable from all known serogroups. Zentralbl Bakteriol 282:35-39.

Lück PC, Helbig JH, Hagedorn HJ, Ehret W (1995). DNA fingerprinting by pulsed-field gel electrophoresis to investigate a nosocomial pneumonia caused by *Legionella bozemanii* serogroup 1, Appl Environ Microbiol 61: 2759-2761.

Maiden MC, Bygraves JA, Feil E, Morelli G, Russell JE, Urwin R, Zhang Q, Zhou J, Zurth K, Caugant DA, Feavers IM, Achtman M, Spratt BG. (1998). Multilocus sequence typing: a portable approach to the identification of clones within populations of pathogenic microorganisms. Proc Natl Acad Sci USA 95: 3140-3145.

Matsui M, Fujii S, Shiroiwa R, Amemura-Maekawa J, Chang B, Kura F, Yamauchi K. (2010). Isolation of *Legionella rubrilucens* from a pneumonia patient co-infected with *Legionella pneumophila*. J Med Microbiol 59: 1242-1246.

McDade JE, Shepard CC, Fraser DW, Tsai TR, Redus MA, Dowdle WR, the Laboratory Investigation Team. (1977). Legionnaires' disease. Isolation of a bacterium and demonstration of its role in other respiratory disease. N Engl J Med 297: 1197-1203

Mentasti M, Fry NK. (2012). Nested Sequence-based Typing (SBT) protocol for epidemiological typing of *Legionella pneumophila* directly from clinical samples, Version 2.0, 08 October 2012. http://www.hpa-bioinformatics.org.uk/legionella/legionella_sbt/php/sbt_homepage.php

Mentasti M, Afshar B, Collins S, Walker J, Harrison TG, Chalker V. (2016). Rapid investigation of cases and clusters of Legionnaires' disease in England and Wales using direct molecular typing. J Med Microbiol 65: 484-493.

Montanaro-Punzengruber JC, Hicks L, Meyer W, Gilbert GL. (1999). Australian isolates of *Legionella longbeachae* are not a clonal population. J Clin Microbiol 37: 3249-3254.

Nederbragt AJ, Balasingham A, Sirevåg R, Utkilen H, Jakobsen KS, Anderson-Glenna MJ. (2008). Multiple-locus variable-number tandem repeat analysis of Legionella pneumophila using multi-colored capillary electrophoresis. J Microbiol Methods 73: 111-117.

Nolte FS, Conlin CA, Roisin AJ, Redmond SR. (1984). Plasmids as epidemiological markers in nosocomial Legionnaires' disease. J Infect Dis. 149: 251-256.

Maher WE, Para MF, Plouffe JF. (1987). Subtyping of *Legionella pneumophila* serogroup 1 isolates by monoclonal antibody and plasmid techniques. J Clin Microbiol 25: 2281-2284.

Maher WE, Plouffe JF, Para MF. (1983). Plasmid profiles of clinical and environmental isolates of *Legionella pneumophila* serogroup 1. J Clin Microbiol 18: 1422-1423.

Mentasti M, Underwood A, Lück C, Kozak-Muiznieks NA, Harrison TG, Fry NK. (2014). Extension of the *Legionella pneumophila* sequence-based typing scheme to include strains carrying a variant of the N-acylneuraminate cytidylyltransferase gene. Clin Microbiol Infect 20: O435-O441.

Pourcel C, Vidgop Y, Ramisse F, Vergnaud G, Tram C. (2003). Characterization of a tandem repeat polymorphism in *Legionella pneumophila* and its use for genotyping. J Clin Microbiol 41: 1819-1826.

Pourcel C, Visca P, Afshar B, D'Arezzo S, Vergnaud G, Fry NK. (2007). Identification of variable-number tandem-repeat (VNTR) sequences in *Legionella pneumophila* and development of an optimized multiple-locus VNTR analysis typing scheme. J Clin Microbiol 45: 1190-1199.

Ratzow S, Gaia V, Helbig JH, Fry NK, Lück PC. (2007). Addition of *neuA*, the gene encoding N-acylneuraminate cytidylyl transferase, increases the discriminatory ability of the consensus sequence-based scheme for typing *Legionella pneumophila* serogroup 1 strains. J Clin Microbiol 45: 1965-1968.

Saunders NA, Harrison TG, Haththotuwa A, Kachwalla N, Taylor AG. (1990). A method for typing strains of *Legionella pneumophila* serogroup 1 by analysis of restriction fragment length polymorphisms. J Med Microbiol 31: 45-55.

Saunders NA, Harrison TG, Haththotuwa A, Taylor AG. (1991). A comparison of probes for restriction fragment length polymorphism (RFLP) typing of *Legionella pneumophila* serogroup 1 strains. J Med Microbiol 35: 152-158.

Schoonmaker D, Heimberger T, Birkhead G. (1992). Comparison of ribotyping and restriction enzyme analysis using pulsed-field gel electrophoresis for distinguishing *Legionella pneumophila* isolates obtained during a nosocomial outbreak. J. Clin. Microbiol 30: 1491-1498.

Schouls LM, Reulen S, Duim B, Wagenaar JA, Willems RJ, Dingle KE, Colles FM, Van Embden JD. Comparative genotyping of *Campylobacter jejuni* by amplified fragment length polymorphism, multilocus sequence typing, and short repeat sequencing: strain diversity, host range, and recombination. (2003). J Clin Microbiol 41: 15-26.

Struelens MJ, Members of the European Study Group on Epidemiological Markers (ESGEM), of the European Society for Clinical Microbiology and Infectious Diseases (ESCMID) (1996). Consensus guidelines for appropriate use and evaluation of microbial epidemiologic typing systems. Clin Microbiol Infect 2: 2-11.

Sobral D, Le Cann P, Gerard A, Jarraud S, Lebeau B, Loisy-Hamon F, Vergnaud G, Pourcel C. (2011). High-throughput typing method to identify a non-outbreak-involved *Legionella pneumophila* strain colonizing the entire water supply system in the town of Rennes, France. Appl Environ Microbiol 77: 6899-6907.

Tompkins LS, Troup NJ, Woods T, Bibb W, McKinney RM. (1987). Molecular epidemiology of Legionella species by restriction endonuclease and alloenzyme analysis. J Clin Microbiol. 25: 1875-1880.

Underwood AP, Bellamy W, Afshar B, Fry NK, Harrison TG. (2006). Chapter 44. Development of an online tool for European Working Group for Legionella Infections Sequence-Based Typing, including automatic quality assessment and data submission, pp. 163-166. *In* NP Cianciotto, KY Abu, PH Edelstein, BS Fields, DF Geary, TG Harrison, CA Joseph, RM Ratcliff,

JE Stout, MS Swanson (eds.), Legionella: State of the Art 30 Years after Its Recognition. ASM Press, Washington, DC.

van Belkum A, Tassios PT, Dijkshoorn L, Haeggman S, Cookson B, Fry NK, Fussing V, Green J, Feil E, Gerner-Smidt P, Brisse S, Struelens M, European Society of Clinical Microbiology and Infectious Diseases (ESCMID) Study Group on Epidemiological Markers (ESGEM). (2007). Guidelines for the validation and application of typing methods for use in bacterial epidemiology. Clin Microbiol Infect 13 S3: 1-46.

van Belkum A, Struelens M, Quint W. (1993). Typing of *Legionella pneumophila* strains by polymerase chain reaction-mediated DNA fingerprinting. J Clin Microbiol 31: 2198-2200.

van Ketel RJ. (1988). Similar DNA restriction endonuclease profiles in strains of *Legionella pneumophila* from different serogroups. (1988). J Clin Microbiol 26: 1838-1841.

van Ketel RJ, ter Schegget J, Zanen HC. (1984). Molecular epidemiology of *Legionella pneumophila* serogroup 1. J Clin Microbiol 20: 362-364.

Valsangiacomo C, Baggi F, Gaia V, Balmelli T, Peduzzi R, Piffaretti JC. (1995). Use of amplified fragment length polymorphism in molecular typing of *Legionella pneumophila* and application to epidemiological studies. J Clin Microbiol 33: 1716-1719.

Visca P, D'Arezzo S, Ramisse F, Gelfand Y, Benson G, Vergnaud G, Fry NK, Pourcel C. (2011). Investigation of the population structure of *Legionella pneumophila* by analysis of tandem repeat copy number and internal sequence variation. Microbiology 157: 2582-2594.

Vos P, Hogers R, Bleeker M, Reijans M, van de Lee T, Hornes M, Frijters A, Pot J, Peleman J, Kuiper M, Zabeau M. (1995). AFLP: a new technique for DNA fingerprinting. Nucleic Acids Res 23: 4407-4414.

Williams JG, Kubelik AR, Livak KJ, Rafalski JA, Tingey SV. (1990). DNA polymorphisms amplified by arbitrary primers are useful as genetic markers. Nucleic Acids Res 18: 6531-6535.

Chapter 11

Typing of *Legionella* Isolates in the Genomic Era

Daniel Wüthrich[1,2,3], Helena M.B. Seth-Smith[1,2,3] and Adrian Egli[1,2*]

[1]Clinical Microbiology, University Hospital Basel, Basel, Switzerland
[2]Applied Microbiology Research, Department of Biomedicine, University of Basel, Basel, Switzerland
[3]Swiss Institute for Bioinformatics, Basel, Switzerland

*Adrian.Egli@usb.ch

DOI: https://doi.org/10.21775/9781913652531.11

Abstract

Precise typing of *Legionella* clinical and environmental isolates is vastly important, as *L. pneumophila* is a highly conserved species with clonal groups such as the sequence type (ST) 1. This chapter discusses the typing of *Legionella* isolates and with emphasis on the new and advantageous genome-based methods. Previously relied on methods such as serotyping, pulsed-field gel electrophoresis (PFGE) and sequence-based typing (SBT) differentiate between *L. pneumophila* isolates. However, these techniques are flawed when evaluating closely related *L. pneumophila* types. This suggests that a new gold-standard is needed for effective surveillance of *Legionella* and prevention of future outbreaks. With the emergence of next-generation sequencing (NGS) and genomics, boundless conclusions can be drawn from the heightened resolution. The four current techniques constantly being applied to novel situations are core genome MLST (cgMLST), whole genome MLST (wgMLST), core genome SNP-tree, and the whole genome SNP-tree. The broad application of WGS is exemplified with mention of prior outbreak investigations utilizing WGS for analysis in conjunction with epidemiological data in varying scenarios.

Introduction

Legionella pneumophila, the cause of legionellosis, has many characteristics of a unique human pathogen. One of the most interesting features of legionellosis is the rarity of human to human transmission (Correia et al., 2016). Therefore, the majority of infections are contracted from the environment. Thus, the normal habitat of this organism appears to be water sources while human or animal infections appear to be only opportunistic. Within nutrient-limited water environments, *Legionella* grows intracellularly inside protozoa within biofilms (Cunha et al., 2016). These factors lead to a low mutation rate and therefore a high conservation of genomic diversity within an environmental pool.

L. pneumophila isolates are found in locations where water does not circulate much, including showers (Laganà et al., 2017; Schjørring et al., 2017), hot tubs, fountains, dishwashers (Yoshida et al., 2018), hot water tanks, larger plumbing systems (Rosendahl Madsen et al., 2017), and air-conditioner cooling towers (ACCT) (Lapierre et al., 2017; Llewellyn et al., 2017; Lucas et al., 2018; Timms et al., 2018). In order to identify the source of infections, it is important to characterise bacterial isolates from the environment and compare to clinical isolates. Still, the source remains unknown in the majority of diagnosed *L. pneumophila* infections.

Several methods have been developed allowing the differentiation between isolates over the last decades. These typing methods allow the characterisation of *L. pneumophila* isolates and help to identify possible sources of *L. pneumophila* infections where environmental isolates are also available. The methods range from traditional serotyping, through molecular methods of pulsed-field gel electrophoresis (PFGE) and sequence-based typing (SBT), to modern whole genome sequencing (WGS).

While the traditional methods are well established, they do not provide sufficient resolution to trace outbreaks to individual sources. The growing accessibility of next generation sequencing (NGS) to clinical laboratories has portended WGS to become the gold standard for high-resolution typing of bacterial species, including *L. pneumophila*. WGS allows access to the complete genomic information of an isolate and transferrable comparison to other strains, either local or from databases (Sabat et al., 2013). A recent publication defined a core genome multi locus sequence typing (cgMLST) scheme based on WGS data which enables fast and reproducible *L. pneumophila* incident investigation (Moran-Gilad et al., 2015).

Legionella species determination

Within the genus *Legionella*, at least 50 species have been described (Burstein et al., 2016) of which *L. pneumophila* is clinically the most important. Several tests have been developed to distinguish the species based on the 16S rRNA gene. This gene is conserved between species to a degree allowing consistent amplification but divergent enough to distinguish between the species. The first molecular test for *Legionella* species determination was developed in the 1990's, and it involved amplification and capillary sequencing of the 16S rRNA gene (Fry et al., 1991). A further assay was developed in which only part of this gene is amplified and sequenced resulting in timely determination of the *Legionella* species (Wilson et al., 2007). Quantitative-PCR (qPCR) tests, targeting the 16S genes and using mip-sequencing, are still the quicker option (Stølhaug and Bergh, 2006).

In recent years, the use of MALDI-TOF mass spectrometry (MS) to determine bacterial species was extended to *Legionella* (Stephan et al., 2014). This method provides direct species identification from a bacterial culture within minutes based on the mass profile of the ribosomal proteins. However, compared to the PCR based method, MALDI-TOF MS requires a pure isolate of the *Legionella* bacterium and cannot directly be applied to clinical or environmental samples.

Serotyping

The cell surface is a feature which varies between bacterial isolates. In the host, variations of the cell surface can cause different immune responses. Serotyping uses characterized antibodies against particular cell surface components to detect variation within a bacterial species. This epidemiologic classification, based on cell surface antigens, allows sub-species level typing. It is also applicable to not only *L. pneumophila*, but also other *Legionella* species as different strains have distinctive cell surfaces (Thacker et al., 1985). A set of monoclonal antibodies was defined to serotype *L. pneumophila* (Brindle et al., 1987) providing resolution into 14 serotypes. However, 95% of clinical case isolates fall within serotype-1. Therefore, while this method can provide valuable information in some situations, the resolution remains insufficient in many cases. Diagnostically, this can be used positively, as *L. pneumophila* serotype-1 strains can be detected using a fast urinary antigen test suggesting possible infection (Murdoch et al., 2003). A PCR-based detection of serotype-1 strains has also been developed (Thürmer et al., 2009).

Classical molecular typing

PFGE is based on restriction enzyme digestion of genomic DNA and separation by electrophoresis within a gel giving specific band patterns for different strains. This method can be used in outbreak investigations providing comparisons of the genomic structure of *Legionella* isolates

(Orsini and Romano-Spica, 2012; Zhou et al., 2010). PFGE has been used in several studies to assess the diversity of *L. pneumophila* in environmental habitats (Furuhata et al., 2014; Sanchez et al., 2008). While this method gives higher resolution than serotyping, the results are centre specific and cannot easily be compared between different laboratories.

The widely used MLST method of bacterial typing is based on the sequence of six to seven conserved household genes comparing different alleles of each gene and often recorded in a database for global comparisons. SBT typing in *L. pneumophila* is similar but combines sequence data from four household genes (*acn*, *groES*, *groEL*, and *recA*) with three more variable genes likely to be under selective pressure (*flaA*, *mompS*, and *proA*) (Gaia et al., 2003). The alleles of these seven genes are determined using PCR and capillary sequencing, and the combination of alleles is then assigned to a specific sequence type. SBT has high resolution and a good reproducibility allowing world-wide comparisons of strains. However, certain *L. pneumophila* clonal complexes (e.g. ST1) are highly diverse, rendering SBT insufficient for public health purposes (David et al., 2016).

Genome based typing

Background

While several of the methods described above access limited parts of the available genomic information, much higher levels of resolution can be obtained from whole genome sequencing (WGS). Next generation sequencing (NGS) technology has revolutionised biological research in the last decade. For the first time, genome sequences can be obtained within a few days on a single NGS device. Such platforms are increasingly installed in clinical settings both for human genetics and for bacterial typing. The major advantages of this technology is that it allows access to the whole genome and provides the highest possible resolution for typing.

As such, it allows highly discriminatory typing of pathogens even for those with conserved genomes such as *Legionella pneumophila, Clostridioides difficle,* or *Mycobacterium tuberculosis*. The significance of NGS technology for *L. pneumophila* outbreak investigations was recently pointed out in an editorial comment in *Clinical Infectious Diseases* (Beatson and Bartley, 2017). Furthermore, NGS provides reproducibility between experiments and labs, high flexibility, and fast turnaround time. The latest developments of the technology allow the typing of large sets of isolates within one to two days enabling almost real-time investigation of outbreaks.

The basic workflow of genome-based typing consists of the four following important steps, DNA extraction, library preparation, sequencing, and data analysis. While the sequencing step is clearly defined by the platform manufacturers and predominantly performed on Illumina devices, the other steps depend on various attributes of the pathogen under study. The DNA extraction protocol must be carefully chosen, as gram classification (cell wall characteristics) and other cellular properties have a high impact on the accessibility, quality, and quantity of DNA. The data analysis also provides challenges, as the genome coverage and read-depth distribution affect the data quality significantly; factors affected by choice of library preparation.

The two main approaches of genome-based typing are expansion of an MLST-like method to compare allelic differences across many more genes (cgMLST / wgMLST), and investigation of the number of base changing mutations/single-nucleotide polymorphisms (SNPs) across most or all of the genome. This results in the four bioinformatic techniques that are usually used in outbreak investigations; core genome MLST (cgMLST), whole genome MLST (wgMLST), core genome SNP-tree, and the whole genome SNP-tree.

cgMLST/ wgMLST

The basic concept of the cgMLST and wgMLST methodologies are very closely related, and both are based on that of MLST. In these methods, the number of different alleles from different genes is used as measure of genetic distance. To perform cgMLST or wgMLST, the whole genome sequence is reconstructed out of NGS data using *de novo* assembly (Bankevich et al., 2012; Zerbino and Birney, 2008). Then, the genes are identified by aligning to the reference core genome (Altschul SF, Gish W, Miller W, Myers EW, 1990). From each genome being studied, the alleles of each gene are determined and compared to those from other strains. In cgMLST analysis, only the alleles of the core genome, i.e. genes conserved between all strains, are compared (Figure 1). In wgMLST analysis, accessory genes are also considered for allele calling and therefore provide an even higher resolution than cgMLST. However, cgMLST generally provides sufficient resolution to determine outbreaks. Results from cgMLST analysis can also be more transferable and comparable when defined genes within established cgMLST schemes are used. This data can then be valuable in situations of wider outbreaks. Both cgMLST and wgMLST methods can be implemented in commercial software packages such as SeqSphere+ (Ridom) or BioNumerics (Applied Maths). Open source solutions are also available (Feijao et al., 2018).

A cgMLST scheme for *L. pneumophila* was recently published, defining 1,521 core genes (Moran-Gilad et al., 2015). The scheme has been deposited on PubMLST (pubmlst.org) where genome sequences can be uploaded for allele designation and comparison with existing data. A centralised maintenance of this database is crucial in order to keep a unified nomenclature of novel alleles. By comparing MLST and cgMLST data from the same isolates, it was found that strains which share the same MLST type are also more closely related in cgMLST analysis and reveal a high diversity within the different sequence type (Moran-Gilad et al., 2015). Thus, cgMLST data correlates with, but also expands MLST

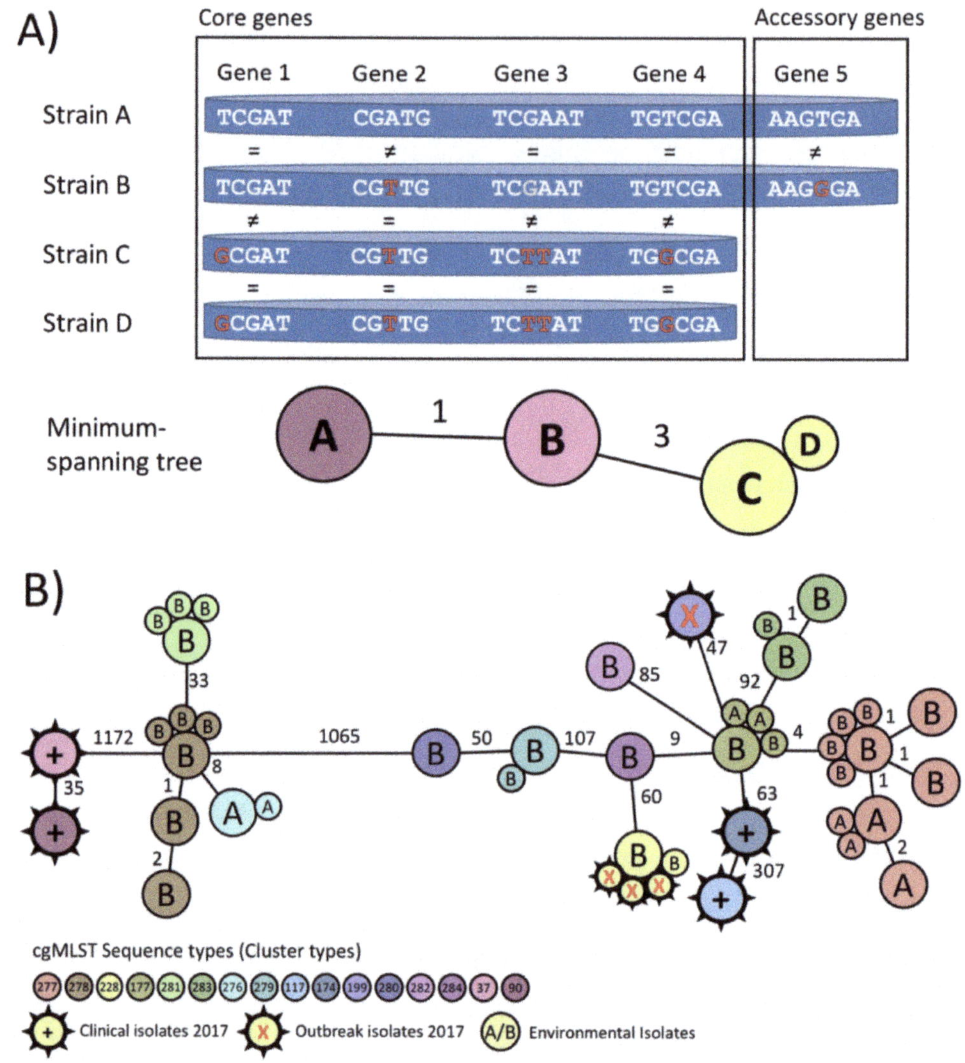

Figure 1. The concept of cgMLST analysis for outbreak investigation. A) A simplified concept of cgMLST. In the figure the number of core genes is reduced for illustrative purposes. To perform cgMLST, first the core genome (genes shared between all isolates) has to be determined from the whole genome sequence. This can either be done by comparing the gene content of the strains to be analysed, or using an established cgMLST scheme, for example those available on PubMLST. The size of a core genome varies greatly with species. For the published *L. pneumophila* scheme, it consists of 1,521 genes. The core genes of the different strains are then compared in order to determine whether they have the same allele ("=") or a different one ("≠"). The total number of different alleles between strains is then use as measure of the genomic distance. A minimum-spanning tree (MST) is used to visualize the differences within the cgMLST. This uses a machine learning algorithm to connect the different strains with the shortest number of allele differences possible. In wgMLST analysis, the alleles of the accessory genome "accessory genes" are also used in the comparison. B) MST of an *L. pneumophila* outbreak in Basel, Switzerland in 2017 (Wüthrich et al., n.d.).

data. Our recent publication studying the correlation of isolates from an outbreak situation with those from air-conditioning cooling towers showed that neither serotyping nor MLST provided the necessary resolution. cgMLST analysis was required to demonstrate a connection between environmental and clinical isolates (Wüthrich et al., n.d.).

SNP-based analysis

In contrast to the allelic differences defined in cgMLST/wgMLST, core genome SNP and whole genome SNP analyses are based on base changes/SNPs between isolate genomes. This means that two mutations within a single gene will count as two differences, rather than one difference, and that insertions and deletions are not taken into account. After determination of the SNPs, a phylogenetic tree is constructed to depict the relatedness of the isolates. To produce a core genome SNP-tree, only SNPs within a defined core genome are analysed. Meanwhile, a whole genome SNP-tree will analyse the complete genome, relative to a reference. Both approaches have advantages including that the core genome is often used if the phylogeny of very diverse strains has to be reconstructed. However, the whole genome approach provides even higher resolution and is used in cases where the studied isolates are closely related. In both cases, non-coding regions can be included which further increases the typing resolution.

SNP-based methods are very sensitive to large scale genomic changes such as recombinations (Didelot and Maiden, 2010), which appear as several to many hundreds of mutations, but are caused by a single evolutionary event. When a recombination has occurred, SNP-based methods overestimate the evolutionary distance between isolates. To address this issue, algorithms that correct for recombinations should be applied before phylogeny construction (Croucher et al., 2015; Didelot and Wilson, 2015).

Table 1. *Legionella* publications based on WGS.		
Topic	Title	Reference
Nosocomial transmission	Seeding and establishment of *L. pneumophila* in hospitals: Implications for genomic investigations of nosocomial Legionnaires' disease	(David et al., 2017)
Basic research	Multiple major disease-associated clones of *L. pneumophila* have emerged recently and independently	(David et al., 2016)
Community acquired	Legionnaires' disease outbreak caused by endemic strain of *L. pneumophila*, New York, New York, USA, 2015	(Lapierre et al., 2017)
Community acquired	Gene flow in environmental *L. pneumophila* leads to genetic and pathogenic heterogeneity within a Legionnaires' disease outbreak	(McAdam et al., 2014)
Nosocomial transmission	Whole-genome sequencing for identification of the source in hospital-acquired Legionnaires' disease	(McAdam et al., 2014)
Community acquired	Genomic investigation of a suspected outbreak of *L. pneumophila* ST82 reveals undetected heterogeneity by the present gold-standard methods, Denmark, July to November 2014	(Schjørring et al., 2017)
Community acquired	Genome sequencing links persistent outbreak of legionellosis in Sydney (New South Wales, Australia) to an emerging clone of *L. pneumophila* sequence type 211	(Timms et al., 2018)
Community acquired	Air-conditioner cooling towers as complex reservoirs and continuous source of *L. pneumophila* infection evidenced by genomic analysis	(Wüthrich et al., n.d.)

SNP-phylogenies were used to study a recent *L. pneumophila* outbreak in Denmark. In this case, isolates which had been found to be identical by other typing methods were found to fall into discrete groups consistent with epidemiological information (Schjørring et al., 2017). In several other cases, SNP analysis has provided insights into outbreaks, transmissions, or environmental populations (Table 1).

Conclusion

Genome based methods provide much higher typing resolution compared to previous methods. The high resolution enables us to draw the correct conclusions in outbreak investigations and to accurately identify sources of infection. High resolution is particularly important for species with high genetic conservation, such as *L. pneumophila*, for which genome-based methods have become the new gold standard for typing. However, even with a high-resolution tool for typing, classical epidemiology information remains key for correct interpretation of data (Petzold et al., 2017).

References

Altschul SF, Gish W, Miller W, Myers EW, L.D. (1990). Basic local alignment search tool. J Mol Biol. Oct 5, 403-10.

Bankevich, A., Nurk, S., Antipov, D., Gurevich, A. a, Dvorkin, M., Kulikov, A.S., Lesin, V.M., Nikolenko, S.I., Pham, S., Prjibelski, A.D., Pyshkin, A. V, Sirotkin, A. V, Vyahhi, N., Tesler, G., Alekseyev, M. A., Pevzner, P. A. (2012). SPAdes: a new genome assembly algorithm and its applications to single-cell sequencing. J. Comput. Biol. 19, 455-77. https://doi.org/10.1089/cmb.2012.0021

Beatson, S.A., Bartley, P.B. (2017). Diving deep into hospital-acquired *Legionella pneumophila* with whole-genome sequencing. Clin. Infect. Dis. 64, 1260-1262. https://doi.org/10.1093/cid/cix156

Brindle, R.J., Stannett, P.J., Tobin, J.O., 1987. *Legionella pneumophila*: monoclonal antibody typing of clinical and environmental isolates. Epidemiol. Infect. 99, 235-239.

Burstein, D., Amaro, F., Zusman, T., Lifshitz, Z., Cohen, O., Gilbert, J.A., Pupko, T., Shuman, H.A., Segal, G., (2016). Genomic analysis of 38 *Legionella* species identifies large and diverse effector repertoires. Nat. Genet. https://doi.org/10.1038/ng.3481

Correia, A.M., Ferreira, J.S., Borges, V., Nunes, A., Gomes, B., Capucho, R., Gonçalves, J., Antunes, D.M., Almeida, S., Mendes, A., Guerreiro, M., Sampaio, D.A., Vieira, L., Machado, J., Simões, M.J., Gonçalves, P., Gomes, J.P. (2016). Probable Person-to-Person Transmission of Legionnaires' Disease. N. Engl. J. Med. 374, 497-498. https://doi.org/10.1056/NEJMc1505356

Croucher, N.J., Page, A.J., Connor, T.R., Delaney, A.J., Keane, J.A., Bentley, S.D., Parkhill, J., Harris, S.R. (2015). Rapid phylogenetic analysis of large samples of recombinant bacterial whole genome sequences using Gubbins. Nucleic Acids Res. https://doi.org/10.1093/nar/gku1196

Cunha, B.A., Burillo, A., Bouza, E. (2016). Legionnaires' disease, in: The Lancet. pp. 376-385. https://doi.org/10.1016/S0140-6736(15)60078-2

David, S., Afshar, B., Mentasti, M., Ginevra, C., Podglajen, I., Harris, S.R., Chalker, V.J., Jarraud, S., Harrison, T.G., Parkhill, J. (2017). Seeding and establishment of *Legionella pneumophila* in hospitals: Implications for genomic investigations of nosocomial legionnaires' disease. Clin. Infect. Dis. 64, 1251-1259. https://doi.org/10.1093/cid/cix153

David, S., Rusniok, C., Mentasti, M., Gomez-Valero, L., Harris, S.R., Lechat, P., Lees, J., Ginevra, C., Glaser, P., Ma, L., Bouchier, C., Underwood, A., Jarraud, S., Harrison, T.G., Parkhill, J., Buchrieser, C. (2016). Multiple major disease-associated clones of Legionella pneumophila have emerged recently and independently. Genome Res. 26, 1555-1564. https://doi.org/10.1101/gr.209536.116

Didelot, X., Maiden, M.C.J. (2010). Impact of recombination on bacterial evolution. Trends Microbiol. https://doi.org/10.1016/j.tim.2010.04.002

Didelot, X., Wilson, D.J. (2015). ClonalFrameML: Efficient Inference of Recombination in Whole Bacterial Genomes. PLoS Comput. Biol. https://doi.org/10.1371/journal.pcbi.1004041

Feijao, P., Yao, H.-T., Fornika, D., Gardy, J., Hsiao, W., Chauve, C., Chindelevitch, L. (2018). MentaLiST - A fast MLST caller for large MLST schemes. Microb. Genomics. https://doi.org/10.1099/mgen.0.000146

Fry, Warwick, Saunders, Embley. (1991). The use of 16S ribosomal RNA analyses to investigate the phylogeny of the family Legionellaceae. J. Gen. Microbiol. 137, 1215-1222. https://doi.org/10.1099/00221287-137-5-1215

Furuhata, K., Ishizaki, N., Umekawa, N., Nishizima, M., Fukuyama, M. (2014). Pulsed-Field Gel Electrophoresis (PFGE) Pattern Analysis and

Chlorine-Resistance of *Legionella pneumophila* Isolated from Hot Spring Water Samples. Biocontrol Sci. https://doi.org/10.4265/bio.19.33

Gaia, V., Fry, N.K., Harrison, T.G., Peduzzi, R. (2003). Sequence-based typing of Legionella pneumophila serogroup 1 offers the potential for true portability in legionellosis outbreak investigation. J. Clin. Microbiol. https://doi.org/10.1128/JCM.41.7.2932-2939.2003

Laganà, P., Gambuzza, M.E., Delia, S. (2017). *Legionella* risk assessment in cruise ships and ferries. Ann. Agric. Environ. Med. 24, 276-282. https://doi.org/10.26444/aaem/74717

Lapierre, P., Nazarian, E., Zhu, Y., Wroblewski, D., Saylors, A., Passaretti, T., Hughes, S., Tran, A., Lin, Y., Kornblum, J., Morrison, S.S., Mercante, J.W., Fitzhenry, R., Weiss, D., Raphael, B.H., Varma, J.K., Zucker, H.A., Rakeman, J.L., Musser, K.A. (2017). Legionnaires' disease outbreak caused by endemic strain of *Legionella pneumophila*, New York, New York, USA, 2015. Emerg. Infect. Dis. 23, 1784-1791. https://doi.org/10.3201/eid2311.170308

Llewellyn, A.C., Lucas, C.E., Roberts, S.E., Brown, E.W., Nayak, B.S., Raphael, B.H., Winchell, J.M. (2017). Distribution of Legionella and bacterial community composition among regionally diverse US cooling towers. PLoS One 12. https://doi.org/10.1371/journal.pone.0189937

Lucas, K.D., Wheeler, C., McLendon, P., Leistikow, B.N., Mohle-Boetani, J.C. (2018). Outbreak of Legionnaires' disease associated with cooling towers at a California state prison, 2015. Epidemiol. Infect. 146, 297-302. https://doi.org/10.1017/S0950268818000110

McAdam, P.R., Vander Broek, C.W., Lindsay, D.S., Ward, M.J., Hanson, M.F., Gillies, M., Watson, M., Stevens, J.M., Edwards, G.F., Fitzgerald, J.R. (2014). Gene flow in environmental *Legionella pneumophila* leads to genetic and pathogenic heterogeneity within a Legionnaires' disease outbreak. Genome Biol. 15, 504. https://doi.org/10.1186/s13059-014-0504-1

Moran-Gilad, J., Prior, K., Yakunin, E., Harrison, T.G., Underwood, A., Lazarovitch, T., Valinsky, L., Lück, C., Krux, F., Agmon, V., Grotto, I.,

Harmsen, D. (2015). Design and application of a core genome multilocus sequence typing scheme for investigation of Legionnaires' disease incidents. Eurosurveillance 20, 1-9. https://doi.org/10.2807/1560-7917.ES2015.20.28.21186

Murdoch, D.R., Weinstein, M.P., Murdoch, D.R. (2003). Diagnosis of Legionella infection. Clin. Infect. Dis. 36, 64-9. https://doi.org/10.1086/345529

Orsini, M., Romano-Spica, V. (2012). A rapid PFGE protocol for typing Legionella isolates from fresh or frozen samples. J. Infect. Chemother. https://doi.org/10.1007/s10156-012-0416-5

Petzold, M., Prior, K., Moran-Gilad, J., Harmsen, D., Lück, C. (2017). Epidemiological information is key when interpreting whole genome sequence data - lessons learned from a large *Legionella pneumophila* outbreak in Warstein, Germany, 2013. Eurosurveillance. https://doi.org/10.2807/1560-7917.ES.2017.22.45.17-00137

Rosendahl Madsen, A.M., Holm, A., Jensen, T.G., Knudsen, E., Lundgaard, H., Skov, M.N., Uldum, S.A., Kemp, M. (2017). Whole-genome sequencing for identification of the source in hospital-acquired Legionnaires' disease. J. Hosp. Infect. 96, 392-395. https://doi.org/10.1016/j.jhin.2017.04.020

Sabat, A.J., Budimir, A., Nashev, D., Sá-Leão, R., van Dijl, J. m, Laurent, F., Grundmann, H., Friedrich, A.W., ESCMID Study Group of Epidemiological Markers (ESGEM). (2013). Overview of molecular typing methods for outbreak detection and epidemiological surveillance. Eur. Commun. Dis. Bull. 18, 20380. https://doi.org/20380 [pii]

Sanchez, I., Garcia-Nuñez, M., Ragull, S., Sopena, N., Pedro-Botet, M.L., Estere, M., Rey-Joly, C., Sabria, M. (2008). Genotypic variability and persistence of *Legionella pneumophila* PFGE patterns in 34 cooling towers from two different areas. Environ. Microbiol. https://doi.org/10.1111/j.1462-2920.2007.01460.x

Schjørring, S., Stegger, M., Kjelsø, C., Lilje, B., Bangsborg, J.M., Petersen, R.F., David, S., Uldum, S.A. (2017). Genomic investigation of a suspected outbreak of legionella pneumophila st82 reveals undetected heterogeneity by the present gold-standard methods, denmark, july to november 2014. Eurosurveillance 22. https://doi.org/10.2807/1560-7917.ES.2017.22.25.30558

Stephan, R., Johler, S., Oesterle, N., Näumann, G., Vogel, G., Valentin, P. (2014). Rapid and reliable species identi fi cation of scallops by MALDI-TOF mass spectrometry 46, 6-9. https://doi.org/10.1016/j.foodcont.2014.04.047

Stølhaug, A., Bergh, K. (2006). Identification and differentiation of *Legionella pneumophila* and Legionella spp. with real-time PCR targeting the 16S rRNA gene and species identification by mip sequencing. Appl. Environ. Microbiol. 72, 6394-6398. https://doi.org/10.1128/AEM.02839-05

Thacker, W.L., Plikaytis, B.B., Wilkinson, H.W. (1985). Identification of 22 Legionella species and 33 serogroups with the slide agglutination test. J. Clin. Microbiol. 21, 779-782.

Thürmer, A., Helbig, J.H., Jacobs, E., Lück, P.C. (2009). PCR-based "serotyping" of *Legionella pneumophila*. J. Med. Microbiol. 58, 588-595. https://doi.org/10.1099/jmm.0.008508-0

Timms, V.J., Rockett, R., Bachmann, N.L., Martinez, E., Wang, Q., Chen, S.C.A., Jeoffreys, N., Howard, P.J., Smith, A., Adamson, S., Gilmour, R., Sheppeard, V., Sintchenko, V. (2018). Genome sequencing links persistent outbreak of legionellosis in Sydney (New South Wales, Australia) to an emerging clone of *Legionella pneumophila* sequence type 211. Appl. Environ. Microbiol. 84. https://doi.org/10.1128/AEM.02020-17

Wilson, D.A., Reischl, U., Hall, G.S., Procop, G.W. (2007). Use of partial 16S rRNA gene sequencing for identification of *Legionella pneumophila* and non-pneumophila *Legionella* spp. J. Clin. Microbiol. 45, 257-258. https://doi.org/10.1128/JCM.01552-06

Wüthrich, D., Gautsch, S., Spieler-Denz, R., Dubuis, O., Gaia, V., Moran-Gilad, J., Hinic, V., Seth-Smith, H.M., Nickel, C., Tschudin-Sutter, S.,

Bassetti, S., Haenggi, M., Brodmann, P., Fuchs, S., Egli, A., n.d. Air-conditioner cooling towers as complex reservoirs and continuous source of *Legionella pneumophila* infection evidenced by genomic analysis. Eurosurveillance Accepted.

Yoshida, M., Furuya, N., Hosokawa, N., Kanamori, H., Kaku, M., Koide, M., Higa, F., Fujita, J. (2018). *Legionella pneumophila* contamination of hospital dishwashers. Am. J. Infect. Control 46, 943-945. https://doi.org/10.1016/j.ajic.2018.01.024

Zerbino, D.R., Birney, E. (2008). Velvet: algorithms for *de novo* short read assembly using de Bruijn graphs. Genome Res. 18, 821-9. https://doi.org/10.1101/gr.074492.107

Zhou, H., Ren, H., Zhu, B., Kan, B., Xu, J., Shao, Z. (2010). Optimization of pulsed-field gel electrophoresis for *Legionella pneumophila* subtyping. Appl. Environ. Microbiol. https://doi.org/10.1128/AEM.01455-09

www.ingramcontent.com/pod-product-compliance
Lightning Source LLC
LaVergne TN
LVHW070848210926
848482LV00016B/893